INTERNATIONAL UNION OF CRYSTALLOGRAPHY
MONOGRAPHS ON CRYSTALLOGRAPHY

IUCr Monographs on Crystallography

IUCr Texts on Crystallography

Polymorphism in Molecular Crystals

JOEL BERNSTEIN

Department of Chemistry
Ben-Gurion University of the Negev

CLARENDON PRESS • OXFORD
2002

OXFORD

UNIVERSITY PRESS

Great Clarendon Street, Oxford OX2 6DP

Oxford University Press is a department of the University of Oxford.
It furthers the University's objective of excellence in research, scholarship,
and education by publishing worldwide in

Oxford New York

Auckland Bangkok Buenos Aires Cape Town Chennai
Dar es Salaam Delhi Hong Kong Istanbul Karachi Kolkata
Kuala Lumpur Madrid Melbourne Mexico City Mumbai Nairobi
São Paulo Shanghai Taipei Tokyo Toronto

with an associated company in Berlin

Oxford is a registered trade mark of Oxford University Press
in the UK and in certain other countries

Published in the United States
by Oxford University Press Inc., New York

A Catalogue record for this title is available from the British Library

Library of Congress Cataloging in Publication Data
Bernstein, Joel.
Polymorphism in molecular crystals / Joel Bernstein.
(IUCr monographs on crystallography; 14)
Includes index.
1. Polymorphism (Crystallography) 2. Molecular crystals. I. Title. II. International
Union of Crystallography monographs on crystallography; 14.
QD951 .B57 2002   548'.3—dc21   2001047556

ISBN 0 19 850605 8

10 9 8 7 6 5 4 3 2 1

Typeset by
Newgen Imaging Systems (P) Ltd., Chennai, India
Printed in Great Britain
on acid-free paper by
T. J. International Ltd, Padstow

To Judy

AT FIRST SIGHT

Once made, this stolid mauve
powder would seem forever;

but people intent on repro-
duction fire up pots next door

or across the sea, and out
of the odd one crystallizes

another, the same, but for
a tell-tale (to X-rays) part

that twists a tad; in a tango
of attractions and absences

molecules nestle in a variant
pattern. Neat, but from here on,

the first will not be made. So
it seems; the ur-makers once

patient hands grow limp—
has desire fled? In all flasks

the second precipitates. Who,
oh who, is to blame? Yes, lay it

to the other coming—as if
seed crystals flew the world.

But the first is the accident,
a small well in a chanced

landscape, a nicked knife edge,
the one parcel of phase space

never to be sampled again,
the vanishing polymorph . . . you.

                              Roald Hoffmann

Preface

Sometime in the middle '60s during an evening stint in the laboratory a fellow graduate student and I were struggling to determine the orientation of a known crystal on a quarter circle manual X-ray diffractometer. When things didn't turn out as expected he raised the possibility that it might be a polymorph (it wasn't). However, I recall being fascinated by the whole idea of a single molecule crystallizing in different structures, and the consequences of such a phenomenon. That fascination has not waned over the intervening years.

In the interim polymorphism has become a much more widely recognized and observed phenomenon, with both fundamental and commercial ramifications. The literature has grown enormously, albeit scattered in a variety of primary sources. In view of the growing interest in the subject there appeared to be the need for a monograph on the subject. Work in polymorphism and on polymorphs is quite interdisciplinary in nature, and as a result there is no single book that provides an introduction and overview of the subject.

The purpose of this book is to summarize and to bring up to date the current knowledge and understanding of polymorphism in molecular crystals, and to con-centrate it in one source. It is meant to serve as a starting point and source book both for those encountering the phenomenon of polymorphism for the first time, and for more seasoned practitioners in any of the disciplines concerned with the organic solid state. It is intended to serve a readership from advanced undergraduate students through experienced professionals. Much of the information in the book does appear in the open literature; however, because of the increasing commercial importance of the phenomenon, a significant portion of the information (for instance, on industrial applications, patents, or previously restricted distribution) is less accessible, and I have attempted to include both the information from those sources as well as full details of their citations. The intention is that even with the passage of time devel-opments in many of the areas covered in the book can be followed by searching for the subsequent citations of the relevant papers cited here. As a point of reference for future readers, an attempt was made to review the relevant literature through December, 2000, although a number of references from 2001 are also included.

I wish to thank the IUCr for its role in fostering this joint series with Oxford University Press and for agreeing to the format of this particular volume.

A work of this type cannot be completed without the help of many other people. This project was initiated during a sabbatical leave (in 1997–98) first at the University of Barcelona with the gracious hospitality of Santiago Alvarez and then at the Cambridge Crystallographic Data Centre. My hosts there and in the contiguous Department of Chemistry at Cambridge University put all their resources at my disposal and simply let me go about my business of reading and writing. I am particularly grateful to

Olga Kennard for encouraging me to spend that time there and to Frank Allen and his colleagues for making it so collegial and so congenial.

Many hours were spent in libraries in a variety of locations, and those at the Royal Society of Chemistry in London were especially enjoyable and fruitful, in particular due to the efforts of Nicola Best.

I have been particularly fortunate to have benefited from the assistance of a small army of bright, enthusiastic students who put up with my changing whims and wishes and managed the logistical aspects of organizing the reprints collection, obtaining reprint permissions, checking and completing the details of references, scanning, modifying and preparing figures, etc.: Megan Fisher, Michal Stark, Avital Furlanger, Margalit Lerner, Noa Zamstein, Shai Allon, Noam Bernstein, and Janice Rubin.

Over the years I have been in touch with countless colleagues—many of whom I have never met—who have willingly, indeed enthusiastically, provided me with preprints, obscure reprints, private documents, observations, and insights on a variety of polymorphic behaviour and systems. To all of them I am grateful, and in the course of this work, a number of them provided exceptional assistance which made my task considerably easier and more enjoyable. They deserve special mention here. Peter Erk at BASF spent many hours helping put the connection between polymorphism and colourants into focus. His colleague Martin Schmidt at Clariant provided almost instantaneous responses and faxes to what must have seemed like an endless stream of questions and requests. The chapter on high energy materials probably could not have been written without the help of Charlotte Lowe-Ma of the Ford Motor Company. Following a brief conversation with her at a scientific meeting a courier showed up in my office with a box of historically important documents and personal notes and summaries on polymorphism of high energy materials that were invaluable. Richard Gilardi from the U.S. Naval Research Laboratories provided similar advice and assistance on many of the newer compounds and systems. Stephen Tarling from Birkbeck College availed himself of his time and experiences in a number of patent litigations involving crystal modifications to lead me to the appropriate cases. Michelle O'Brien and Howard Levine of the firm of Finnegan, Henderson, Farabow, Garrett and Dunner, Washington, D. C. managed to get hold of every legal document I requested. Angelo Gavezzotti of the University of Milan was ready with wise counsel and support whenever I needed it.

When it came to finding examples, systems and references among the pharmaceuticals Jan-Olav Henck of Bayer and Ulrich Griesser of the University of Innsbruck were always ready and willing with immediate detailed answers, and faxed reprints if necessary.

Many graduate students and associates in my laboratory carried out the examples taken from our own work described here. I am grateful for their dedication and their contribution to the contents: Ilana Bar, Ehud Goldstein (Chosen), Leah Shahal, Liat Shimoni, Sharona Zamir, Arkady Ellern, Oshrit Navon, and the same Jan-Olav Henck mentioned above.

The exchanges with Roald Hoffmann of Cornell University on disappearing polymorphs in song and story were particularly memorable, and I am grateful for his permission to reprint his poem on the subject as the frontispiece of this tome. As he has been for a couple of generations of chemical crystallographers, Jack Dunitz was a constant inspiration and standard of excellence.

As a postdoctoral fellow myself I was very fortunate to have worked with two inspiring scientists whose scientific integrity and talent for precision in writing and expression have served as models throughout my career: K. N. Trueblood at UCLA and Gerhard Schmidt at the Weizmann Institute. Countless times in the course of preparing this work I found myself asking if they would have passed a sentence, a phrase or a scientific judgement or opinion that had just been written. I hope they would, but as they also taught me, I alone am responsible for what follows.

My late wife Judy was a source of constant encouragement and support for nearly 35 wonderful years together, especially those during which this book developed and took shape. Dedicating it to her is but a minor recognition of her contribution and the life we shared.

Beer Sheva   J. B.
June 2001

Contents

Colour plates can be found between pp. 114 and 115

1

Introduction and historical background

With the accumulation of data, there is developing a gradual realization of the generality of polymorphic behavior, but to many chemists polymorphism is still a strange and unusual phenomenon. (Buerger and Bloom 1937)

In spite of the fact that different polymorphs of a given compound are, in general, as different in structure and properties as the crystals of two different compounds, most chemists are almost completely unaware of the nature of polymorphism and the potential usefulness of knowledge of this phenomenon in research. (W. C. McCrone 1965)

1.1 Introduction

In spite of chemists' occupation and fascination with structure and the connection between structure and properties, in McCrone's view in the nearly three decades following the observation of Buerger and Bloom there had not been any serious change in their awareness of polymorphism, its importance to chemistry and its potential usefulness. More than thirty-five additional years have passed, and that awareness is now increasing. As analytical methods have become more sophisticated, more precise and more rapidly carried out, the proliferation of data has revealed differences in structure and behaviour which can be attributed to polymorphism. The increasingly rapid accumulation and archiving of structural data allow for the systematic search and retrieval of those data for the purpose of correlating structure with properties. In short, polymorphism in chemistry is moving from a 'strange and unusual phenomenon' to one which is a legitimate and important area of research in and of itself that can also be utilized by chemists in unique and efficient ways for the study, understanding and development of structure–property relations in solids.

Structural diversity surfaces in almost every facet of nature. Chemistry in general is no exception, nor, in particular is structural chemistry, and crystal polymorphism is one manifestation of that diversity. The emphasis here will be on molecular crystals for a number of reasons. Inorganic compounds and minerals traditionally have been the purview of geologists and inorganic chemists and their innate interest in structure–property relationships led naturally to more organization and more awareness of polymorphism than in other pursuits. Monographs such as Wells' (1989) 'Structural Inorganic Chemistry' and Verma and Krishna's (1966) 'Polytypism and Polymorphism' are typical examples. On the other hand, organic solid state chemistry is a relatively new discipline (or multidiscipline), founded (or refounded) in the 1960s by the schools of Schmidt (1971) in Israel and Paul and Curtin (1973, 1975) in Urbana, Illinois, so that information and knowledge of polymorphism in this area is scattered through a wider variety of literature. Our aim is to provide within the framework of a single volume an introduction to the fundamental physical principles upon which polymorphism is based, together with a variety of examples from the literature which demonstrate the importance of understanding polymorphism and, in McCrone's words (1965), the 'potential usefulness of knowledge of this phenomenon in research'. This can then be used as a reference and source book for those encountering polymorphism for the first time, those embarking on polymorphism-related research, or those already involved in such endeavours who wish to find additional examples and an entrance to the related literature. The diversity of the field as well as its rapid development in the past few years makes a comprehensive survey prohibitive in terms of space and almost immediately out of date. As a necessary compromise we have attempted to choose examples which are meant to be representative of the phenomena they exhibit, as well as to provide leading references that can be updated with subsequent citations.

1.2   Definitions

1.2.1   *Polymorphism*

Polymorphism (Greek: *poly* = many, *morph* = form) specifying the diversity of nature, is a term used in many disciplines.[1]

According to the Oxford English Dictionary the term first appears in 1656 in relation to the diversity of fashion. In the context of crystallography, the first use is generally credited to Mitscherlich (1822, 1823), who recognized different crystal structures of the same compound in a number of arsenate and phosphate salts. The historical development of polymorphism is discussed in Section 1.4.

As in many terms of chemistry, an all-encompassing definition of polymorphism is elusive. The problem has been discussed by McCrone (1965), whose working definition and accompanying caveats are as relevant today as when they were first enunciated. McCrone defines a polymorph as 'a solid crystalline phase of a given compound resulting from the possibility of at least two different arrangements of the molecules of that compound in the solid state'.

At first glance this definition seems straightforward. What are some of the complications? For flexible molecules McCrone would include *conformational polymorphs*, wherein the molecule can adopt different conformations in the different crystal structures (Corradini 1973; Panagiotopoulis *et al.* 1974; Bernstein and Hagler 1978; Bernstein 1987). But this is a matter of degree: dynamic isomerism or tautomerism

---

[1]   In a recent internet search Threlfall (2000), found 1.5 million references to the term, of which 90 per cent refer to video games in which creatures change shape. 90 per cent of the remainder refer to *genetic* polymorphism, which involves minor change of protein or DNA sequences that may lead, for instance to particular sensitivity to drugs. Of references that refer to crystallographic polymorphism, approximately 90 per cent are devoted to inorganic structures, which are not covered here. The remainder deal with molecular crystals.

would be excluded, because they involve the formation of different molecules. The 'safe' criterion for classification of a system as polymorphic would be if the crystal structures are different but lead to identical liquid and vapour states. For dynamically converting isomers, this criterion invokes a time factor (Dunitz 1995). As with polymorphs, dynamic isomers will melt at different temperatures. However, the composition of the melt will differ. That composition can change with time until equilibrium is reached, however, and the equilibrium composition will be temperature dependent. Using these criteria, a system in which the isomers (or in the limit conformers) were rapidly interconverting would be considered a polymorphic one, while a slowly interconverting system would not be characteristic of polymorphic solids. As Dunitz (1995) has pointed out, such a definition would lead to the situation in which a racemate and a conglomerate would be determined to be polymorphic when the rate of interconversion of enantiomers in the melt or in solution is fast, but would be classified as three different compounds when that rate of interconversion is slow. Since no time frame is defined for slow or fast, the borderline is indeed fuzzy. Dunitz has also noted that the distinction has important ramifications when considering the Phase Rule (see Section 2.2.1), since the number of components must be defined. In general, components are 'chemically distinct constituents' whose concentrations may be varied independently at the temperature concerned. McCrone (1965) has attempted to clarify the distinction between polymorphism and dynamic isomerism. The latter involves chemically different molecules 'more or less readily convertible in the melt state. The basic difference between two polymorphs can occur in the solid state, and the difference between any two polymorphs disappears in the melt state.' A number of examples of these phenomena are described in Section 3.5.2. Chemists may certainly differ on precisely what comprises 'chemically distinct molecules' and 'more or less readily convertible' which can lead to the lack of precision in the definition of polymorphism.

Some additional aspects of the definition deserve mention here. Since polymorphism involves different states of matter with potentially different properties, debates about definitions of the phenomenon have centred alternatively on differences in thermodynamic, structural, or other physical properties. For instance, Buerger and Bloom (1937) cited Goldschmidt's use of 'building blocks', 'polarization properties', and 'thermodynamic environment' to describe the state of the art and understanding of polymorphism at that time:

... if a member of an isomorphous series is constructed of building blocks whose size and polarization properties lie near the limit which the structure of this series can accommodate, changes in the thermodynamic environment may cause this limit to be exceeded and a new structure to be developed. This is polymorphism.

On the other hand McCrone's definition appears to have been simplified by Rosenstein and Lamy (1969) as 'when a substance can exist in more than one crystalline state it is said to exhibit polymorphism.' This simplified definition was apparently adopted by Burger (1983), 'If these [solids composed of only one component] can exist in different crystal lattices, then we speak of polymorphism', which unfortunately confuses the concept of crystal lattice and the concept of crystal structure. Some of these misconceptions have been carried through to more recent publications (Wood 1997).

There has also been an ongoing debate about the use and misuse of the terms allotropy and polymorphism (Jensen 1998). The former was originally introduced by Berzelius (1844) to describe the existence of different crystal structures of elements, as opposed to different structures of compounds. Findlay (1951) opposed the use of two terms for essentially the same phenomenon, and even proposed that polymorphism be abandoned in favour of allotropism as a description for the general phenomenon. The distinction between the two terms was debated by Sharma (1987) and Reinke *et al.* (1993). Sharma suggested that polymorphs be denoted as 'different crystal forms, belonging to the same or different crystal systems, in which the identical units of the same element or the identical units of the same compound or the identical formulas or identical repeating units are packed differently.' Reinke *et al.* invoked the modern language of supramolecular chemistry, by proposing 'an extended and modified definition' for polymorphism as 'the phenomenon where supermolecular structures with different, well-defined physical properties can be formed by chemically uniform species both in the liquid and solid state.' This line of thought has apparently come full circle, with Dunitz's (1991, 1996) description of the crystal as the 'supermolecule par excellence' and on that basis, 'If a crystal is a supermolecule, then polymorphic modifications are superisomers and polymorphism is a kind of superisomerism . . .'.

As with many other concepts in chemistry, in a room full of chemists there is general agreement about the meaning, consequences, and relevance of polymorphism. Although the language of chemistry is constantly developing, McCrone's working definition noted at the beginning of this section appears to have stood the test of time, and is the one that would be recognized and used by most chemists today.

1.2.2 *Pseudopolymorphism, solvates, and hydrates*

The literature on polymorphism and related phenomena has spawned a number of additional definitions and terms that potentially lead to confusion rather than clarification. One of these is *pseudopolymorphism*, and it is of interest that authors (McCrone 1965; Haleblian and McCrone 1969; Dunitz 1991; Threfall 1995) who have given serious thought to the definition of polymorphism and its ramifications almost unanimously argued, strenuously in a number of cases, against the use of the term pseudopolymorphism. Even in arguing against its use, for the sake of completeness and to define some phenomena which are not to be considered as polymorphic behaviour, it is impossible to ignore the term and how it has been used and perhaps misused in the past—*caveat emptor*!

McCrone (1965) and Haleblian and McCrone (1969) pointed out that pseudopolymorphism has been used to describe a number of phenomena that are related to polymorphism: among them are desolvation, second-order transitions (some of which may be considered examples of polymorphism), dynamic isomerism, mesomorphism, grain growth, boundary migration, recrystallization in the solid state, and lattice strain effects.

Probably the most common use, particularly prevalent in the pharmaceutical industry (David and Giron 1994; Henck *et al.* 1997), involves the confusion between solvates (including hydrates) and crystalline materials that do not contain solvent (anhydrates in the case of water). As Byrn (1982) and Byrn *et al.* 1999) have pointed out, crystal solvates exhibit a wide range of behaviour. At one extreme, the solvent is tightly bound, and vigorous conditions are required for the desolvation process. In many of these cases the solvent is an integral part of the original crystal structure, and its elimination leads to the collapse of the structure and the generation of a new structure. At the other extreme are solvates in which the solvent is very loosely bound, and desolvation does not lead to the collapse of the original structure (Van der Sluis and Kroon 1989). Anything between the two extremes is also possible. Threlfall (1995) has noted that since a solvate and an unsolvated crystalline form are constitutionally distinct, they cannot be defined as polymorphs by any definition.

McCrone (1965) and Haleblian and McCrone (1969) have proposed a simple experimental test to distinguish between a desolvation phenomenon and a true polymorphic transformation, using the microscope hot stage (Section 4.2). During heating of a crystalline sample, both a true polymorphic phase transition and a desolvation process will often lead to loss of transmission and/or crystal darkening (due to formation of polycrystallites of the product phase). However, if the original sample is placed in a drop of solvent which is immiscible with the (suspected) solvent of crystallization, then upon heating the liberated solvent will form an easily observable bubble in the surrounding droplet. No such observation can be made for a true polymorphic transformation. A more sophisticated technique, involving very much the same principle, is the measurement of the TGA, which involves following the change in mass (in this case a loss in mass due to loss in solvent) corresponding to the heating process (Gruno *et al.* 1993; Perrenot and Widman 1994) (see Section 4.3).

In spite of the objections to the use of pseudopolymorphism to describe solvated structures of a material, the term seems to have gained quite a general acceptance in this context, especially in the pharmaceutical industry, both in the characterization (Kitamura *et al.* 1994; Nguyen *et al.* 1994; Kiaoka and Ohya 1995; Brittain *et al.* 1995; Kitaoka *et al.* 1995; Caira *et al.* 1996; Gao 1996; Kalinkova and Hristov 1996; Kritl *et al.* 1996; de Ilarduya *et al.* 1997; Ito *et al.* 1997; deMatas *et al.* 1998) and production/processing aspects (Adyeeye *et al.* 1995; Hendrickson *et al.* 1995; Joachim *et al.* 1995).

In light of the variety of behaviour exhibited by solvates, Byrn (1982) has suggested a classification scheme for crystal solvates based on that behaviour, rather than on stability. He proposed that the solvates for which the solvent can be removed from the crystal and added back to the crystal reversibly 'without greatly changing the X-ray powder diffraction pattern' (Section 4.4) would be considered *pseudopolymorphic solvates*. Those which undergo a change in structure, as evidenced by a different powder diffraction pattern, would be described as polymorphic solvates. The appellation does not seem to have been adopted by many other workers.

McCrone (1965) also noted that second-order phase transitions have been termed as pseudopolymorphic. Such transitions are difficult to detect by optical methods, because of the small structural changes that occur; hence, the origin of the prefix *pseudo* sometimes used to describe them. However, the birefringence of the crystals changes during such phase changes (see Section 4.2), so the use of crossed polarizers makes the phase change readily detectable.

A third phenomenon which has been described as pseudopolymorphism is dynamic isomerism (McCrone 1965). This takes us back to the problems of defining polymorphism in general, where the questions of degree and time are raised. Dynamic isomers (including tautomers as well as geometric isomers) are generally considered to be chemically different. However, it is not always simple to make a distinction between geometric isomers and conformationally different species. Dynamic isomers exist in both the solid and the molten state, and are in equilibrium over a wide temperature range. Over that range, both isomers are stable in varying amounts depending on the temperature, and in solution, on the solvent. Equilibrium between two polymorphs, on the other hand, can occur in the solid state, but upon melting the difference between the two polymorphs disappears. At any particular temperature only one polymorph is the thermodynamically stable one, except at a transition point where two polymorphs are in equilibrium (Section 2.2.2).

(E)   (Z)

1-I

In principle, the distinction appears rather straightforward. However, a practical example will serve to demonstrate the difficulty. Matthews *et al.* (1991) described the crystal structures of three crystalline forms of 4-methyl-N-(4-nitro-α-phenylbenzylidene)-aniline **1-I**. In solution the material exists as a mixture of rapidly interconverting stereoisomers with Z and E configurations, hence dynamic isomers. In the solid state it is trimorphic. The so-called A crystal form has three molecules in the asymmetric unit, all exhibiting the Z configuration. The B form, which can be crystallized simultaneously with the A form at 0 °C from ethanol or hexane-ether, has two molecules in the unit cell, both exhibiting the E configuration. A third C form, obtained at room temperature from ethanol, also exhibits the E configuration. At ambient temperature the latter two forms are converted to A, with the appropriate molecular configurational change from E to Z.

While this system falls somewhere on the fuzzy line between polymorphism and dynamic isomerism we agree with McCrone (1965) and Threlfall (1995) that this phenomenon should not be described as pseudopolymorphism.

McCrone (1965) attempted to summarize the distinction using a number of important criteria, and again suggested some rather simple thermomicroscopic tests to determine it. They are worth noting here, since systems of this type have received little experimental attention, and the example cited demonstrates the problems well.

McCrone (1965) notes that polymorphs, existing only in the solid, can convert at least in one direction without going through the melt. On the other hand Curtin and Engelman (1972) observed that the equilibrium in melt or in solution between the two configurational isomers may be shifted by crystallization or by chemical reaction to form a derivative of one of the isomers. In solution, the two isomers will have different solubilities, in the same way that different polymorphs can have different solubilities (Sections 3.2 and 7.3.1). The solubility curves may cross, and with a change in temperature the solution can become saturated with one form. This is apparently what happens in the case of **1-I**, as the C form is obtained from the room temperature crystallization, while at lower temperatures, a mixture of A and B is obtained. Dynamic isomers exist in both the melt and the solid state. Each isomer can exist in polymorphic forms, which is true for forms B and C. Details on the experimental techniques and observations are given in Chapter 4.

The thermomicroscopic differentiation between two phases that are known to be related either by polymorphism or dynamic isomerism is elegantly straightforward. The two phases should be melted side by side between a microscope slide and a cover slip, and then allowed to crystallize. Two possibilities exist for the subsequent crystallization events. In the case of polymorphism, the crystal fronts from the two melts will grow at a constant velocity until they come into contact, at which point one phase will grow through the other, due to a solid–solid transformation to the stable phase at that temperature. In the case of dynamic isomerism, the two crystal fronts would slow down as they approach each other, and in the so-called 'zone of mixing' (McCrone 1965) a eutectic could appear.

Another suggestion for making the distinction between polymorphs and dynamic isomers is to melt each sample by the equilibrium melting procedure (McCrone 1957), and observe the melt as a function of time. For a polymorphic system, the melting point will not change unless a solid–solid transformation takes place. Such transformations are usually sudden, and the resulting melting point will not change. For the case of dynamic isomers, the melting point of each will decrease gradually with the attainment of equilibrium. The final melting point should be the same for each, since the same equilibrium composition will be attained for both. As the melting point is followed through the eutectic composition, one of the isomers should show an apparent phase transformation. In another test suggested by McCrone two crystals of the same compound suspected of being polymorphs are placed side by side on a microscope slide in a mutually suitable solvent. If they are polymorphs of different thermodynamic stability the more stable one will grow at the expense of the less stable one.

McCrone (1957, 1965) has also given detailed descriptions of the microscopic examinations and phenomena that can be used to distinguish polymorphism from other phenomena that sometimes have been mistakenly labelled as pseudopolymorphism: mesomorphism (i.e. liquid crystals), grain growth (boundary migration and recrystallization), and lattice strain.

1.2.3   *Conventions for naming polymorphs*

Part of the difficulty encountered in searching and interpreting the literature on polymorphic behaviour of materials is due to the inconsistent labelling of polymorphs. In many cases, the inconsistency arises from lack of an accepted standard notation. However, often, and perhaps more important, it is due to the lack of various authors' awareness of previous work or lack of attempts to reconcile their own work with earlier studies (see, for instance, Bar and Bernstein 1985). While many polymorphic minerals and inorganic compounds actually have different names (e.g. calcite, aragonite and vaterite for calcium carbonate or rutile, brookite, and anatase for titanium dioxide) this has not been the practice for molecular crystals, which have been labelled with Arabic (1, 2, 3, . . .) or Roman (I, II, III, . . .) numerals, lower or upper case Latin (a, b, c, . . . or A, B, C, . . .) or lower case Greek (α, β, γ, . . .) letters, or by names descriptive of properties (red form, low-temperature polymorph, metastable modification, etc.).

As Threlfall (1995) and Whitaker (1995) have commented, arbitrary systems for naming polymorphs should be discouraged to avoid confusion surrounding the number and identity of polymorphs for any compound. Relative stability and/or order of melting point, as well as a specification of the monotropic or enantiotropic nature of the polymorphic form (see Section 2.2.4) have also been suggested as a basis for labelling (Herbstein 2001), but these do not allow for the discovery of forms with intermediate values, in addition to the fact that small differences in stability or melting point might lead to different order and different labelling by different workers. McCrone (1965) proposed using Roman numerals for the polymorphs in the order of their discovery, with the numeral I specifying the most stable form at room temperature. By Ostwald's Rule (Ostwald 1897) (Section 2.3) the order of discovery should in general follow the order of stability the least stable appearing first. McCrone also supported the suggestion by the Koflers (Kofler and Kofler, 1954) that the Roman numeral be followed by the melting point in parentheses. In fact, the successors of the Koflers at the Innsbruck school have very much followed this practice (Kuhnert-Brandstätter 1971), although in general it has not been adopted by others. The use of melting points is complicated by the fact that while this datum has a clear thermodynamic definition, a number of techniques are employed to determine the melting point (or melting point range, in many cases) so that real or apparent inconsistencies may arise from such a designation (see Sections 4.2 and 4.3).

In view of the body of literature already extant and the questions surrounding the definition of a polymorph it does not appear to be practical to define hard and fast rules for labelling polymorphs. The Kofler method has clear advantages, since the melting point designation may eliminate some questions of identity; hence its use should be encouraged. For those studying (and naming) polymorphic systems it is important to be fully aware of previous work, to try to identify the correspondence between their own polymorphic discoveries and those of earlier workers, and to avoid flippancy in the use of nomenclature in the naming of truly new polymorphs.

1.3 Is this material polymorphic?

1.3.1 Occurrence of polymorphism

Perhaps the most well-known statement about the occurrence of polymorphism is that of McCrone (1965): 'It is at least this author's opinion that every compound has different polymorphic forms and that, in general, the number of forms known for a given compound is proportional to the time and money spent in research on that compound.' As a corollary to this rather sweeping, even provocative, statement, McCrone noted that 'all the common compounds (and elements) show polymorphism', and he cited many common organic and inorganic examples.

These echo similar statements by Findlay (1951) p. 35, '[polymorphism] is now recognized as a very frequent occurrence indeed', Buerger and Bloom (1937), 'polymorphism is an inherent property of the solid state and that it fails to appear only under special conditions', and Sirota (1982), '[polymorphism] is now believed to be characteristic of all substances, its actual non-occurrence arising from the fact that a polymorphic transition lies above the melting point of the substance or in the area of yet unattainable values of external equilibrium factors or other conditions providing for the transition.'

Such statements tend to give the impression that polymorphism is the rule rather than the exception. The body of literature in fact indicates that caution should be exercised in making them. It appears to be true that instances of polymorphism are not uncommon in those industries where the preparation and characterization of solid materials are integral aspects of the development and manufacturing of products (i.e. those on which a great deal of time and money is spent): silica, iron, calcium silicate, sulphur, soap, pharmaceutical products, dyes, and explosives. Such materials, unlike the vast majority of compounds that are isolated, are prepared not just once, but repeatedly, under conditions that may vary slightly (even unintentionally) from time to time. Similarly, in the attempt to grow crystals of biomolecular compounds, much time and effort is invested in attempts to crystallize proteins under carefully controlled and slightly varying conditions, and polymorphism is frequently observed (Bernstein *et al.* 1977; McPherson 1982). Even with the growing awareness and economic importance of polymorphism, most documented cases have been discovered by serendipity rather than through systematic searches. Some very common materials, such as sucrose and naphthalene, which certainly have been crystallized innumerable times, have not been reported to be polymorphic. The *possibility* of polymorphism may exist for any particular compound, but the conditions required to prepare as yet unknown polymorphs are by no means obvious. There are as yet no comprehensive systematic methods for feasibly determining those conditions. Moreover, we are almost totally ignorant about the properties to be expected from any new polymorphs that might be obtained.

With the growing awareness among chemists of the phenomenon of polymorphism its actual occurrence in any particular system may not be as great a surprise as a generation or two ago. The predicted existence of any particular polymorphic structure for a single compound, the conditions and methods required to obtain it, and the properties it will exhibit are still problems that will challenge researchers for many years to come.[2]

1.3.2   *Literature sources of polymorphic compounds*

As noted in the previous section, the phenomenon of polymorphism is not new to chemistry. Nineteenth century chemists were very much aware of the properties of solids, and in the decades preceding the development of spectroscopic and X-ray crystallographic methods, the characterization of solids was a crucial aspect of the identification of materials. Chemists grew crystals carefully in order to obtain characteristic morphologies and then determined physical properties such as colour, interfacial angle, indices of refraction, melting point, and even taste (e.g. Schorlemmer 1874; Senechal 1990; Kahr and McBride 1992). Being critically observant was essential, for there was little other information to rely on.

A great deal of information on crystalline properties, including polymorphism was summarized in the five-volume compendium covering over 10 000 compounds by Groth, published between 1906 and 1919. The first two of these volumes (Groth 1906*b*, 1908) deal with elements and inorganic compounds, while the last three (Groth 1910, 1917, 1919) are concerned with organic materials. The genesis of this opus is vividly described by Ewald (1962):

'Groth's most stupendous work was the *Chemische Kristallographie*, five volumes which appeared between 1906 and 1919, comprising *in toto* 4208 pages and 3342 drawings and diagrams of crystals. The manuscript was written entirely by Groth in his fine hand and corrected over and over again by him until there was hardly a white spot left on the manuscript and again on the galley proofs. Oh for the admirable compositors in the Leipzig printing centres in the days before the general use of typewriters! The volumes contain a review of all crystallographic measurements . . . Each section is preceded by a survey of the crystal–chemical relations and includes many hints of gaps which should be filled by further work. In many instances Groth doubted the correctness of the work reported in the literature, and wherever possible, he got his pupils, assistants, or visiting colleagues to prepare the same substances again, and to recrystallize and remeasure them . . . Altogether measurements on between 9000 and 10 000

---

[2] While perhaps anecdotal, the following appears to be a good measure of the state of our knowledge about the 'pervasiveness' of polymorphism. In assigning a research project to a graduate student, a research advisor assumes a certain risk that the project will not succeed. One could imagine as a perfectly reasonable project the assignment to prepare and characterize the polymorphic forms of a *single* compound of interest, which is, of course, the practical manifestation of all four quotations at the start of this section. Unexpected results can constitute the basis for a Ph.D. thesis, but the absence of results, that is the inability to obtain *any* polymorphs, would constitute a total failure of project. This author has yet to encounter an academic research advisor who would be prepared to take the responsibility of assigning such a research project to a Ph.D. student. That is, in spite of the hyperbole of McCrone's statement and the notoriety it has received, and the increasing importance of polymorphism in the market place, there is not sufficient confidence in its veracity to risk the career of a student on any single particular compound.

are critically discussed in *Chemische Kristallographie*, an astounding feat considering the small number of the team . . .'

The work thus contains a thorough, checked survey of the physical properties of many of the crystals that had been studied up to its publication. Typical pages of the 'crystal–chemical relations' for dimorphic diphenyl malonic anhydride are shown in Fig. 1.1, in which the methods for obtaining both structures are described. A few pages on appear the entries for the description of crystal habit, melting point, solvent, appropriate reference(s), interfacial angles, and indices of refraction, if reported in the literature. Many of the substances had been reported to be polymorphic, and Groth recorded those facts, along with methods for preparing the polymorphs and the original literature references. It is a remarkable work, and one which should be consulted to check for the existence of polymorphism in a specific material, as well as for the source of physical phenomena, once observed, but since forgotten.

A second rich collection of references on the polymorphic behaviour of organic materials is the compilation by Deffet (1942). This contains information and references to primary sources on 1188 substances that exhibit polymorphism at atmospheric pressure and another 32 that exhibit polymorphic behaviour at elevated pressures. A typical entry contains the number of reported polymorphic forms, their melting points, temperature(s) of transition, crystal system, some physical properties, and literature references, of which there are nearly 1000. Substances are organized by empirical formula with an index organized by compound name (in French).

A third compilation intended to be devoted to polymorphic materials is that of Kuhnert-Brandstätter (1971). The body of this book is an identification table for hot stage studies of pharmaceutical materials (see Sections 4.2 and 7.2), in which materials are arranged by increasing melting point, with eutectic data for mixtures with azobenzene and benzil. There is considerable descriptive detail on the melting behaviour and identification and description of polymorphic forms, albeit only microscopic determinations, for approximately 1000 pharmaceutically important

das **Diphenylmaleïnsäureanhydrid** $= C_6H_5 . C{=}{=}{=}C . C_6H_5$ und
$$CO . O . CO$$

in der Tat zeigen nun beide Körper eine ähnliche Verwandtschaft in krystallographischer Beziehung wie jene, indem sie die gleiche Symmetrie und sehr nahe übereinstimmende Werte zweier Axen (a und b) besitzen; die ungesättigte Verbindung existiert aber außerdem noch in einer metastabilen Modification, deren monokline Krystalle sich neben der stabilen in wässeriger Acetonlösung bilden, aber sich sehr bald umwandeln, wenn sie mit Krystallen der stabilen Form in Berührung sind; durch Erwärmung kann die Umwandlung bei jeder Temperatur bewirkt werden, niemals die umgekehrte (Monotropie); die metastabile Modification entsteht außerdem durch Unterkühlen der Schmelze, wenn keine Spur der stabilen vorhanden ist.

Fig. 1a

Diphenylmaleïnsäureanhydrid $= C_6H_5 \cdot \overset{\displaystyle C}{\underset{\displaystyle CO}{\big\|}}\!\!\!=\!\!\!\overset{\displaystyle C}{\underset{\displaystyle CO}{\big\|}} \cdot C_6H_5 .$

$$C_6H_5 \cdot C = C \cdot C_6H_5 .$$
$$CO \, . \, O \, . \, CO$$

Stabile Modification.

Schmelzpunkt 155°.

Spec. Gew. 1,340 Drugman[44]).

Rhombisch bipyramidal.

$$a : b : c = 0,5176 : 1 : 0,7024 \quad \text{Drugman[44]).}$$

Aus Aceton entsteht die Combination (Fig. 2575): $m\{110\}$, $o\{111\}$, $q\{011\}$, ebenso aus Benzol, Chloroform, Äther und Alkohol, aus letzterem nach der c-Axe dünn nadelförmig. Einmal wurden aus Aceton kleine oktaëderähnliche Krystalle erhalten, die nur $m\{110\}$ und $q\{011\}$ zeigten; aus etwas harzhaltiger Lösung wurden Combinationen mit untergeordneten Flächen von $x\{112\}$, $y\{122\}$, $c\{001\}$ und $b\{010\}$ beobachtet. Die Krystalle aus Toluol zeigen die Formen (Fig. 2576): $m\{110\}$, $k\{021\}$, $b\{010\}$, untergeordnet: $q\{011\}$, $c\{001\}$, seltener $o\{111\}$; die hier vorhandene Verlängerung nach der a-Axe tritt noch mehr hervor an den Krystallen aus Xylol, welche dieselben Formen, aber mit besser ausgebildetem $o\{111\}$, zeigen. Eine solche nach der a-Axe prismatische Combination mit untergeordnetem $a\{100\}$ hatte früher bereits Jenssen[43]) (l. c. 64) an den von Anschütz und Bendix aus Äther erhaltenen Krystallen beobachtet, ihr aber eine andere Aufstellung gegeben.

Fig. 2575. Fig. 2576.

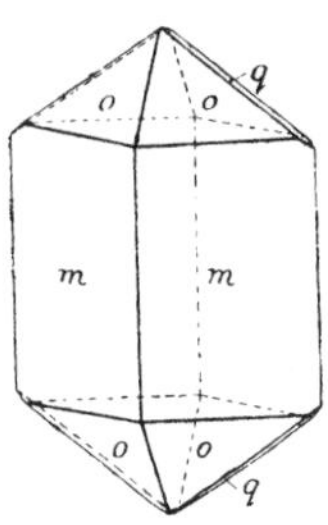

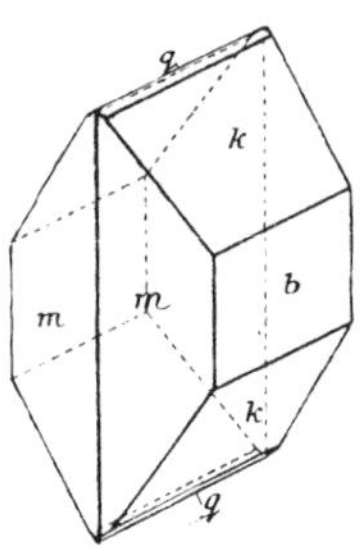

		Berechnet:		Beobachtet: Drugman:	Jenssen:
$m:m=$	$(110):(1\bar{1}0)=$	—		*54° 44'	54° 42'
$q:q=$	$(011):(0\bar{1}1)=$	—		*70 10	—
$o:o=$	$(111):(1\bar{1}1)=$	45°	14'	45 13	—
$o:q=$	$(111):(011)=$	48	0	48 1	—
$o:m=$	$(111):(110)=$	33	12	33 15	—
$o:m=$	$(111):(1\bar{1}0)=$	61	6½	61 7	—
$q:m=$	$(011):(110)=$	74	41	74 42	—
$k:b=$	$(021):(010)=$	35	27	35 5 ca.	34 43
$k:m=$	$(021):(110)=$	68	0½	67 55 »	67 49
$x:o=$	$(112):(111)=$	19	25	19 13	—
$x:q=$	$(112):(011)=$	37	23	37 24	—
$y:q=$	$(122):(011)=$	29	2½	29 2½	—

Keine deutliche Spaltbarkeit.

Doppelbrechung positiv; Axenebene $a\{100\}$, 1. Mittellinie Axe c; Axenwinkel klein.

$\alpha = 1,505$ $\;Li,$ $\quad 1,511$ $Na,$ $\quad 1,517$ $\quad Tl$ (alle optischen Angaben von
$\beta = 1,505$ ca. » $\quad 1,5115$ » $\quad 1,518$ ca. » $\qquad$ Drugman).
$\gamma = 1,811$ $\quad$ » $\quad 1,836$ » $\quad 1,865$ »

44) Drugman, Zeitschr. f. Krystall. 1912, **50**, 576.

Fig. 1b

Metastabile Modification.

Schmelzpunkt 146°.

Spec. Gew. 1,345 Drugman[44]).

Monoklin prismatisch.

$a : b : c = 2,5615 : 1 : 2,3275;$ $\beta = 101°33'$ Drugman[44]).

Diese Modification erhielt Drugman neben der stabilen aus wässeriger Acetonlösung mit den Formen: $c\{001\}$ (oft sehr stark vorherrschend), $a\{100\}$, $m\{110\}$, $o\{111\}$; aus Xylol bilden sich kleine Krystalle der gleichen Form mit $n\{210\}$ (Fig. 2577). Einmal wurde ein Zwilling nach $a\{100\}$ beobachtet.

Fig. 2577.

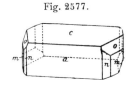

	Berechnet:	Beobachtet:
$m : a = (110):(100) =$	—	*68° 15′
$a : c = (100):(001) =$	—	*78 27
$o : c = (111):(001) =$	—	*64 13
$m : c = (110):(001) =$	85° 45′	85 45
$n : c = (210):(001) =$	82 50	—
$n : a = (210):(100) =$	51 25	51 21
$o : n = (111):(210) =$	24 28	24 17
$o : o_{,} = (111):(\bar{1}\bar{1}) =$	65 58	65 57
$o : m' = (111):(\bar{1}10) =$	51 5	51 7
$o_{,}: n = (\bar{1}\bar{1}):(210) =$	66 21	66 21

Spaltbarkeit nach $a\{100\}$ und $c\{001\}$ vollkommen.

Ätzfiguren auf $c\{001\}$ nach $b\{010\}$ symmetrisch.

Ebene der optischen Axen $b\{010\}$, durch $a\{100\}$ und $c\{001\}$ je ein Axenbild, durch eine Schlifffläche $\parallel \{101\}$ beide sichtbar; starke Dispersion.

Fig. 1c

Fig. 1.1   A typical entry from Groth's *Chemische Kristallographie*. (a) Textual description of the dimorphic diphenyl maleic anhydride. (b) Physical data for the stable modification melting at 155 °C. (c) Physical data for the metastable modification melting at 146 °C.

compounds. There is no formula index, and the subject index contains only a partial listing of the compounds included. Nevertheless, the book contains some very useful information about the existence of polymorphism and the characterization of its behaviour in many of these commercially important materials. In this context, it is perhaps noteworthy that the 1997 edition of the Merck Index describes polymorphic behaviour for only 140 of over 10 000 entries, many of which appear in the Kuhnert-Brandstätter compilation.

There are a number of additional sources for consultation on information on polymorphism of particular compounds. From 1948 to 1961, McCrone edited a regular column in *Analytical Chemistry* entitled 'Crystallographic Data', in which were published the details on crystal growth, physical properties, and polymorphic behaviour of approximately 200 compounds. The series was undertaken at the time 'because optical crystallography is neglected as an analytical tool because too few compounds have been described', and with the desire to '...initiate a process which [would] enable a group of crystallographers to complete the tabulation of crystal data for most of the common everyday compounds.' (Grabar and McCrone 1950). About 140 of these were organic compounds, and 25 per cent of these exhibited polymorphism. Even in the cases where there is no evidence of polymorphism, these reports contain detailed descriptions of conditions for growth of crystals with well-defined faces, and the characterization of crystal habit very much in the tradition of Groth. It is information that future investigators will be able to utilize for a variety of studies. The need for recording the detailed description of crystal growth, crystal habit, and crystal properties was recently echoed in an appeal by Dunitz (1995) to authors of crystallographic structure analyses:

... please give the color (easy to observe) and melting point of crystals studied (easy to measure); if possible, also the heat of fusion and of any observed phase transitions (only slightly more difficult to measure): report also any 'unusual' behavior, any observed change of physical properties or of the diffraction pattern.

The short reports solicited and edited by McCrone are models of the kind of data which should be required and included in descriptions of crystals and crystal structure reports, even if only in deposited form (Section 1.3.3).

Some additional literature sources should also be consulted to check for earlier reports of polymorphism. The Barker Index (Porter and Spiller 1951, 1956; Porter and Codd 1963) made use of the characteristic interfacial angles for purposes of identification of crystals. The Index is based on Groth's earlier compilation (which is organized by chemical composition) and is arranged by increasing interfacial angle within a crystal system. There are some additional compounds, with totals of 2991 in tetragonal, trigonal, and orthorhombic space groups (Volume I) (Porter and Spiller 1951), 3572 in monoclinic (Volume II) (Porter and Spiller 1956), and 871 in triclinic (Volume III) (Porter and Codd 1963), space groups. However, the method of arrangement means that polymorphs of a compound crystallizing, say, in monoclinic and orthorhombic space groups, requires that the compound be checked in all three volumes.

Another approach was taken by Winchell (1943, 1987), who prepared a compilation of 'all organic compounds whose optical properties are sufficiently well known to permit identification by optical methods'. The compilation is arranged in the same fashion as the fourth edition of *Beilstein's Handbuch der organischen Chemie* (Beilstein 1978), and at the time of its first publication was meant to include all organic compounds whose indices of refraction had been measured. Since indices of refraction differ among them, polymorphs could be easily recognized by different optical properties. The book does contain references to primary sources and drawings of crystals, as illustrated in a typical entry Fig. 1.2.

Another useful compilation of crystallographic data as a source of examples of polymorphic systems, NIST *Crystal Data* (NIST 2001; Mighell and Stalick 1983), which contains the principal crystallographic data on over 237 000 organic and organometallic entries. Each entry contains cell constants, space group, and other crystallographic information and bibliographic citations. In some cases the fact that a crystalline compound is one of a polymorphic system is specifically noted. In other cases the

p-Methylbenzophenone or phenyl p-tolyl ketone $[C_6H_5 \cdot CO \cdot C_6H_4(CH_3)]$ has two phases. The stable phase is monoclinic with $a:b:c = 1.012:1:0.412$, $\beta = 95°7'$. Crystals {010} tablets or equant with {110}, {210}, {100}, {011}, {001}, etc. Figs. 60, 61. No distinct cleavage. M.P. 60°. The optic plane is 010 for red to green and normal thereto for blue and violet. X $\wedge$ c = +37°. (-)2E = 49°11' Li, 35°15' Na, 6°55' Tl, 49°32' blue. The matastable phase is ditrigonal pyramidal with $c/a = 1.225$. Crystals show both trigonal prisms, {10$\bar{1}$0} and {1$\bar{1}$00}, etc. Fig. 62. M.P. 55°. Uniaxial negative with N_O = 1.7067 Li, 1.7170 Na, 1.7250 Tl; N_E = 1.5564 Li, 1.5629 Na, 1.5685 Tl; $N_O - N_E$ = 0.1541 Na.

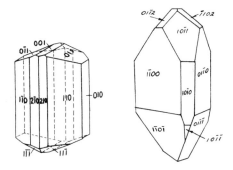

Figs. 60, 61. Phenyl-β-tolyl ketone.

Fig. 1.2 Typical entry from Winchell's *Optical Properties of Organic Crystals* for dimorphic p-methylbenzophenone (reproduced, with permission).

polymorphism may be recognized by the fact that a compound has more than one entry either in the formula index or the compound name index.

In addition to these compilations of crystal data in which instances of polymorphism may be recorded, a number of texts on the subject of the solid state properties of organic compounds contain many examples of polymorphism. Since these books are based in part, at least, on work by the authors not published elsewhere, they may be considered as primary literature sources. Particularly noteworthy in this regard are the books by Pfeiffer (1922), Kofler and Kofler (1954), and McCrone (1957).

The usual search strategies for information on the preparation and properties, such as *Chemical Abstracts* and Beilstein can also be useful for determining if a particular compound has been reported to be polymorphic. However, reference to the primary sources on the preparation and the characterization of the compound may reveal unusual behaviour (e.g. melting points or colours which differed from one crystallization to the next) which testifies to the possible existence of polymorphic forms, behaviour that is not specifically noted in the abstracted material.

1.3.3 *Polymorphic compounds in the Cambridge Structural Database*

The Cambridge Structural Database (CSD) is the repository for the results obtained from the X-ray crystal structure analysis of organic and organometallic compounds

(Allen *et al.* 1991; Allen and Kennard 1993; Kennard 1993; Allen *et al.* 1994). The October 2000 release of the data base contains over 240 000 entries, and as of this date approximately 20 000 structures are added annually. It is now also serving as a depository for crystallographic data that may not be published elsewhere. In the past three decades the database has increasingly influenced the way structural chemists carry out their trade. An enormous amount of geometric and structural information is available in a very short time for searches, correlations, model compounds, packing arrangements, reaction coordinates, hydrogen-bonding patterns, and a variety of studies. The rapid increase in the data availability which has been accompanied by increasingly sophisticated software has opened opportunities that could not have been imagined even a quarter of a century ago. Formerly accessible only on mainframes or work stations it has recently become available in a PC version on a CD-ROM.

As the repository for all organic and organometallic crystal structures, the CSD naturally contains entries for polymorphic materials. As of the October 2000 release, approximately 5000–6000 compounds may possibly be classified as polymorphic (for details see Section 7.2.1). However, some words of caution are necessary in the use of these data. Each entry in the CSD contains 1D, 2D and 3D information. The 2D information is used to generate the structural formula and chemical connectivity, which clearly will be the same for polymorphs. The 3D information contains the results of the X-ray structure determination: cell constants, space group, atomic coordinates, and atomic attributes needed to generate the three-dimensional molecular and crystal structures. The 1D data contain bibliographical and chemical information (name and empirical formula), including qualifying phrase(s) such as 'neutron study', 'absolute configuration', etc. It is here that the CSD notes that the material is polymorphic with a qualifying phrase such as 'red phase', 'metastable polymorph', 'Form II' *if the author of the primary publication noted this feature or if the abstractors recognized that the structure was one of a polymorphic system.* In many cases note is taken of the fact that this is some special crystal form only when a second (or third, etc.) structure of a polymorphic series is being reported. The first report may not contain such a notation, since the author may not have been aware that the material is polymorphic. (This may be the case for subsequent structure determinations as well. In the early days of the CSD some polymorphic structures were archived with different REFCODEs—the unique identifier for each chemical species. The more sophisticated archiving software used now prevents such duplication and has eliminated many of the older 'orphans', but some may still exist.) Once one member of a polymorphic set of structures has been identified care should be taken to extract all entries of that compound, including those not identified by appropriate descriptors. In short, the absence of a descriptor indicating that a material belongs to a polymorphic system is not a foolproof indication that the material is not polymorphic. Other literature sources should be consulted to make that determination.

An example of the caution which must be exercised in performing such searches and the numbers obtained was given by Gavezzotti and Filippini (1995). The search was defined for organic compounds (containing only C, H, N, O, F, Cl, or S) and for which the crystal structures of more than one polymorphic form had been determined.

A total of 163 'clusters' were obtained, where a cluster is a group of polymorphic crystal structures of the same compound. Of the 163 clusters, 147 contained two structures, 13 had three, and three had four structures. The authors note that these numbers are 'first evidence of the high frequency of polymorphism in organic crystals', although the number of clusters is a relatively small percentage of the entries in the database. The number of these clusters is probably more a measure of certain authors' interest in the particular polymorphic system in question. A more realistic measure (although certainly not precise because of the caveats mentioned above) of the frequency of polymorphism in these compounds would be the fraction of compounds in the database known to be polymorphic, whether multiple structures have been done or not.

1.3.4 *Powder Diffraction File*

The second crystallographic database which can serve as a source of examples of polymorphic structures is the Powder Diffraction File (PDF) (ICDD 2001; Jenkins and Snyder 1996). This is the depository for over 130 000 powder diffraction patterns of solids (2000 release), roughly divided into organic, inorganic, and metallic compounds, of which organics are about 25 per cent. Bibliographic searches may be run on compound name or formula, and again, the existence of polymorphism for a particular compound may be recognized by the presence of more than one entry for a compound. An example of identifying polymorphism from the bibliographic entries (formula index and compound name index) of the PDF is shown in Fig. 1.3.

1.3.5 *Patent literature*

As polymorphism has become an increasingly important factor in the commercial aspects of many solid materials, the number of patents relating to the discovery and use of particular polymorphic forms has increased. This is particularly important for pharmaceuticals, pigments and dyes, and explosive materials, which are discussed in Chapters 7–9. Some examples of the role of polymorphism in patent litigation are described in detail in Chapter 10. The patent literature is readily searchable using terms such as 'crystal form', 'polymorph' etc., and since polymorphic behaviour often forms the basis of a patent (as opposed to many journal publications, where it may be peripheral to the main point of the paper) instances of polymorphism are relatively straightforward to locate.

1.3.6 *Polymorphism of elements and inorganic compounds*

Berzelius (1844) introduced the term 'allotropy' as the phenomenon of polymorphism in elements. There has been some debate about the necessity of a special term to designate the polymorphism of elements, as opposed to compounds (Sharma 1987; Reinke *et al.* 1993), but the term is still introduced in first year chemistry texts, so it has become part of the chemical language. Sharma (1987) has given some examples

	Sulphamethylthiazole	$C_{10}H_{11}N_3O_2S_2$	7.80_x	4.34_x	6.80_6	8–521
i	Sulphamidochrisoidine	$C_{12}H_{13}N_5O_2S$	3.86_x	5.13_9	3.27_8	39–1610
o	Sulphamidochrysoidine	$C_{12}H_{13}N_5O_2S$	4.51_x	3.96_8	13.9_5	39–1611
*	β-Sulphanilamide	$C_5H_8N_2O_2S$	6.12_x	3.90_8	4.91_6	41–1909
*	Sulphanilamide	$C_6H_8N_2O_2S$	4.49_g	3.78_x	6.57_g	38–1710
*	Sulphanilamide	$C_5H_8N_2O_2S$	4.47_g	3.70_g	7.82_g	38–1709
o	α-Sulphanilamide	$C_5H_8N_2O_2S$	4.23_x	3.36_x	3.57_7	30–1944
	2-Sulphanilamidopyrimidine Sodium	$C_{10}H_9N_4NaO_2S$	9.16_x	5.17_7	4.06_7	5–112
*	Sulphanilic Acid	$C_6H_7NO_3S$	4.91_x	6.96_5	3.48_2	30–1945
o	Sulphaphenazole	$C_{15}H_{14}N_4O_2S$	4.37_x	7.29_5	3.90_2	30–1946
	Sulphapyrazine	$C_{10}H_{10}N_4O_2S$	5.59_x	7.21_7	4.79_7	5–213
*	Sulphapyridine	$C_{11}H_{11}N_3O_2S$	5.48_x	3.57_5	4.01_5	37–1695
*	Sulphapyridine	$C_{11}H_{11}N_3O_2S$	4.77_x	4.13_5	3.81_8	37–1698
i	Sulphapyridine	$C_{11}H_{11}N_3O_2S$	3.81_x	4.76_7	6.49_5	37–1700
	Sulphasalazine	$C_{18}H_{14}N_4O_5S$	3.77_x	5.73_8	4.28_7	29–1928
	Sulphathiazole	$C_9H_9N_3O_2S_2$	5.81_x	4.12_x	4.02_x	5–206
	Sulphathiazole	$C_9H_9N_3O_2S_2$	5.77_x	4.03_x	4.33_x	29–1930
	Sulphathiazole	$C_9H_9N_3O_2S_2$	5.59_x	5.06_x	4.75_x	29–1931
	Sulphathiazole Sodium Hydrate	$C_9H_8N_3O_2S_2 \cdot 1.5H_2O$	6.85_x	4.50_x	3.77_x	8–684
	Sulphathiazole Sodium Hydrate	$C_9H_8N_3NaO_2S_2 \cdot 1.5H_2O$	6.80_x	12.3_3	3.96_8	8–802

Fig. 1.3 Example of the bibliographic entries in the PDF for substances listed by compound name. Each name is followed by the formula and the d-spacings of the three strongest diffraction lines, with the relative intensity as a subscript. The last column on the right is the card number in the PDF. Multiple entries with different principle lines are indications of polymorphic systems, for instance the three entries for sulphapyridine and sulphathiazole, the four entries for sulphanilamide and the two entries for sulphathiazole sodium hydrate, but additional bibliographic and crystallographic information should be obtained from the entries themselves.

of allotropism, and Sirota (1982) has noted that '54–55 elements' exhibit the property (Samsonov 1976; Smithells 1976). More complete descriptions can be found in the texts by Donohue (1974) and Wells (1989).

The inorganic equivalent of the CSD is the Inorganic Crystal Structure Database (ICSD) (FIZ 2001; Bergerhoff *et al.* 1983). This currently contains over 53 000 entries (August 2000) with two updates per year, and may be searched in a manner similar to that used for the CSD. There are currently efforts under way to unify the searching software for these two important data bases, a move which would considerably facilitate and widen their use. Another useful source is the inorganic section of the PDF (ICDD 2001; Jenkins and Snyder 1996). For older references, the first two volumes of Groth (1906, 1908) are particularly valuable.

1.3.7   *Polymorphism in macromolecular crystals*

Protein crystal structures are archived in the Protein Data Bank (PDB) (Bernstein *et al.* 1977; Berman *et al.* 2000). About 5 per cent of the approximately 14 000 (December 2000) entries (~12 500 proteins, peptides, and viruses, ~900 nucleic acids, ~600 protein/nucleic acid complexes, ~20 carbohydrates) contain the qualifier 'form' in the compound name/descriptor field, and most of those refer to polymorphic varieties. In biomolecular crystallography great efforts are expended varying crystallization conditions in the attempts to obtain single crystals suitable for structural investigations

(McPherson 1982, 1989, 1998). These myriad attempts and the variety of conditions have lead to the acquisition of many polymorphic forms, especially for those compounds on which a great deal of work has been done. For instance, the extensively studied lysozyme has entries in the PDB for triclinic, monoclinic, orthorhombic, trigonal, tetragonal, and hexagonal modifications; human haemoglobin has been studied in monoclinic, orthorhombic, and tetragonal modifications. The amount of effort expended in a typical protein crystal structure analysis means that the isolation of crystals and the determination of cell constant and space group is an accomplishment worthy of publication in and of itself. Thus, much of the information on polymorphism in macromolecular structures can be found in the primary literature (King *et al.* 1956, 1962; Cramer *et al.* 1974; McClure and Craven 1974; Kim *et al.* 1973; Falini *et al.* 1996). One secondary source, which should be of increasing importance as the number of proteins studied increases, is the Biological Macromolecule Crystallization Database and the NASA Archive for Protein Crystal Growth Data (Gilliland *et al.* 1994). In late 2000 this database contained nearly 3300 crystal entries from about 2300 biological macromolecules. McPherson (1982) summarized the crystallization procedures for 331 proteins. Of these, 23 (or about 7 per cent) were listed as being polymorphic. Another primary source is the citations of the McPherson book (1982); of the nearly 700 citations by early 1998, 20 were for polymorphic systems. For smaller proteins, at least some of the incidents of polymorphism have been included in the above-mentioned NIST Crystal Data Compilation.

1.4 Historical perspective

Following the historical development of a particular scientific concept or discipline helps to recall the way certain modes of thinking developed, were debated and accepted as new facts came to light, and perhaps were abandoned. Tracing that development serves as a reminder that the field is dynamic, with new techniques and new findings changing our ideas and the problems we are seeking to solve. As in any human activity, knowing where we have come from and where we are helps to define where we have to go, and that is certainly true for the field of polymorphism. An early historical account may be found in Hartley (1902) and a later ones in Verma and Krishna (1966) and Leonidov (2000).

Mitscherlich is generally credited with the first recognition of the phenomenon of polymorphism (e.g. Tutton 1911*a,b*). Early in his career in 1818 he discovered that crystals of certain phosphates and arsenates were very similar. He termed this phenomenon *isomorphism*, and pursued further investigations with Berzelius in Stockholm on the pairs of salts $NaH_2PO_4 \cdot H_2O/NaH_2AsO_4 \cdot H_2O$ and $Na_2HPO_4 \cdot H_2O/Na_2HAsO_4 \cdot H_2O$ and the corresponding ammonium and potassium salts. Among the measurements he carried out were the interfacial angles of the crystals, then a standard technique for characterizing solids (Romé de l'Isle 1783; Lima-de-Faria 1990). Mitscherlich (1822) found that the members of the first pair of compounds usually have different crystals, but that the phosphate sometimes crystallizes in the same form as the arsenate. Typical of so many other subsequent discoveries of polymorphism, this one also appears to have been serendipitous:

Whilst I was still seeking a difference in chemical composition [in the different crystals of the phosphate] I succeeded several times, in the recrystallization of the phosphate, in obtaining crystals having the same form as the acid arsenate. Since I knew definitely that there was no difference between the two salts I proceeded with the investigation of this phenomenon, and the whole solution of the acid phosphate crystallized several times in the form of the arsenate.

Hence it is established that one and the same body, composed of the same substances in the same proportions, can assume two different forms. This is easily understood from the atomic theory: different forms can result according as the position of the atoms with respect to one another is changed, but the number of different forms remains quite restricted.

Mitscherlich's mentor, Berzelius, considered the discoveries of isomorphism and *dimorphism*, as it was initially called, 'the most important made since the doctrine of chemical proportions, which depends on them of necessity for its further development'.

Mitscherlich followed this paper shortly thereafter with another one on the dimorphism of sulphur (1822–1823). Actually, others had earlier identified more than one crystal form for a number of materials. Klaproth (1798) had recognized that calcite and aragonite have the same chemical composition and Davey had recognized that diamond was a form of carbon (Encyclopedia Brittanica 1798). This prompted Thenard and Biot (1809) to reach nearly the same conclusion as Mitscherlich, in stating that

the same chemical elements combined in the same proportions can form compounds differing in their physical properties either because the molecules of these elements have the intrinsic faculty of combining in different ways or because they acquire this faculty through the temporary influence of a foreign agent which afterwards disappears without destroying itself (Webb and Anderson 1979).[3]

Monoclinic sulphur (in addition to the more common orthorhombic form) had also been recognized and documented by a number of other people (see, e.g. Partington 1952 which also contains many early references to polymorphism and polymorphic materials).

The microscope played a crucial role in research on polymorphism, and as this analytical tool became of wider and more sophisticated use, so polymorphism became the subject of increasing interest and study (Lima-de-Faria 1990). Frankenheim's early 1839 investigation of the polymorphism of potassium nitrate is one of the classic studies of that period. He demonstrated that phase changes could be brought about by solvent mediation and by physical perturbations of a crystal, such as scratching or physical contact with another polymorph. With a detailed study of the mercuric iodide system he also established many of the principles still recognized today regarding the

---

[3] The controversy that arose about the nature of these discoveries and who should get credit for them, prompted correspondence, among others, between Berzelius and the pioneering French crystallographer Haüy. Detailed accounts have been given by Amorós (1959, 1978).

nature of polymorphism. Some of these are as follows:

- polymorphs have different melting and boiling points (*sic!*) and their vapours have different densities.
- the transition from a low-temperature form (A) to a high-temperature form (B) is distinguished by a specific temperature of transition.
- the low-temperature form (A) cannot exist at a temperature above the transition point to form B, but B can exist below the transition point; below the transition point it is a metastable form.
- at temperatures below the transition point B will transform to A upon contact with A, the transition proceeding in all directions, but with differing velocities.
- in some cases, B can be converted without contact with A by mechanical shock or by scratching.
- heat is absorbed upon the transition from A to B.

As early as 1835, Frankenheim was particularly concerned with cohesive forces in different states of aggregation, and suggested that in the various solid states of a material the attractions which lead to the aggregation in different solids are different, and are characterized by different special symmetry relations.

The first *polarizing* microscope, an instrument that was destined to play such an important role in the development of chemical crystallography in general and polymorphism in particular, was invented by Amici in 1844 (see also Wrede 1841). It was also at about this point that Berzelius (1844), Mitscherlich's early mentor, suggested that the pyrite–marcasite polymorphism of FeS_2 was due to the polymorphism of the sulphur in the two solids, while the iron was the same in the two, although the concept of structure, *per se*, had not yet really crept into the lexicon of chemical crystallography. As Hartley (1902) pointed out, in spite of the investigation of many polymorphic modifications, the middle decades of the nineteenth century were not noted for any new generalizations in terms of the characterization and understanding of the phenomenon itself.

In the 1870s things started to change rapidly. Mallard (1876, 1879) had been concerned with geometrical crystallography and had considered the structural basis for polymorphism in an 1876 paper. He considered crystals as being built up of minute elementary crystallites that can pack in a number of ways giving rise to different crystal forms. The ideal form is that with the closest packing thereby being the most dense, and different forms have different packing which results in different physical properties such as optical properties and density. He attributed the differences in physical properties to differences in the arrangement of these elementary crystallites. In general, though, he still saw a great deal of similarity in the structures of two forms of the same substance:

It has been known for a long time that when the same substance displays two fundamentally incompatible forms, often belonging to two different chemical systems, these two forms are always only slightly different and the symmetry of the less symmetrical is very similar to that of the other.

As an early pioneer of chemical crystallography (particularly of organic compounds) Lehmann's Ph.D. thesis, much of which was published in the first issues of *Zeitschriff für Kristallographie* (founded by Groth) (1877a,b) already contained some new concepts for polymorphic systems (Lehman 1891). He characterized two different types of polymorphism. The first, which he termed *monotropic*, involves two forms in which one undergoes an irreversible phase change to the second form; the second form is termed *enantiotropic*, in which the two phases can undergo a reversible phase transition (see Chapter 2). An increase in temperature tends to lead to the transformation to the more stable form.[4] Lehman also showed that many organic compounds crystallize from the melt as monotropic forms, and that these tend to be the less-stable form with a lower melting point.[5]

Lehmann further reduced Mallard's 'structural crystallites' to be aggregates of 'physical molecules'. Then the structural crystallites could differ in the number or in the arrangement of the physical molecules of which they were composed, thereby constituting the difference between two polymorphs. These distinctions were then related to the transformation phenomena: an enantiotropic transformation was characterized by Lehmann as a reversible polymerization that is, with an increase in temperature, elementary particles of a large size were transformed into elementary particles of a smaller size. In a monotropic transition, according to Lehmann, there is no such relationship between temperature and the mode of rearrangement.

The problem of distinguishing between molecular isomerism and polymorphism (Section 1.2) arose in this period as well. For instance, in a manner similar to Berzelius' arguments about the pyrite–marcasite system, Geuther (1883) postulated that the calcite–aragonite polymorphism arose from the existence of two carbonic acids. Wyrouboff (1890) differed with this view, claiming that polymorphs differ only in their physical properties. Crystals with different molecular isomers would give different products upon reaction, whereas true polymorphs would give the same reaction products. Polymorphic products, according to Wyrouboff, are distinguishable only by their physical properties.[6] He also differed with Lehmann's classification of polymorphs based on monotropic and enantiotropic phase transformations, choosing

---

[4] It is remarkable how particular systems attain the status of 'classics'. Hartley (1902) noted α and β sulphur (transition temperature 95.6 °C), red and yellow mercuric iodide (transition temperature 126 °C) (Hostettler *et al.* 2001) and the four modifications of ammonium nitrate as examples of enantiotropic behaviour. These three systems are given as archetypical experiments in Chamot and Mason's book (1973) on chemical microscopy.

[5] It is of interest to note Tutton's optimistic assesment of Lehmann's definition of monotropism and enantiotropism, published just prior to the dawn of the age of structural crystallography:

It thus appears that any general acceptance of Lehmann's ideas will only tend to amplify and further explain the nature of polymorphism on the lines here laid down, the temperature conversion of one form into another being merely that at which either a different homogeneous packing is possible, or that at which the stereometric relations of the atoms in the molecule are so altered as to produce a new form of point-system without forming a new chemical compound. Tutton (1911a,b)

[6] On first glance this seems consistent with our definition above. However, the topochemical principles, first enunciated by Cohen and Schmidt (1964a); Cohen *et al.* (1964e) were actually developed from the fact that different polymorphs of a substance (*trans* cinnamic acid) undergo different photo*chemical* reactions, leading to different products Schmidt (1964) (Section 6.4).

a scheme based essentially on the physical manifestations of the phase changes. For most materials, labelled heteroaxial by Wyrouboff, the starting crystal loses homogeneity upon transformation, becoming optically clouded, and the transformation results in the breaking up of the crystal into many smaller crystallites. The heteroaxial designation results from the lack of any correspondence between axes of the initial and product phases. In the second class, labelled isoaxial, the phase transformation takes place without the crystal losing its optical transparency. If it does break up into smaller crystals, they remain parallel to each other and to the axes of the parent crystal.

Following the elaboration of many of the principles of thermodynamics in the latter three decades of the nineteenth century, a major development in polymorphism came with the work by Ostwald (1897) on the relative stability of different polymorphs, and the reason for the mere existence of less-stable forms. Among the findings were the fact that unstable polymorphic forms have a greater solubility than the more-stable forms in a particular solvent, and that monotropic forms have a lower melting point than enantiotropic forms. Ostwald related these findings to the phenomena of supersaturation and supercooling. The result is Ostwald's so-called 'Rule of Steps' or 'Law of Successive Reactions', although as Findlay (1951) has pointed out, the designation 'law' is not justified since many exceptions are known, but as a guideline or rule of thumb, it is still a useful concept. In Ostwald's words (1897), '. . . that on leaving any state, and passing into a more stable one, that which is selected is not the most stable one under the existing conditions, but the nearest' (i.e. that which can be reached with the minimum loss of free energy). Groth (1906a) provided an explanation for the phenomenon, which is discussed in detail in Chapters 2 and 3.This phenomena described by Ostwald is, in fact often (unknowingly) observed by synthetic chemists. The first synthesis of a new material with a melting point above room temperature may result in a metastable form, which eventually (either spontaneously or through an intentional recrystallization) will yield a more-stable form. The metastable form may not always be recognised or the stable form may not appear immediately—it may take years until the appropriate constellation of conditions exists. However, once seeds of the more stable form exist in a particular environment, it may be difficult to obtain the metastable form (Dunitz and Bernstein 1995) (see Section 3.5). An example of the stable form crystallizing out of the metastable one over a period of days is shown in Fig. 1.4.

Ostwald (1897) was aware of the fact that his 'rule' was tenuous, since it was not based on a very large set of observations. In addition, if the metastable region were to shrink to a vanishingly small value, then sufficient time would not be allowed for crystallization of Form I to appear, and the 'rule' would be invalidated. In fact, this does happen in many cases. Nevertheless, Ostwald's Rule has remained in the lexicon of crystal chemists, probably because it is generally observed that if a succession of polymorphic forms is obtained, those which appear later are generally more stable than those which appear earlier.

The turn of the twentieth century brought to play a convergence of many experimental techniques and theoretical developments in the investigation and understanding of polymorphism. Experimentally, hot stage microscopy (Lehmann 1891), dilatometry,

Fig. 1.4   Example of Ostwald's Rule of Successive Reactions. 2,4-dibromoacetanilide initially crystallizes from alcoholic solutions as small needle-shaped crystals, forming the voluminous mass in (1). Successive photos (2,3,4) of the same crystallization vessel, taken at two-day intervals show the transformation to the more stable chunky rhombic crystals (from Findlay 1951, with permission).

precise vapour pressure and solubility measurements, heat capacity and transition point determinations, all served to provide data by which models and theories could be tested. Theoretically, thermodynamic relations, in particular the Gibbs Phase Rule and the Clapeyron equation applied to the solid state, established the equilibrium relationships that exist among polymorphic forms. There appears to have been a real symbiotic relationship between theory and experiment here, as improved theories required more precise measurements and data, which in turn provided the impetus generated for theoretical refinements. Most of Frankenheim's conclusions were shown to be correct, and Lehmann's assertions about monotropism and enantiotropism also were validated (Tutton 1911a,b).

The study of organic crystals had gained considerable momentum during this period, particularly in Germany. This was a period of intense activity in organic chemistry. As noted earlier, in those days preceding the invention and application of today's armoury of spectroscopic methods for the characterization of physical properties of a solid material, chemists had to resort to other simpler techniques, such as the melting point, but often involving their own sensual perception for such characteristics as color, shape, smell, and even taste. Chemists employed somewhat simpler techniques. By the end of the nineteenth century, one of the principal tools of the organic chemist for characterizing and identifying materials was the polarizing microscope, which by 1879 had essentially developed to the state we recognize it today (Lima-de-Faria 1990). Many of the observations on organic crystals are summarized in the last three volumes of Groth's five-volume compendium (Groth 1910, 1917, 1919). By the time the last volume was published X-ray crystallography was well on the way to overtaking polarized light microscopy as a method for examining and characterizing solids. The view of polymorphism turned from phenomenological to structural.

Gustav Tamman from Göttingen was one who bridged this period, publishing a book on crystallization and melting prior to the advent of X-ray crystallography (Tamman 1903) and one on the structural aspects of crystals well into the age of X-ray crystallography (Tamman 1926). Tamman considered two polymorphic molecular crystals as being identical molecular species being arranged on different lattices, and he made a point of distinguishing between the outer form of the crystal and the inner structure of the crystal (habit and form in modern terms, see Section 2.4.1). He revisited Lehmann's division of enantiotropic and monotropic forms, noting that this distinction in principle is really not sufficient, but in practice works quite well, since most investigators do not operate outside of the domain of one atmosphere. For a universal criterion for distinction among polymorphs, he proposed the relative thermodynamic stability, namely the thermodynamic potential per unit of mass. This is the ς-surface of Gibbs for which the less-stable forms have higher values than the stable form.

Tamman demonstrated the use of this thermodynamic measure with three-dimensional $P/T/\varsigma$ plots. If the surfaces for two polymorphs intersect then each is partly stable and partly unstable, depending on the P–T domain. If they don't intersect, then one form may be considered 'totally stable' while the other form is 'totally unstable'. Tamman (1926) reviews the thermodynamic details of a number of possible cases of relative stability, but he summarizes a still-used rule of thumb in stating that the 'relative magnitudes of the volumes and of the heats of fusion of two forms differing in stability are to be regarded as indications of total or partial stability'. In fact, these generalizations are echoed in the so-called 'density rule' (Section 2.2.10) and 'heat-of-fusion rule' (Section 2.2.7) enunciated more than half a century later on the basis of a large body of experimental data (Burger and Ramburger 1979a). Tamman also proposed models for nucleation and the relationship between molecular structure and the possibility of polymorphism.

By 1925, Niggli (1924) had proposed a model for enantiotropism that was based on structural *changes*. If two forms are similar in structure, then the structural change upon a phase transformation is not large, and the two forms were considered to be enantiotropically related. A second type of relationship exists for polymorphs in which the structural differences are large (e.g. calcite–aragonite or graphite–diamond) and the phase transformation perforce results in much greater structural change.

The emerging detailed knowledge of crystal structures in the 1920s changed the direction from thermodynamic to structural, and stimulated interest in the relationship between the structure and properties of materials, as enunciated by Goldschmidt in 1929: 'The task of crystal chemistry is to find systematic relationships between chemical composition and physical properties of crystalline substances . . .'. But there still was an abiding interest '. . . in relating crystal structure to chemical composition especially to find how crystal structure, the arrangement of atoms in crystals, depends on chemical composition.' After noting 'the very extensive wealth of observations [of an earlier] epoch, which has been treasured by v. Groth . . .,' Goldschmidt considered polymorphism to be the result of 'thermodynamic alteration. The substance, under different conditions, may no longer be isomorphous with itself . . . The amount of thermal energy, involved in polymorphic changes are mostly rather small, compared for instance with the heats of chemical reactions.'

By the 1930s, though, interest in polymorphism had very much waned, as the chemical and physical aspects of crystallography were relegated to a relatively minor role in the shadows of the rapidly developing discipline of structural crystallography, with its capability of revealing molecular structure at the atomic level. For about thirty years interest and activity in polymorphism of organic materials was limited to a few devoted (and probably isolated) practitioners, notably L. and A. Kofler and their successors in Innsbruck and McCrone (1957)[7] in the US. In both of these centres of activity the optical microscope often equipped with a hot stage was a principal tool of investigation (Kuhnert-Brandstätter 1971; McCrone 1957), which may have played an important role in saving it from extinction in the search for and investigation of organic polymorphic materials.

After roughly thirty years' hiatus awareness of and interest in polymorphism was aroused again in the 1960s by McCrone's (1965) comprehensive chapter on the subject and his later review on its growing pharmaceutical importance (Haleblian and McCrone 1969). These have set the stage for subsequent developments. In summarizing their 1937 account of the historical development of polymorphism Buerger and Bloom (1937) raised two 'major questions that demand answer':

1. What fundamental property of crystalline matter calls for different forms of the same chemical substance in compounds of all types and causes the appearance of discontinuous jumps in physical properties at definite transformation points?

---

[7] In 1956 McCrone founded McCrone Associates, a private analytical laboratory in which the principal analytical technique employed was polarized light spectroscopy. Over the years he and his staff learned to visually identify over 30 000 particles (McLafferty 1990). McCrone Associates specialized in the identification of polymorphs, asbestos samples, airborne impurities, among others. McCrone recently endowed a chair of chemical microscopy to Cornell University, his Alma Mater.

2. What are the factors that determine what particular crystalline form will be generated from vapour liquid or solution when the thermodynamic phase region of the solid state is entered?

In the interim a great deal has been learnt about the nature and consequences of polymorphism, but the fact that over sixty years later both of these questions are still largely unanswered is a challenge to chemists, physicists, and crystallographers. What has been learnt is the subject of the remainder of this book and is meant to provide a jumping-off point for the investigation of these questions.

1.5 Commercial/industrial importance of polymorphism— some additional comments

The discovery, whether accidental or intended, of polymorphs is unlikely to be greeted with enthusiasm by senior management, and the situation is better treated as an opportunity rather than as a problem. Opportunities are likely to exist for increasing patent cover, for retaining a competitive edge through unpublished knowledge and in formulating pharmaceutical products. A metastable polymorph can be used in capsules for tabletting, and the thermodynamically stable one for suspensions (Bavin 1989)

As we document further in a number of chapters towards the end of the book, polymorphism can and does play an important role in a number of industrial and commercial applications. Again, the overriding reason for this is the variation in properties that often accompany differences in structure. Obviously, if two polymorphs exhibited the same properties required for a certain product specification, there would be no concern about which polymorph (or what mixture of polymorphs) was actually present. A number of examples will serve to demonstrate the scope of this role.

As Bavin (1989) has succinctly noted, in the real world of chemical processing 'few compounds reach development and fewer still are marketed'. Since in many ways a new polymorph is a new material, the characterization and control of the polymorphic behaviour is an integral part of its development and marketing.

In the photographic industry, differences in solubility of polymorphs can pose problems during manufacture, and can lead to impaired performance (Nass 1991). The transformation to undesired structures may take place through a solvent-mediated phase transformation (Section 3.3) during a manufacturing process. On the other hand, recognition of such a process may lead to its utlization to obtain a desired polymorph. In an example from the food additives industry, for L-glutamic acid, later converted to the monosodium salt (MSG) used for taste enhancement, it is crucial to obtain the α polymorph rather than the β form. The latter can lead to a situation in which the crystallizing slurry coagulates into a gel and can no longer be processed (Sugita 1988). A recent study (Garti and Zour 1997) indicates that the addition of selected surface active agents can lead to the preferential crystallization of the α polymorph (Section 3.4).

Polymorphism plays an important role in the huge industries of fat-based food products, for instance ice cream, chocolate, and margarine (Garti and Sato 1988;

O'Connor 1960; Loisel *et al.* 1998), as well as in the chemistry involved with long alkyl chains (Robles *et al.* 1998). The melting point and melting behaviour are clearly important physical properties of such materials, but the appearance as well as the properties perceived by the consumer are to a great extent determined by the structure of the solid fat phase. The solid fats provide structure to the marketed product, with other additives supplying taste, stability, and colour. One of the key ingredients of these fats is the triacylglycerols, and the different polymorphic forms play an important role in both the designing of properties, and the processing of the product (Hagemann and Rothfus 1993; Johansson *et al.* 1995; Jovanovic *et al.* 1995; Herrera and Rocha 1996). Hence considerable effort is often expended in characterizing the polymorphic forms of many of these materials, their properties, and the methods required to control which polymorph is obtained (Sato *et al.* 1985; Sato 1999; Suzuki and Ogaki 1986; Ng 1990; Kellens *et al.* 1992; Sato and Suzuki 1986).

Similarly the chain-like structure of polymers also results in a proliferation of polymorphic structures (e.g. Keller and Cheng 1998; Lotz 2000; Rastogi and Kurelec 2000). The differences in structure lead to a variety of properties (e.g. Calleja *et al.* 1993; Chunwachirasiri *et al.* 2000), which of course is one of the driving forces for the development of new polymeric materials. Although not referenced specifically, many of the principles and examples presented in subsequent chapters apply equally well to many of the polymorphic molecular systems.

2

Fundamentals

Of course, one cannot ignore the physics, and, in particular, one has to pay due respect to thermodynamic considerations ... (Dunitz 1991)

... a major deficiency in our current knowledge and understanding concerns the relationship among the members of a polymorph cluster—what is their relative mutual stability, how do they transform, one into another, what are the thermodynamic factors governing their mutual stability, what are the kinetics of the transitions. Answers to the last of these questions are very important to users of, and sufferers from, polymorphism. (Herbstein 2001)

2.1 Introduction

Polymorphic structures of molecular crystals are different phases of a particular molecular entity. To understand the formation of those phases and relationships between them we make use of the classic tools of the Phase Rule, and of thermo-dynamics and kinetics. In this chapter we will review the thermodynamics in the context of its relevance to polymorphism and explore a number of areas in which it has proved useful in understanding the relationship between polymorphs and poly-morphic behaviour. This will be followed by a summary of the role of kinetic factors in detecting the growth of polymorphic forms. We will then provide some guidelines for presenting and comparing the structural aspects of different polymorphic structures, with particular emphasis on those that are dominated by hydrogen bonds.

2.2 The thermodynamics of polymorphic molecular crystals

It is beyond the scope of this book to provide a comprehensive review of the thermody-namics of molecular crystals (Stall *et al.* 1969). The field was very adequately covered in the comprehensive chapter by Westrum and McCullough (1963), a work which has stood the test of time remarkably well, and can serve as an excellent resource on this general subject. An earlier useful reference is the chapter on polymorphism in the classic book by Tammann (1926).

2.2.1 The Phase Rule

The Phase Rule was first formulated by Gibbs (1876, 1878) on the basis of thermo-dynamic principles and then applied to physical chemistry by Roozeboom (1911). As with so much of chemistry, the apparently absolute physical principles stated in the Phase Rule must be tempered by real chemical situations (Dunitz 1991; Brittain

1999*c*; Alper 1999; Jensen 2001). However, in order to establish a working language, it is necessary to define terms and then indicate the difficulties that may arise in the practical use of those definitions.

The Phase Rule is simply stated as

$$F = C - P + 2$$

where F is the number of degrees of freedom of the system, C the number of components, and P the number of phases.

A phase is defined (Glasstone 1940; Findlay 1951) as *any homogeneous and physically distinct part of a system which is separated from other parts of the system by definite bounding surfaces.* By this definition for any substance there is one gaseous phase and one liquid phase, since these must be physically and chemically homogeneous. Each crystalline form constitutes an individual phase, for example, the different forms of ice. A mixture of two polymorphs contains two solid phases, but a homogeneous solid solution or an alloy of two totally miscible metals is only one phase. Problems arise when one must consider the dimensions of the structural realm for the definition of 'homogeneous'—uniform throughout. As both Findlay (1951) and Dunitz (1991) have pointed out, at the molecular level such a definition certainly breaks down, and even some molecular substances that exhibit, say, the X-ray diffraction patterns expected from single crystals (i.e. a single phase), have been shown on closer inspection to be inhomogeneous mixed crystals (Weissbuch *et al.* 1995) or inhomogeneous racemic mixtures (conglomerates) (Green *et al.* 1981; Ramdas *et al.* 1981). In other cases, it has been claimed that a single crystal was actually a hybrid in which two polymorphs coexist (Freer and Kraut 1965; Coppens *et al.* 1998; Fomitchev *et al.* 2000).

The number of *components* is the minimum number of independent species required to define the composition of all of the phases in the system. The simplest example usually cited to demonstrate the concept of components is that of water, which can exist in various equilibria involving the solid, liquid, and gas. In such a system there is one component. Likewise for acetic acid, even though it associates into dimers in the solid, liquid, and gaseous state, the composition of each phase can be expressed in terms of the acetic acid molecule and this is the only component. The important point for such a system is that the monomer–dimer equilibrium is established very rapidly, that is, faster than the time required to determine, say, the vapour pressure. In the cases in which the equilbrium between molecular species is established more slowly than the time required for a physical measurement, the vapour pressure, for example, will no longer be a function only of temperature, but also of the composition of the mixture, and the definition of a component acquires a kinetic aspect.

The *number of degrees of freedom* (sometimes also referred to as the *variance*) is the number of variable factors, such as temperature, pressure, and concentration that must be fixed in order to define the condition of a system at equilibrium. Thus, a one-component system in one phase, say a gas, would have two degrees of freedom; a one component system in two phases (liquid and gas) would have one degree of freedom. A system of one component and three phases would have no degrees of freedom. For the water system this would correspond to the familiar triple point. These relationships are often described, respectively as bivariant, monovariant, and invariant (Trevor 1902).

For polymorphic systems of a particular material we are interested in the relationship between polymorphs of one component. A maximum of three polymorphs can coexist in equilibrium in an invariant system, since the system cannot have a negative number of degrees of freedom. This will also correspond to a triple point. For the more usual case of interest of two polymorphs the system is monovariant, which means that the two can coexist in equilibrium with either the vapour or the liquid phases, but not both. In either of these instances there will be another invariant triple point for the two solid phases and the vapour on the one hand, or for the two solid phases and the liquid on the other hand. These are best understood in terms of phase diagrams, which are discussed below, following a review of some fundamental thermodynamic relationships that are important in the treatment of polymorphic systems.

2.2.2 *Thermodynamic relations in polymorphs*

In terms of thermodynamics, one of the key questions regarding polymorphic systems is the relative stability of the various crystal modifications and the changes in thermodynamic relationships accompanying phase changes and different domains of temperature, pressure, and other conditions. Buerger's (1951) treatment of these questions provides the fundamentals upon which to base further discussion.

For simplicity, and to demonstrate the principles of these considerations, we will generally limit ourselves to a discussion of two polymorphic solids, although the extension to more complex systems is based on precisely the same principles. The relative stability of the two polymorphs depends on their free energies, the most stable form having the lowest free energy. Because of this energy relationship those forms which are less stable will be energetically driven to transform into the most stable form, although kinetic factors may prevent the transformation. Since we are dealing with solids the differences in volume between polymorphs are small fractions of the volumes of the solids themselves; we can then neglect volume and pressure changes with energy. Under these conditions (of essentially constant temperature and pressure) the free energy of a solid phase may be represented by the Helmholz relationship

$$A = E - TS$$

where E is the internal energy, T the absolute temperature, and S the entropy.

At absolute zero TS vanishes and the Helmholz free energy equals the internal energy. As a consequence, at absolute zero the most stable polymorphic modification should have the lowest internal energy.[1] Above absolute zero the entropy term will

---

[1] Buerger (1951) defines the internal energy as being composed of the 'structural energy' and the zero-point energy. He objects to non-crystallographers equating the term 'lattice energy' with 'structure energy'. In the 50 years since his chapter was written, 'lattice energy' appears to have become accepted usage and will be further used here as well.

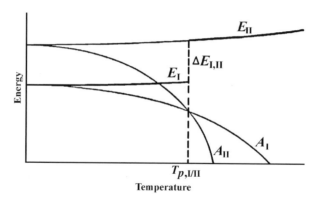

Fig. 2.1   Energy vs temperature curves for two polymorphs I and II. A is the Helmholz free energy and E is the internal energy. Consistent with the labelling scheme proposed by McCrone (see Chapter 1) Form I is assumed to be the stable form at room temperature. (From Buerger 1951, with permission.)

play a role which may differ for the two polymorphs. Hence the behaviour of the free energy as a function of temperature can differ for the two polymorphs, as presented by the curves A_I and A_{II} in Fig. 2.1. Form I is more stable at absolute zero and the two curves behave differently, crossing at the transition temperature $T_{p,I/II}$. Above the transition temperature Form II is more stable. At the transition temperature the free energy of the two forms is identical, but since the internal energy of Form I is less than that of Form II a quantity of energy ΔE is required to be input for the phase transition, which must be endothermic. Buerger (1951) has also demonstrated the endothermic nature of any transformation which takes place upon raising the temperature.

Moving from low temperature to high temperature for a trimorphic system the positive heat of transformation at each step (ΔE) corresponding to a crossing point of the corresponding free energy curves leads to the behaviour in Fig. 2.2. If the entropy (which is the slope of the free energy curve) increases at a uniform rate, then each pair of curves crosses only once. To put these diagrams into perspective it is perhaps useful at this point to get some idea of the magnitudes of some of the differences in lattice energy and vibrational entropy between polymorphic forms. ΔEs for real polymorphs often are in the range or $0–10\,kJ\,mol^{-1}$ (Kitaigorodskii 1973a,b; Kuhnert-Brandstätter and Sollinger 1989; Chickos *et al.* 1991) and probably do not exceed $25\,kJ\,mol^{-1}$, while ΔSs are less than $15\,J/K\cdot mol$ (Gavezzotti and Filippini 1995).

2.2.3   *Energy vs temperature diagrams—the Gibbs free energy*

Diagrams qualitatively very similar to Figs 2.1 and 2.2 may be prepared on the basis of the Gibbs free energy,

$$G = H - TS$$

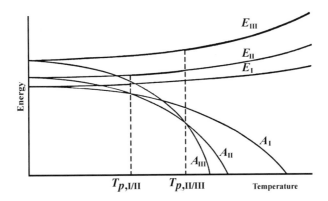

Fig. 2.2 Energy vs temperature curves for three polymorphs I, II, and III. A is the Helmholz free energy and E is the internal energy. (Adapted from Buerger 1951, with permission.)

rather than the Helmholz free energy. In such a case E is replaced by the enthalpy H, and A is replaced by G, the Gibbs free energy. Since the data required to produce these are experimentally more readily accessible, the diagrams based on the Gibbs free energy are more commonly in use, and will be described in some detail here (Grunenberg *et al.* 1996). The utility of these diagrams is that they contain a great deal of information in a compact form, and provide a one page visual and readily interpretable summary of what can be complex interrelationships among polymorphic modifications. As their utility is recognized these diagrams should become more widely used for the characterization of polymorphic systems and the phase relationships among various polymorphs (Henck and Kuhnert-Brandstätter 1999).

A typical energy vs temperature diagram using the Gibbs relationship is given in Fig. 2.3. Compared to the analogous Fig. 2.1 there are two additional isobars: the H_{liq} curve (above the two (H_{I} and H_{II}) solid curves), and the G_{liq} curve. The H vs temperature curves may be constructed experimentally by determination of the heat capacity C_p, from

$$\left(\frac{\delta H}{\delta T}\right)_p = C_p$$

the fundamental relationship between the enthalpy and the heat capacity, as demonstrated in Fig. 2.4. A number of points concerning this plot are worthy of note. It can be shown from the third law of thermodynamics that the heat capacity of an ideal crystal is zero at 0 K. Therefore the slope of the curve in Fig. 2.4 (and those of H in Fig. 2.3) at 0 K must also be zero. From the expression for the Gibbs free energy the partial derivative with respect to temperature is

$$\left(\frac{\delta G}{\delta T}\right)_p = -S$$

Since S is always positive, G is a constantly decreasing function, as seen in Fig. 2.3. The G isobars can follow different paths, and their intersections represent transition

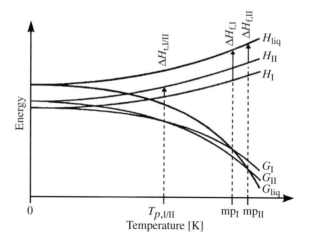

Fig. 2.3 Energy vs temperature (E/T) diagram of a dimorphic system. G is the Gibbs free energy and H is the enthalpy. This diagram represents the situation for an enantiotropic system, in which Form I is the stable form below the transition point, and presumably at room temperature, consistent with the labelling scheme for polymorphs proposed by McCrone (see Chapter 1). (Adapted from Grunenberg *et al.* 1996, with permission.)

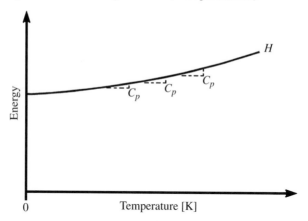

Fig. 2.4 Plot of enthalpy H vs temperature indicating the relationship with the heat capacity, C_p (Adapted from Grunenberg *et al.* 1996, with permission.)

points between phases. Buerger (1951) suggested the geometric possibility for two of the G isobars to cross twice, but Burger and Ramburger (1979*a*,*b*) have shown by statistical–mechanical arguments that only one crossing is physically possible.

2.2.4 *Enantiotropism and monotropism*

Enantiotorpism and monotropism were referred to in Chapter 1. We now provide a thermodynamic basis for these two important descriptors of polymorphic behaviour.

In Fig. 2.3 it is seen that the thermodynamic transition point $T_{p,\mathrm{I/II}}$, defined by the point at which G_I and G_II cross, falls at a temperature below the melting point of the lower melting form, mp$_\mathrm{II}$. This is the thermodynamic definition of an enantiotropic polymorphic system. The melting point itself is defined by the crossing of the G curves for Form II and the liquid (and similarly for the melting point of Form I). The enthalpies of transition (ΔH_t) or fusion (ΔH_f) appear at the corresponding temperatures as the vertical differences between the appropriate H curves. The energy vs temperature diagram for a monotropic relationship between two polymorphs is shown in Fig. 2.5. In this case, however, the free energy curves do not cross at a temperature below the two melting points.

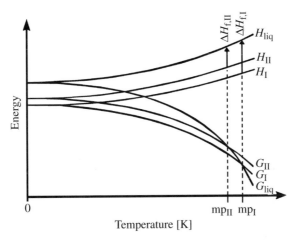

Fig. 2.5 Energy vs temperature (E/T) diagram for a monotropic dimorphic system. The symbols have the same meaning as in Fig. 2.3. Form I is more stable at all temperatures; the crossing of the G_I and G_II curves (not shown) will be above the melting point for Form I and Form II. (From Grunenberg *et al.* 1996, with permission.)

2.2.5 *Phase diagrams in terms of pressure and temperature*

Because pressure and temperature are two readily measured experimental quantities (e.g. Griesser *et al.* 1999), the relationships among the vapour, liquid, and polymorphs of a substance are often represented on diagrams of pressure vs temperature. These are also very useful in summarizing the polymorphic behaviour of a system.

Figure 2.6 shows the prototypical plots of pressure vs temperature for the enantiotropic and monotropic cases. These are best understood by proceeding along various curves, which represent equilibrium situations between two phases. The l./v. line in the high-temperature region of Fig. 2.6(a) is the boiling point curve. Moving to lower temperatures along that line one encounters the II/v. line, which is the sublimation curve for Form II. The intersection of the two curves is the melting point

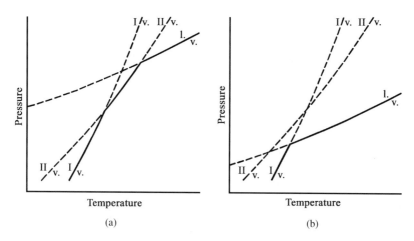

Fig. 2.6 Pressure vs temperature plots. I/v. and II/v. represent sublimation curves; I.v. is the boiling point curve. Broken lines represent regions which are thermodynamically unstable or inaccessible. (a) enantiotropic system; (b) monotropic system. The labelling corresponds to earlier figures to indicate that Form I is stable at room temperature which is below the transition point in the enantiotropic case. (From McCrone 1965, with permission.)

of Form II. Under thermodynamic conditions Form II would crystallize out at this point and the solid part of the II/v. line would govern the behaviour. However, if kinetic conditions prevail (e.g. if the temperature is lowered rapidly) the system may proceed along the broken l./v. line to the intersection of with the I/v. line, at which point Form I would crystallize. Continuing downward along the solid part of the II/v. sublimation curve, the crossing point with the II/v. sublimation curve is the transition point between the two polymorphic phases. Once again, if thermodynamic conditions prevail Form II will be transformed to Form I. Under kinetic conditions Form II may continue to exist (even indefinitely in some cases) along the II/v. sublimation curve. Figure 2.6(a) represents the enantiotropic case because the transition point between the two solid phases is at a temperature below melting point of Form II and Fig. 2.6(b) is the diagram for the monotropic case.

As McCrone (1965) has pointed out, the complete pressure/temperature diagram must also contain the curves representing the phase boundaries between the two solid forms (transition temperature curve) and between each form and the liquid (melting point curves). These are shown included in Fig. 2.7 for the same system as in Fig. 2.6. The transition temperature curve rises from the intersection of the two solid–vapour curves and goes through the intersection of the two solid–liquid curves. For the enantiotropic case the curve is a solid line, indicating the existence of true thermodynamic equilibrium between the two phases. The II<->l. line is also solid, since it originates from a thermodynamically accessible point. The I<->l. line is a broken one, however, since it originates from a thermodynamically inaccessible, but kinetically accessible point. Similar arguments describe the three additional lines for the monotropic diagram, Fig. 2.7(b).

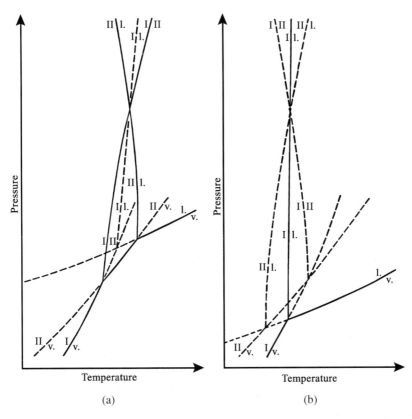

Fig. 2.7 Transition temperature (I<–>II) and melting point (I<–>melt, II<–>melt) curves added to Fig. 2.6. (a) enantiotropic; (b) monotropic. (From McCrone 1965, with permission.)

A review of some of the features of these diagrams in terms of the Phase Rule is enlightening (Findlay 1951). A system composed of two different solid forms of a substance will have one component and two solid phases. In the absence of a further definition of the system there will be one degree of freedom. In Fig. 2.7 this is either the temperature or the pressure along the I<–>II line. Choosing either variable fixes a point and defines the system. However, suppose that we are interested in the situation when the two phases are in equilibrium with the liquid or the vapour. Each one of those is an additional phase, making three in total and, by virtue of the Phase Rule, rendering the system invariant. Invariance results in a triple point for each case, defined by the intersection of the I<–>II curve with the I<–>v. and II<–>v. curves on the one hand and the intersection of the I<–>II curve with the I<–>l. and II<–>l. curves on the other hand.

Recalling the classic definition of the triple point, say, for water as the intersection of the solid–vapour and the liquid–vapour curves, the analogy in Fig. 2.7 is the intersection of the II<–>v. and I<–>v. curves. Below the triple point only one of the solid phases (I) can exist in stable equilibrium with the vapour; above the triple point only II can exist in equilibrium with the vapour. The triple point I–II–v. may be looked upon as a point at which there is a change in the relative stability of the two phases. In a general way, then, phase changes may be viewed as what can transpire at the triple point.

Which of the two representations (energy/temperature or pressure/temperature) is preferred? Westrum (1963) has pointed out that most textbooks use the pressure/ temperature representation because of the availability of such data for inorganic systems and examples which can be given using those data. On the other hand for organic systems, energy data are more abundant and more convenient to use (Chickos 1987; Chickos *et al.* 1991). The sources of those data and their use in generating (semi-quantitative) energy/temperature diagrams are given in Chapters 4 and 5. Examples of enantiotropic and monotropic energy/temperature diagrams for two real systems are given in Fig. 2.8.

Consideration of the details of these diagrams leads to a number of 'rules' which are helpful in characterizing, understanding and predicting the behaviour of poly-morphic systems, including monotropism and enantiotropism as well as generating the energy/temperature diagrams from experimental measurements and observations. A number of these rules were originally developed by Tammann (1926) and then expanded by Burger and Ramberger (1979*a,b*) and by Grunenberg *et al.* (1996).

2.2.6   *Heat-of-transition rule*

As indicated in Figs 2.3–2.5, ΔH and ΔG are usually positive, and it is assumed that the H curves do not intersect. Also as discussed above, the G curves intersect only once. This set of circumstances leads to a statement of the heat-of-transition rule (Grunenberg *et al.* 1996; Burger and Ramberger 1979*a*):

> If an endothermic phase change is observed at a particular temperature, the transition point lies below that temperature, and the two polymorphs are enantiotropically related. If an exothermic transition is observed, then there is no thermodynamic transition point below that transition temperature. This can occur when the two modifications are monotropically related or when they are enantiotropically related and the thermodynamic transition point is higher than the measured transition temperature.

Burger and Ramberger (1979*b*) claim that this rule is observed in at least 99 per cent of the cases examined.

2.2.7   *Heat-of-fusion rule*

The heat-of-fusion rule states that in an enantiotropic system the higher-melting poly-morph will have the lower heat of fusion. If the higher-melting polymorph has a higher heat of fusion the two are related monotropically. This is a direct consequence of the relationship between the H curves, and the rule will be valid so long as the thermody-namic behaviour for the two cases can be represented by Figs 2.3 and 2.5. Deviations will arise when the H curves diverge significantly, or the melting points are not close

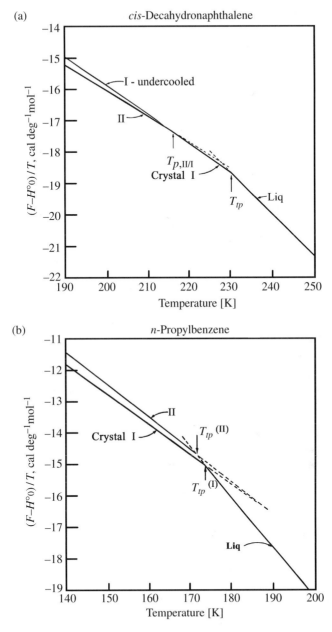

Fig. 2.8 Diagrams illustrating (a) enantiotropic and (b) monotropic phase relationships for two organic compounds, *cis*-decahydronaphthalene and *n*-propylbenzene, respectively. Note that the *y* scale is actually given in units of entropy calculated from the energy terms. $T_{p,II/I}$ represents the transition point between phases I and II, being above the melting point in (a) and (by extrapolation) below the melting point in (b). (From Westrum and McCullough 1963, with permission.)

together (Δmp $\approx$30 K), or both. In such cases it may be preferred to use the entropy-of-fusion rule or the heat-capacity rule as guidelines, rather than the heat-of-fusion rule.[2] Nevertheless the success rate of this rule is essentially as high as that of the heat-of-transition rule (Burger and Ramberger 1979b).

2.2.8   *Entropy-of-fusion rule*

The melting point is defined as the temperature at which the liquid is in equilibrium with the solid so that the difference in Gibbs free energy between the two phases is zero. The entropy of fusion can then be expressed as,

$$\Delta S_f = \frac{\Delta H_f}{T_f}$$

According to the rule, if the polymorph with the higher melting point has the lower entropy of fusion, the two modifications are enantiotropically related. If the lower-melting form has the lower entropy of fusion then the two forms are monotropically related (Burger 1982a,b).

2.2.9   *Heat-capacity rule*

For a pair of polymorphs, if the modification with the higher melting point also has a higher heat capacity at a given temperature than the second polymorph, then there exists an enantiotropic relationship between them. Otherwise, the system is monotropic.

2.2.10   *Density rule*

Kitaigorodskii (1961) enunciated the principle of closest packing for molecular crystals. Briefly this principle states that the mutual 'orientation of molecules in a crystal is conditioned by the shortest distances between the atoms of adjacent molecules', and because the periphery of molecules is often dominated by hydrogen atoms that these distances will usually 'be determined by the interactions between hydrogen atoms or the interaction of hydrogen atoms with other atoms of other elements'. What determines the existence or non-existence of a crystal structure is the free energy, and the energetic manifestation of these distance arguments is that the most stable structure energetically should be expected to correspond to the one that has the most efficient packing. In other words, on the multidimensional energy surface the lowest lying among all the deep minima at zero degrees should also correspond to the structure with the highest density. Additional polymorphic structures, each located at a minimum with higher free energy than the minimum energy structure at zero degrees, will be expected to have less efficient packing and correspondingly lower density.

---

[2] Burger and Ramberger (1979b) discuss the error in calculation of the difference of heats of fusion of two polymorphs in detail.

This rule is quite general for ordered molecular solids that are dominated by van der Waals interactions. Exceptions are not unexpected when other interactions, such as hydrogen bonds, dominate the packing, since some energetically favourable hydrogen-bond dominated packing arrangements can lead to large voids in the crystal structure with correspondingly lower density.

While the density is arguably the easiest to obtain of the physical properties noted for deriving these rules, an increasing portion of crystallographic investigations no longer include the *experimental* determination of density. The *calculated* crystal density is routinely obtained from the unit cell dimensions and contents, but at the very minimum this requires an indexed X-ray powder diffraction pattern. While the density of polymorphs can be very useful in ranking the relative stability of polymorphs for which no X-ray data are available, the determination of experimental densities is becoming a lost art (Tutton 1922; Reilly and Rae 1954; Richards and Lindley 1999). Although accuracy as good as 0.02 per cent can be obtained for the experimental determinations, Stout and Jensen (1989) have estimated that the agreement between calculated and experimental values of density is normally about 1–1.5 per cent. Since densities of polymorphs often differ by very close to that amount, a *caveat* must be associated with the use of this rule, which Burger and Bamberger (1979b) indicate is correct 90 per cent of the time, excluding cases of density differences of less than 1 per cent.

2.2.11 *Infrared rule*

Burger and Ramberger (1979a) have also proposed an 'infrared rule' for the highest frequency infrared absorption band in polymorphic structures containing strong hydrogen bonds. The formation of strong hydrogen bonds is associated with a reduction in entropy and an increase in the frequency of the vibrational modes of those same hydrogen bonds. The intramolecular N–H or O–H bond is correspondingly weaker (if the hydrogen participates in a hydrogen bond) with a reduction in the frequency of the associated bond stretching modes. The assumption is that these vibrations are only weakly coupled to the rest of the molecule, which is usually true for the O–H and N–H stretching vibrations and the NH_2 symmetric stretch. Under these conditions, the infrared rule says that the hydrogen-bonded polymorphic structure with the higher frequency in the bond stretching modes may be assumed to have the larger entropy.

Many cases could not be tested by Burger and Ramberger using this rule, since there was no difference in the highest frequency infrared absorption on going from the solid to the melt. Also, compounds containing the strong hydrogen-bonding group –CO–NH were prominent among the exceptions. However, after eliminating these approximately 10 per cent exceptional cases all the remaining 113 cases examined behaved according to the rule (Burger and Ramberger 1979b). The successful application of the infrared rule requires detailed information on nature of the hydrogen bonds in the solid state. If a molecule is able to form more than one type of hydrogen bond it may be difficult to correlate the frequencies observed with the polymorphic form. More studies on these correlations would be helpful in demonstrating the potential utility of this rule.

None of these rules is foolproof. However, they are useful guidelines, and the combination of relatively simple techniques can often be used to get a good estimate of the relative stability of polymorphs under a variety of conditions, information which is useful in understanding polymorphic systems, the properties of different polymorphs and the methods to be used to selectively obtain any particular polymorph (see Section 3.2). As noted above, much of that information can be included in the energy/temperature diagram, and the actual preparation of that diagram from experimentally determined quantities is described in Sections 4.2 and 4.3 following the description of the techniques used to obtain those physical data.

2.3   Kinetic factors determining the formation of polymorphic modifications

The starting point for a discussion of the kinetic factors is the traditional energy–reaction coordinate diagram, Fig. 2.9. This shows G_0, the free energy per mole of a solute in a supersaturated fluid which transforms by crystallization into one of two crystalline products, I or II, in which I is the more stable ($G_{II} > G_I$). Associated with each reaction pathway is a transition state and an activation free energy which is implicated in the relative rates of formation of the two structures. Unlike a chemical reaction, crystallization is complicated by the nature of the activated state since it is not a simple bi- or trimolecular complex as would be expected for a process in which a covalent bond is formed; rather the activated state relates to a collection of self-assembled molecules having not only a precise packing arrangement but also existing as a new separate solid phase.

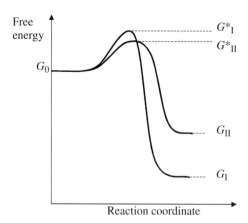

Fig. 2.9   Schematic of the reaction coordinate for crystallization in a dimorphic system, showing the activation, barriers for the formation of polymorphs I and II. (Adapted from Bernstein *et al.* 1999, with permission.)

The existence of the phase boundary between the solid and liquid phase complicates matters, since a phase boundary is associated with an increase in free energy of the system which must be offset by the overall loss of free energy. For this reason the magnitudes of the activated barriers are dependent on the size (i.e. the surface to volume ratio of the new phase) of the supramolecular assembly (crystal nucleus). This was recognized in 1939 by Volmer in his development of the kinetic theory of nucleation from homogeneous solutions and remains our best model today (Volmer 1939).

One of the key outcomes of this theory is the concept of critical size which must be achieved by an assembly of molecules in order to be stabilized by further growth. The higher the operating level of supersaturation the smaller is this size (typically a few tens of molecules). Now, in Fig. 2.9 the supersaturation with respect to II is simply $G_0 - G_{II}$ and is lower than $G_0 - G_I$ for structure I. However it can now be seen that if for a particular solution composition the critical size is lower for II than for I then the activation free energy for nucleation is lower and kinetics will favour form II. Ultimately form II will have to transform to form I, a process that we discuss later. Overall we can say that the probability that a particular form i will appear is given by

$$P(i) = f(\Delta G, R) \tag{2.1}$$

in which ΔG is the free energy for forming the ith polymorph and R is the rate of some kinetic process associated with the formation of a crystal by molecular aggregation. Thus, for example, if we follow the above reasoning we could equate the rate process with J, the rate of nucleation of the form. If all polymorphs had the same rates of nucleation, then their appearance probability would be dominated by the relative free energies of the possible crystal structures.

The rates of nucleation as expressed by the classical expression of Volmer are related to various thermodynamic and physical properties of the system such as surface free energy (γ), temperature (T), degree of supersaturation (σ), solubility (hidden in the

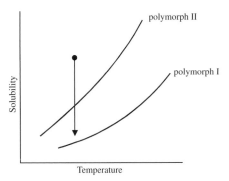

Fig. 2.10   Schematic solubility diagram for a dimorphic system (polymorphs I and II) showing a hypothetical crystallization pathway (vertical arrow) at constant temperature. (Adapted from Bernstein *et al.* 1999, with permission.)

pre-exponential factor A_n), which will not be the same for each structure but will correctly reflect the balance between changes in bulk and surface free energies during nucleation. This is seen in eqn (2.2) which relates the rate of nucleation to the above parameters (v is the molecular volume and k is Boltzmann's constant):

$$J = A_n \exp(-16\pi\gamma^3 v^2/3k^3 T^3 \sigma^2) \qquad (2.2)$$

From this analysis it is clear that the trade-off between kinetics and thermodynamics is not at all obvious. Consider a monotropic, dimorphic system (for simplicity) whose solubility diagram is shown schematically in Fig. 2.10. It is quite clear that for the occurrence domain given by solution compositions and temperatures that lie between the form II and I solubility curves only polymorph I can crystallize. However, the outcome of an isothermal crystallization that follows the crystallization pathway indicated by the vector in Fig. 2.10 is not so obvious since the initial solution is now supersaturated with respect to both polymorphic structures, with thermodynamics favouring form I and kinetics (i.e. supersaturation) form II.

Experimentally, the reality of this overall scenario of kinetic vs thermodynamic control was known long before the development of nucleation theory and is encompassed by Ostwald (1897) in his Rule of Stages (Davey 1993; Cardew and Davey 1982). The German scientific literature between 1870 and 1914 contains many organic and inorganic examples in which crystallization from melts and solutions initially yield a metastable form that was ultimately replaced by a stable structure. On the basis of such a phenomenon Ostwald was led to conclude that 'when leaving a metastable state, a given chemical system does not seek out the most stable state, rather the nearest metastable one that can be reached without loss of free energy' (Ciechanowicz *et al.* 1976).

There are significant flaws in Ostwald's conclusion that led to his rule. For instance, when a crystallization experiment yields only a single form there is the question of whether it contradicts the Rule or whether the material is simply not polymorphic. Moreover, there is no way of answering this question. However, a sufficient number of cases of successively crystallizing polymorphic forms have been observed to warrant considering the principles behind Ostwald's Rule as guidelines for understanding the phenomenon of the successive crystallization of different polymorphic phases.

By making use of Volmer's equations some attempts have been made by Becker and Döring (1935), Stranski and Totomanov (1933), and Davey (1993) to explain the Rule in kinetic terms. In doing this, it becomes apparent that the situation is by no means as clear cut as Ostwald might have us believe. Figure 2.11 shows the three possible simultaneous solutions of the nucleation equations which indicate that by careful control of the occurrence domain there may be conditions in which the nucleation rates of two polymorphic forms are equal, and hence their appearance probabilities are nearly equal. Under such conditions we might expect the polymorphs to crystallize concomitantly (see Section 3.3). In other cases, there is a clearer distinction between kinetic and thermodynamic crystallization conditions, and that distinction may be utilized to selectively obtain or prevent the crystallization of a particular polymorph.

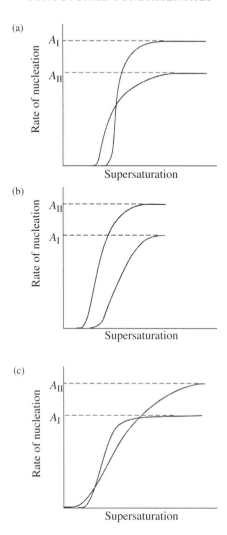

Fig. 2.11 The rates of nucleation as functions of supersaturation for the dimorphic system defined in Fig. 2.10. The three diagrams a, b and c represent the three possible solutions for the simultaneous nucleation of two polymorphs each of which follows a rate equation of the form of eqn (2.2). Note that solutions a and c both allow for simultaneous nucleation of the forms at supersaturations corresponding to the crossover of the curves. (From Bernstein *et al.* 1999, with permission.)

2.4 Structural fundamentals

Structure is the fundamental property of polymorphs. In this section we will deal with the definition of that structure, how it may be best viewed and understood, and how polymorphic structures may be compared. The determination of crystal structure, by

X-ray methods is the subject of many fine texts (Dunitz 1979; Giacovazzo 1992; Stout and Jensen 1989; Glusker *et al.* 1994), which should be referred to, if necessary, as background for this section.

2.4.1   *Form vs habit*

In the description of crystals and crystal structures the two terms *form* and *habit* have very specific and very different meanings. *Form* refers to the internal *crystal structure* and etymologically is the descendant of the Greek *morph.* Hence, *polymorph* refers to a number of different crystal modifications or different crystal structures, and the naming of different structures as 'Form I' or '*α* Form' follows directly from this definition and usage. As we have seen above, the difference in crystal structure is very much, although not exclusively, a function of thermodynamics. Certainly, only the structures which are thermodynamically accessible can ever exist, but there often is a question of thermodynamic vs kinetic control over which particular structure may be obtained under any particular set of crystal growth conditions.

Habit, on the other hand, derives from the Latin and Old French word for mode of growth, and describes the shape of a particular crystal.[3] That shape is influenced greatly by the environment. It is essentially a manifestation of kinetic factors determining the relative rate of growth along various directions of the crystal, and hence the preferential growth or inhibition of the development of the different crystal faces that ultimately define the shape of the crystal. Examples of definitions of different habits and the variation in habit resulting from changes in crystal growth conditions are given in Fig. 2.12.

Unfortunately, the distinction between *form* and *habit* is often blurred, and consequently some confusion has crept into the literature. Especially in the older crystallographic literature, *form* was used to describe a set of crystal faces which are alike or symmetry related, and the *habit* then described the collection of the forms that are exhibited (Chamot and Mason 1973). Ideally, the external shape of the crystal reflects internal symmetry of the crystal. Crystallographers refer to this symmetry as the *crystal class*, while chemists traditionally refer to it as the point group (Hahn and Klapper 1992). The study of the external shape and symmetry of crystals is called crystal morphology. An excellent collection of crystal habits according to crystal class, their descriptors and drawings of both mineral and organic examples, can be found in Buerger's (1956) text.

The important point is that differences in external crystal shape, *habit*, or crystal *morphology* may not necessarily indicate a change in the polymorphic *form* or polymorphic crystal structure. An example of this distinction is given in Fig. 2.13.

---

[3] According to the Oxford English Dictionary (OED), the initial use for *habit* was in zoology and botany, indicating the characteristic mode of growth and general appearance of an animal part. The usage was transferred to crystallography to indicate the characteristic mode of formation of a crystal. For instance, the OED cites an 1895 usage by M. H. N. Story-Maskelyne: 'Such differences, then, may generally be held to indicate a mero-symmetrical habit'.

Control of crystal habit and habit modification is an important aspect of the preparation and ultimate use of solids (Wood 1997; Davey *et al.* 1994), and a great deal of
work has been done in this area. Some early leading references are those of Tipson
(1956) and Buckley (1951), and there have been important theoretical (Clydesdale
et al. 1997) and experimental developments, for example, in the more recent work
of Hartman and Hartman (1973), Bennema and Hartman (1980), Black *et al.* (1990),
Clydesdale *et al.* (1997), Lahav and Leiserowitz (1993) and Addadi *et al.* (1985).
Even with the significant developments in learning to control of both crystal habit
and crystal form, a single modification may crystallize in a number of habits, one
of which has more desirable processing and packaging characteristics, other chemical and physical properties being equal, for example, aspartame (see Section 10.5).
Hence, these two intimately related characteristics of internal structure and external
shape must often be considered together, which has been the approach of most of the
modern researchers in this area.

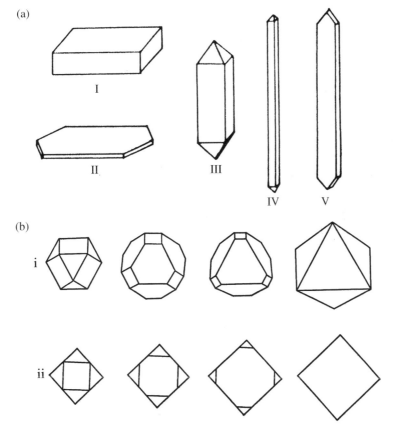

Fig. 2.12   (*Continued*)

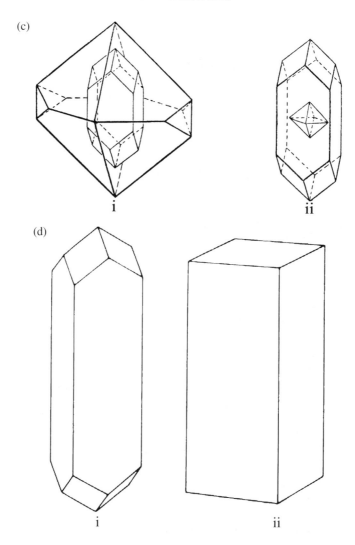

Fig. 2.12 (a) Some different crystal habits of crystals and their descriptions: I, tabular; II, platy; III, prismatic; IV, acicular; V, bladed (from Hartshorne and Stuart (1964), with permission. (b) Demonstration of the difference of rate of growth of cubic or octahedral faces of a crystal as governed by the rate of deposition on different crystal faces. Starting on left hand side, both crystals have equal development of the cubic and octahedral faces. Deposition is faster on the cubic faces in i and the octahedral faces in ii, leading eventually to the crystal habit on the right (from Chamot and Mason (1973), with permission). (c) Demonstration of the effect of solvent on the crystal habit of anthranilic acid. i, crystal initially grown in water, then in ethanol; ii, crystal from i transferred to acetic acid and allowed to grow (from Wells (1946), with permission). (d) Demonstration of the use of *form* to describe the family of faces bounding a crystal. In the commonly used terminology both orthorhombic crystals exhibit the same prismatic *habit*, but different combinations of *forms*. i is bounded by (110), (101) and (011) faces, while ii is bounded by (100), (010) and (001).

Fig. 2.13 Demonstration of the difference between crystal habit and polymorphic form for the three polymorphs of naphthazarin. Upper left, polymorph A; upper right, polymorph B; lower, polymorph C in two different habits. (Crystals prepared and photos taken by Kress and Etter 1989.)

2.4.2   *Structural characterization and comparison of polymorphic systems*

As Herbstein (2001) has noted, consideration of polymorphism in terms of classical thermodynamics ignores the structural aspects of the phenomenon, an approach he terms as purely physical. At the other extreme of his proposed physics–chemistry scale for polymorphism, molecules of a substance would exhibit appreciable chemical differences between the chemical entities in the two polymorphs. At the chemical end of the scale one is dealing with structural characterization and comparison, and a number of typical examples along the physics–chemistry scale are discussed in detail by Herbstein.

In dealing with a polymorphic system perhaps the fundamental question is how similar or how different are the various crystal structures. Although Gavezzotti and Fillippini (1995) have attempted to provide a quantitative measure to this comparison, we believe that investigation of this question should include a close visual examination of the crystal structures as outlined in the next section.

2.4.3   *Presentation of polymorphic structures for comparison*

Even for the trained and practised eye, a single crystal structure of a molecular solid is rarely understood with ease, so that *comparison* of two or more crystal structures, even involving the same molecule in polymorphic structures, can be an exercise in frustration. Very often, the best means for examining and comparing polymorphic crystal structures is to plot them *on the same molecular reference plane*, and orient that plane in the same way for all of the structures. Such a strategy allows a ready comparison of the immediate environment about the reference molecule and the intermolecular interactions that dominate the structures. When the molecule in question can exhibit conformational flexibility then the molecular reference plane chosen for projection of the structure should be a rigid part of the molecule (e.g. a phenyl ring) or an appropriate group of three atoms. If the purpose of the figure is to demonstrate the differences in

Table 2.1   Comparison of cell constants for the two polymorphs of L-glutamic acid

	α Form	β Form
a (Å)	7.068	5.519
b (Å)	10.277	17.30
c (Å)	8.755	6.948

conformation as well as the differences in crystal structure, then the reference plane should be chosen such that the torsion or dihedral angles that lead to the largest differences in molecular conformation are immediately adjacent to the reference plane. However, such figures can be confusing, simply because the projection may contain many overlapped atoms, in which case an alternative is to choose a reference plane which is distant by a number of bonds from the ones that are the major source of the conformational difference. If these views are given in stereo, then views for the different structures to be compared should appear above and below each other to facilitate comparison. Exemplary samples of this presentation strategy are shown in Fig. 2.14.

This strategy unfortunately is contrary to the usual practice in preparing packing diagrams, most of which are given as the view down a particular crystallographic axis or on a particular crystallographic plane. Such views carry very little information with regard to similarities or differences in polymorphic structures. However, in some cases, polymorphs have some similar cell constants or symmetry elements in common, and the similarity may be manifested in the crystal packing. In such a case, plotting the structures in such a way as to view and compare those ostensibly common features is essential for the analysis. An example is the α- and β-modifications of L-glutamic acid (Lehmann *et al.* 1972). The two structures crystallize in the same orthorhombic space group, $P2_12_12_1$. Comparison of the cell constants in Table 2.1 suggests a possible axial relationship as follows:

$$a_\alpha \cong c_\beta \qquad b_\alpha \cong 2a_\beta \qquad 2c_\alpha \cong b_\beta$$

Therefore, plotting the structures so that the possibly common $a_\alpha \cong c_\beta$ axis is horizontal and the possibly doubled $2c_\alpha \cong b_\beta$ axis is vertical greatly aids in making this comparison. In fact, it is seen in Fig. 2.15 that in spite of the identity in space group and the apparent possible relationship between some cell constants of the two polymorphs, there is no similarity in the crystal structures (see also Section 5.9, Fig. 5.17).

An additional example will serve to demonstrate the similarities in structures that can be revealed by taking care in the way they are plotted. The cell constants for the two forms of 5-methyl-1-thia-5-azacyclooctane 1-oxide perchlorate (Paul and Go 1969) suggested the following axial relationships:

$$2a_\alpha \cong a_\beta \qquad b_\alpha \cong b_\beta \qquad c_\alpha \cong 2c_\beta$$

Table 2.2 Comparison of cell constants for the two polymorphs of 5-methyl-1-thia-5-azacyclooctane 1-oxide perchlorate

	α Form	β Form
a (Å)	9.87	20.10
b (Å)	8.78	8.89
c (Å)	13.26	6.77
β (°)	97.90	97.80
Space Group	$P2_1/c$	$P2_1/a$

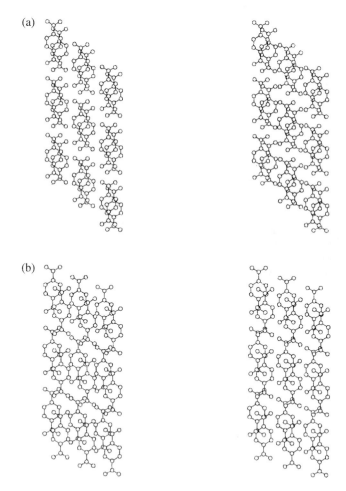

Fig. 2.14 (*Continued*)

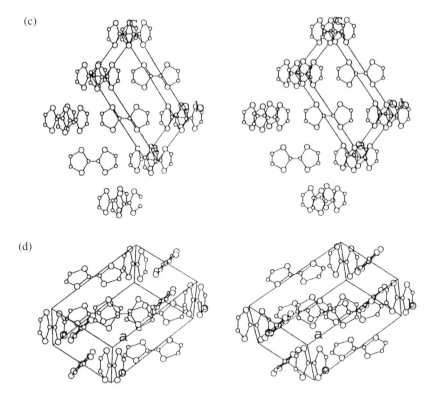

Fig. 2.14 Examples of comparative stereoplots of the packing of polymorphic structures, with the view chosen to convey similarities and differences in packing. (a) and (b) Terephthalic acid (Bailey and Brown 1967, 1984; Berkovich-Yellin and Leiserowitz 1982), plotted on the plane of the phenyl ring. Both polymorphs are triclinic, $P\bar{1}$, and are composed of very similar layers (which is essentially parallel to the plane of the paper) comprising linear chains of hydrogen-bonded molecules. The lateral offset of chains within a layer is also very similar for the two structures. The structures do differ in the manner in which subsequent layers are offset from each other. (after Davey *et al.* 1994.) (c) and (d) Tetrathiafulvalene (Coppens *et al.* 1971; Cooper *et al.* 1974; Ellern *et al.* 1994.) (c) Form 1 is monoclinic, $P2_1/c$; (d) Form 2 is triclinic $P\bar{1}$. In (c) and (d) the two structures are plotted on the best plane of the central molecule and oriented so that the long axis of that molecule is horizontal. Especially in mono view (c) suggests a layered structure for Form 1; however, a rotation of slightly less than 90 degrees about the horizontal axis (to better enable viewing of all of the atoms of the molecule) would indicate that the packing is better described as a herring-bone motif. Note that the views chosen here, especially that for Form 2, may not be the best one for interpreting or discussing that structure *individually*. The views chosen do, however, greatly facilitate *comparison* of the polymorphic structures.

and plotting both structures with a view down the *b* axis (Fig. 2.16) reveals the close similarity between them, as discussed in detail in the original paper.

The two examples given in Figs 2.15 and 2.16 demonstrate another point that deserves emphasis, especially to non-crystallographers. The identity or non-identity

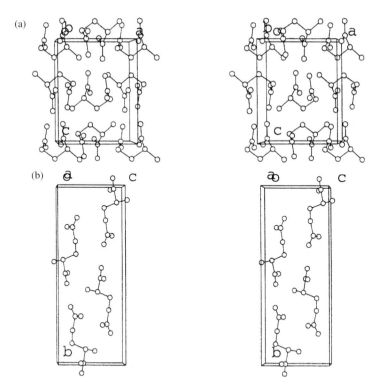

Fig. 2.15 Stereoplots of the packing of the two forms of L-glutamic acid. The figures are oriented to facilitate comparison of the axes of the two forms, which are apparently either similar or related by whole number ratios (Table 2.1 and text). Upper, α Form; lower, β Form (adapted from Bernstein 1991*a*, with permission).

of space groups for polymorphic structures is purely coincidental. No structural similarity can be associated with polymorphic structures on the basis of space group identity. Also, the lack of structural similarity between polymorphs cannot be assumed simply on the basis of a comparison of the space groups. Space group symbols provide information on the symmetry elements present in the crystal structure. They do not provide details on the intermolecular relationships or on the environment of any particular molecule in a crystal structure. For the non-crystallographer it should be noted that $P2_1/c$, $P2_1/a$, and $P2_1/n$ are equivalent space groups (No. 14 in the *International Tables*) (1987).

2.4.3.1 *Characterization of hydrogen-bond structures*

Because hydrogen bonds are the strongest and most directional of intermolecular interactions, they have been the subject of study since the 1920s (Latimer and Rodebush 1920; Pimentel and McClellan 1960). The strength and directionality of these bonds might suggest a high incidence of polymorphism for those compounds which contain

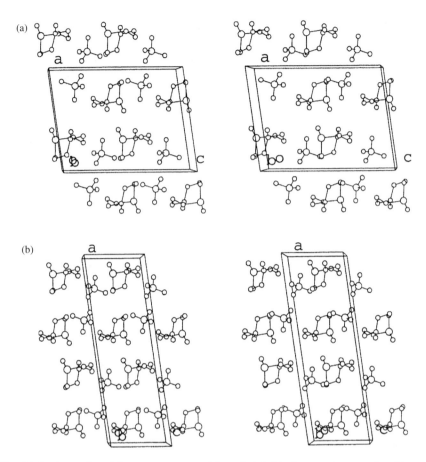

Fig. 2.16 Stereoplot of the two structures of 5-methyl-1-thia-5-azacyclooctane 1-oxide perchlorate. In both cases the view is along the *b* crystallographic axis. Note the similarity in the *arrangement* of the molecules, although a casual glance at the unit cell dimensions might suggest that the structures are very different. Upper, *α* Form; lower, *β* Form. (Adopted from Paul and Go 1969, with permission.)

a number of potential hydrogen-bond donors and hydrogen-bond acceptors, since they might be combined in a number of different ways. However, Etter (1990, 1991) pointed out that in the formation of hydrogen bonds not all combinations of donor and acceptor are equally likely, since strong hydrogen donors (strongly acidic hydrogens) will tend to form hydrogen bonds preferentially with strong hydrogen acceptors (atoms with electron pairs, etc.). Hence, the preference for certain hydrogen bonds to form might tend to lead to the preservation of hydrogen-bond patterns among polymorphs rather than the proliferation of polymorphs for molecules with many potential hydrogen bonds. The traditional approach for comparing the hydrogen bonding among polymorphs is to list the geometry (bond lengths and angles) of the individual hydrogen bonds for each structure, sometimes including the crystallographic symmetry elements which generated each of the hydrogen bonds (e.g. Bernstein 1979). This approach carries very little information which can serve as a basis for *comparing* the *patterns* of hydrogen bonds among the polymorphs. A much more useful method involves the utilization of graph sets (Wells 1989; Hamilton and Ibers 1968; Kuleshova and Zorky 1980; Zorky and Kuleshova 1980; Chang *et al.* 1984), also originally proposed and demonstrated in its current context by Etter (1985, 1990, 1991; Etter *et al.* 1990*a,b*), with subsequent modifications and refinements (Bernstein *et al.* 1995; Grell *et al.* 1999). This method can be quite useful and informative in comparing polymorphs dominated by hydrogen bonds (e.g. Griesser *et al.* 1997); hence we review the fundamental principles of the graph set method in the next section and follow that with a section describing a number of typical examples of polymorphic systems.

Although the Etter approach was developed and applied to hydrogen-bonded networks in molecular crystals, earlier application of graphs in inorganic systems Wells (1989) and subsequent developments (Navon *et al.* 1997; Grell *et al.* 1999) suggest the potential widespread use of graphs for characterizing crystal structures in general (e.g. Moers *et al.* 2000), and comparing polymorphic structures in particular.

2.4.3.1.1 *Graph set notation for the description of hydrogen bonds:*

1. *The designator:* A remarkable feature of the graph set approach to analysis of hydrogen-bond patterns is the fact that most complicated networks can be reduced to combinations of four simple patterns, each specified by a designator: chains (**C**), rings (**R**), intramolecular hydrogen-bonded patterns (**S**), and other finite patterns (**D**). Specification of a pattern is augmented by a subscript **d** designating the number of hydrogen-bond donors (most commonly covalently bonded hydrogens, but certainly not limited to them), and a superscript **a** indicating the number of hydrogen-bond acceptors. When no subscript and superscript are given, one donor and one acceptor are implied. In addition, the number of bonds **n** in the pattern is called the degree of the pattern and is specified in parentheses. The general graph set descriptor is then given as $\mathbf{G_d^a(n)}$ where **G** is one of the four possible designators.

These four patterns and their descriptors are best illustrated by examples. A chain whose 'link' is composed of four atoms as in **2-I** is specified as **C(4)**. Similarly, the intramolecular hydrogen bond in **2-II** would be specified as **S(6)**, for the six atoms comprising the intramolecular pattern. When the donor and acceptor are from two (or more) *discrete* entities (molecules or ions), as in **2-III**, the designation of the hydrogen bond is **D**. The entities may differ on chemical grounds (different ions or molecules) or on crystallographic grounds (chemically identical but not related by a crystallographic symmetry operation). In **2-III** there is only one donor and one acceptor, and the pattern involves only one hydrogen bond. The fourth possible pattern is the ring **2-IV**. In the example shown, the two hydrogen bonds in the ring could be different but in this case they are related by a crystallographic inversion centre. The

2-I   C(4)   2-II   S(6)   2-III   D

2-IV $R_2^2(8)$

pattern contains a total of eight atoms, two of them donors and two acceptors, and hence is designated $\mathbf{R_2^2(8)}$.

2. *Motifs and levels:* A motif is a pattern containing only one type of hydrogen bond (Etter *et al.* 1990a,b). Specifying the motif for each different hydrogen bond in a network according to one of the four pattern descriptors above leads to a description of the network in the form of a list of the motifs. This is the *unitary*, or *first level*, graph set, noted as $\mathbf{N_1}$.

The chemically interesting or topologically characteristic patterns of a system often appear when more than one type of hydrogen bond is included in the description, that is in higher-level graph sets (Etter *et al.* 1990a; Bernstein *et al.* 1995; Bernstein 1991a,b; Bernstein *et al.* 1990). This will be true for **S**, **D**, and **C** patterns. Suppose a structure contains three distinct hydrogen bonds, designated *a*, *b*, and *c*. There will now be several possible *binary* (or *second level*) graph sets, each one describing a pattern formed by two of these H-bonds—that is $\mathbf{N_2(ab)}$, $\mathbf{N_2(ac)}$, and $\mathbf{N_2(bc)}$. These concepts are illustrated in Fig. 2.17. The *ternary* or *third level* graph sets are those that involve three hydrogen bonds (Bernstein *et al.* 1995).

Different pathways might also be found that include the same set of H-bonds, but with different degrees ($\mathbf{n}$). To describe this situation, the term *basic* was suggested to describe the graph set of the *lowest degree* and the term *complex* to describe ones of *higher degree*. Consideration of some of the choices for the binary graph set for α-glycine (Power *et al.* 1976; Legros and Kvick 1980) illustrates this point (Fig. 2.18). The shortest path involving H-bonds *a* and *b* gives the binary graph set $\mathbf{C_2^2(6)}$; this is the *basic* binary graph set for *a* and *b*. A longer chain, $\mathbf{C_2^2(10)}$, represents a *complex* binary graph set for the two H-bonds. Neither of these, however, describes the most obvious feature of this array, the ring structure, which is denoted by another complex binary graph set $\mathbf{R_4^4(16)}$. In addition, in this structure it is possible to define an infinite number of increasingly larger ring systems. In such a case it remains for the chemist or crystallographer to choose a ring or those rings that best characterize the particular structure in question (Bernstein *et al.* 1995).

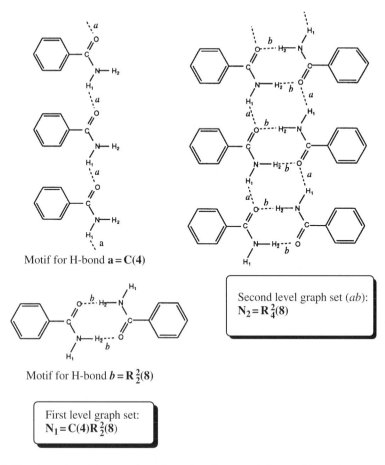

Motif for H-bond **a = C(4)**

Motif for H-bond **$b = R_2^2(8)$**

Second level graph set (*ab*):
$N_2 = R_4^2(8)$

First level graph set:
$N_1 = C(4)R_2^2(8)$

Fig. 2.17 Examples of the use of graph set descriptors to define motifs and first and second level graph sets for schematic representations of the hydrogen-bond patterns in the crystal structure of benzamide. (From Bernstein and Davis 1999, with permission.)

3. *Practicalities of assigning graph sets*: The early work on graph sets required a fairly tedious process of plotting the structures, identifying the patterns and counting the number of atoms, donors and acceptors in order to determine the descriptor. It was also a process fraught with possibilities for errors and inconsistencies. A significant advance has been achieved recently with the automatic identification and visualization of hydrogen-bonding patterns and the assignment of the graph set in the standard software of the CSD (Grell *et al.* 1999; Motherwell *et al.* 1999). These developments will greatly facilitate the use of graph sets in the study and comparison of polymorphic systems.

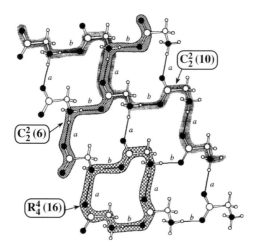

Fig. 2.18   Graph set assignments for the binary level of α-glycine. As in Fig. 2.17, different types of hydrogen bonds (solid lines) are distinguished by labelling with lower case bold letters; carbon and hydrogen atoms are shown as open circles, oxygen atoms as solid circles, and nitrogen atoms as shaded circles. (From Bernstein and Davis 1999, with permission.)

2.4.3.1.2   *Differences and similarities in hydrogen-bond patterns among poly-morphs:*   When hydrogen bonding (and in principle any other pattern-definable intermolecular interaction (Grell *et al.* 2002)) is involved in the packing it is possible to use the graph set notation to describe the similarities and differences among polymorphic structures. This application of graph sets to polymorphic systems was first demonstrated with iminodiacetic acid (Bernstein *et al.* 1990) and glutamic acid (Bernstein 1991*a*), and has subsequently been employed with increasing frequency (MacDonald 1994; Bernstein *et al.* 1995; Nakano *et al.* 1996; Ceolin *et al.* 1996; Griesser *et al.* 1997). Perhaps not surprisingly, it is generally found that the hydrogen-bond patterns do differ among polymorphs, although the analysis may have to take into account higher complex or higher-level graph sets in order to recognize the distinction (Bernstein 1991*a*; MacDonald 1993).

The bookkeeping and the comparison of the graph sets for different crystal structures are considerably simplified by the preparation of a matrix-like table that summarizes the graph set assignments (Bernstein *et al.* 1995; Grell *et al.* 1999). The columns and rows are designated in identical order from top to bottom and left to right, respectively, each column and each row being labelled by one of the identified hydrogen bonds. The 'diagonal elements' of the resulting matrix then contain first level graph set assignments (motifs), while the off-diagonal ones contain the binary level graph set assignments. For purposes of comparison it is important that chemically identical (or lacking identity, chemically similar) hydrogen bonds be given identical literal

designation. Then, ordering the columns and rows the same way for, say, polymorphic structures or a family of compounds with similar hydrogen-bonding capabilities
greatly facilitates the comparison of the pattern of interactions in these structures.

2.4.3.1.3 *A worked example—anthranilic acid:*   These principles can be demonstrated with the structures of anthranilic acid **2-V**, a trimorphic material whose
structural chemistry has been widely studied (Anonymous 1949; Brown 1968*c*;
Arnold and Jones 1972; Boone *et al.* 1977; d'Avignon and Brown 1981; Hardy
et al. 1981; Brown and Ehrenberg 1985; Takezawa *et al.* 1986; Ojala and Etter 1992).
The molecule may exist in either the neutral **2-Va** or zwitterionic **2-Vb** form. In either
case each molecule has three hydrogens that can participate in hydrogen bonds. For
the neutral molecule the carbonyl oxygen is the obvious candidate as the best hydrogen acceptor, with the hydroxyl oxygen considerably less so. For the zwitterionic
case, both oxygens are equally good acceptors. These structures and their graph sets
are presented in Figs 2.19 and 2.20. In Form II and in Form III there is one neutral
molecule per asymmetric unit. In Form I there are two molecules per asymmetric unit,
one neutral and one zwitterionic. In Form II (Fig. 2.19(a)) the hydrogen of the carboxyl group forms the ring structure $\mathbf{R}_2^2(\mathbf{8})$ about a crystallographic inversion center.
One of the amino-hydrogens participates in an intramolecular motif $\mathbf{S(6)}$. The closest
acceptor to the second amino-hydrogen is 2.38 Å. Hence by 'normal' standards it
might not be considered as participating in a hydrogen bond (Jeffrey and Saenger
1991). However, that hydrogen clearly points in the direction of a carbonyl oxygen
(O$\cdots$H $= 2.49$ Å), and although the latter must then be counted as participating in
three hydrogen bonds we believe that the pattern exhibited here warrants including
this in the hydrogen-bond network as a chain, $\mathbf{C(6)}$. Thus the first level graph set is
$\mathbf{N_1 = C(6)S(6)R_2^2(8)}$(diagonal terms in Table 2.3).

2-V

In Form III (Fig. 2.19(b)) once again the carboxylic acid forms the $\mathbf{R}_2^2(\mathbf{8})$ ring
about a crystallographic inversion center. One amino-hydrogen is intramolecularly
hydrogen bonded to the carboxyl carbonyl to again yield $\mathbf{S(6)}$, and the second
amino-hydrogen is hydrogen bonded this time to the carboxy-hydroxyl oxygen

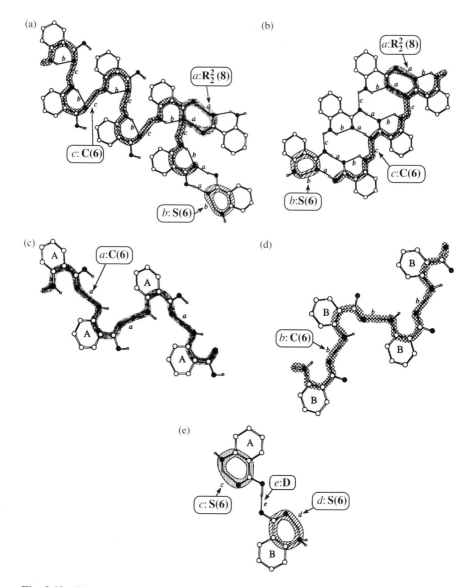

Fig. 2.19 First level graph sets for the three polymorphs of anthranilic acid **2-V**. The two molecules in the asymmetric unit of Form I are shown in their true crystallographic relationship to each other (see text). In (d) and (e) the third ammoniacal hydrogen, which does not participate in a hydrogen bond, is hidden by the nitrogen to which it is attached. (a) Form II; (b) Form III; (c) Form I, **C(6)** of A type molecules; (d) Form I, **C(6)** of B type molecules; (e) Form I, intramolecular motifs **S(6)** for molecules A and B and the dimer **D** relationship connecting them. (Adapted from Bernstein *et al.* 1995, with permission.)

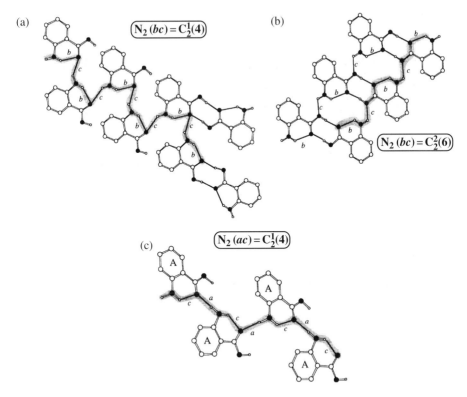

$$\boxed{N_2\,(bc) = C_2^1(4)}$$

$$\boxed{N_2\,(bc) = C_2^2(6)}$$

$$\boxed{N_2\,(ac) = C_2^1(4)}$$

Fig. 2.20 Selected members of the second level graph set for anthranilic acid **2-V**. (General comments in the caption in Fig. 2.19 also apply here.) (a) Form II; (b) Form III; (c) Form I. (Adapted from Bernstein *et al.* 1995, with permission)

$(\mathrm{O}\cdots\mathrm{H} = 3.06\,\text{Å})$ to form the chain $\mathbf{C(6)}$.[4] Thus at the first level the graph sets of Forms II and III are identical $N_1 = \mathbf{R_2^2(8)S(6)C(6)}$.

The two molecules in the asymmetric unit of Form I can be distinguished chemically. Neutral molecule A forms a $\mathbf{C(6)}$ using an amino-hydrogen and a carboxyl carbonyl (Fig. 2.19(c)); the zwitterion B has the same motif using an ammoniacal

---

[4] The designator for the *c* hydrogen bond for both Forms II and III deserves some additional comment. For Form II that hydrogen bond is recognized automatically by the CSD software, due to a default threshold value of 2.65 Å for O—H···O hydrogen bonds. The distance for a similar hydrogen bond in Form III is beyond the threshold value. Etter (1990, 1991) pointed out the importance of recognizing *patterns* of hydrogen bonds rather than identifying them solely on the basis of some geometrical criteria. Noting that the potential energy function for the hydrogen bond (and for other interactions) is a continuous function Jeffrey and Saenger (1991) also indicated the dangers of using an arbitrary geometrical parameter as a criterion for defining a hydrogen bond. Those caveats have been heeded here resulting in the recognition of hydrogen bond *c* in Form III and the comparison of the two forms that can then be readily distinguished at the binary level. We are not proposing here abandoning geometric criteria for the automatic definition of hydrogen bonds, but we do suggest using caution in choosing the threshold values for those geometric parameters, and in cases such as this one, examining the structures carefully for patterns that suggest the presence of hydrogen bonds beyond the threshold values.

hydrogen and a carboxylate oxygen (Fig. 2.19(d)). As in Forms II and III, a second hydrogen on the nitrogen leads to an **S(6)** motif for each of the molecules in the asymmetric unit (Fig. 2.19(e)). Molecules A and B are linked together via an O–H$\cdots$O (carboxylate) bond, a **D** motif. The sixth hydrogen, that is, the remaining ammoniacal one on molecule B, is not involved in hydrogen bonding, at least for O$\cdots$H distances less than 3 Å. Hence the first order graph set for Form I is $\mathbf{N_1 = C(6)C(6)S(6)S(6)D}$.

To distinguish between Form II and Form III we turn to the binary graph sets. In principle it is possible to combine the three hydrogen bonds pairwise in three different ways. One of the combinations, $\mathbf{C_2^1(4)}$ for Form II is shown in Fig. 2.20(a), which reflects the fact that both hydrogen bonds in the chain are to the same carbonyl oxygen acceptor. For Form III (Fig. 2.20(b)) the chain $\mathbf{C_2^2(6)}$ contains both oxygens of the carboxyl group; hence there are two acceptors and two additional atoms in each chain 'link'. Also, note that an equal number of atoms in the chain formed by *ca* in both Forms II and III, but the former has one acceptor and the latter has two acceptors. The

Table 2.3 Unitary motifs (diagonal) and binary graph sets (off-diagonal) for anthranilic acid **2-Va**, Form II[a]

Designator H-bond	*a*	*b*	*c*
a	$R_2^2(8)$		
b	$R_2^2(16)$	S(6)	
c	$C_2^1(8)$	$C_2^1(4)$	C(6)

[a] See Figs 2.19 and 2.20.

Table 2.4 Unitary motifs (on-diagonal) and binary graph sets (off-diagonal) for anthranilic acid **2-Va**, Form III[a]

Type of H-bond	*a*	*b*	*c*
a	$R_2^2(8)$		
b	$R_2^2(16)$	S(6)	
c	$C_2^2(8)$	$C_2^2(6)$	C(6)

[a] See Figs 2.19 and 2.20.

Table 2.5 Unitary motifs (on-diagonal) and binary graph sets (off-diagonal) for anthranilic acid **2-Va**, Form I[a]

Type of H-bond	*a*	*b*	*c*	*d*	*e*
a	C(6)				
b	—	C(6)			
c	$C_2^1(4)$	—	S(6)		
d	—	—	—	S(6)	
e	$D_3^3(11)$	—	—	—	D

[a] See Figs 2.19 and 2.20.

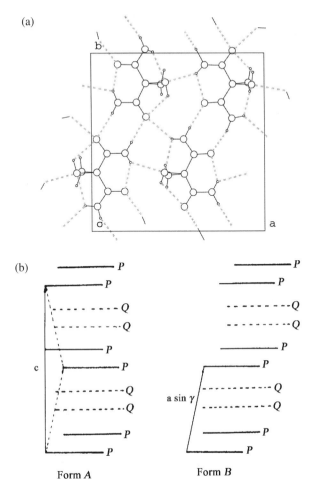

Fig. 2.21 (a) The chain structure (running horizontally) of 2,2-aziridinecarboxamide **2-VII** common to the two polymorphic structures. (b) The schematic relationship between stacking of the layers composed of chains in the two forms. The views are along the *b* axis in the tetragonal Form A and along $c \times a^*$ in the triclinic Form B. (Adopted from Brückner 1982, with permission.)

binary graph set for Form I, $C_2^1(4)$(Fig. 2.20(c)) is identical to that of Form II, but it is important to note that only A type molecules are present in this pattern. These similarities and differences are conveniently summarized and are readily apparent in the summary 'matrix tables' 2.3–2.5.

2.4.3.1.4 *Additional examples:* There are, of course, cases where the hydrogen bonding is essentially identical among the polymorphs of a system. A simple, and simply understood, case is that of terephthalic acid **2-VI** (Davey *et al.* 1994) (Fig. 2.14). The presence of two identical carboxylate groups in the *para* position

(a)

Form I

Form II

(b)

Form I

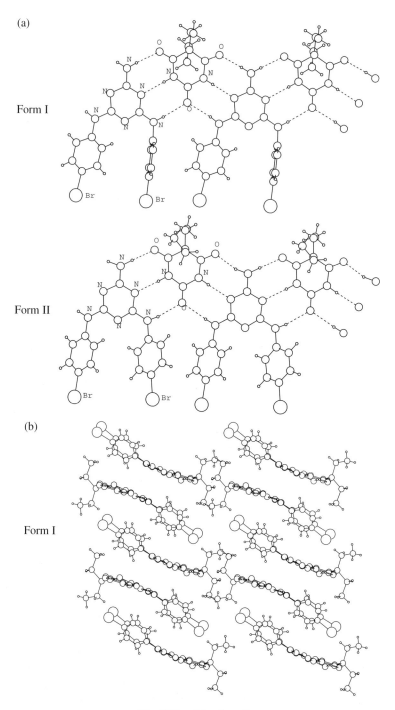

Fig. 2.22 (*Continued*)

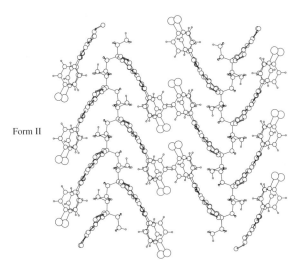

Form II

Fig. 2.22 Comparison of the crystal structures in the two polymorphs of the cocrystal of **2-VIII** and **2-IX**. (a) The hydrogen-bonded tapes. Note the difference in the orientation of the bromophenyl substituents along the chain. (b) End-on view of the packing of the tapes in the two polymorphs. (From Zerkowski *et al.* 1997, with permission.)

of the benzene ring essentially dictates a chain motif, and the two polymorphs differ slightly (but measurably) in the relationship of neighbouring chains to each other (to form layers) and the subsequent relationship of layers to each other in the two polymorphs (See Section 2.4.3). A second example may be found in the two polymorphs of 2,2-aziridinecarboxamide **2-VII** (Brückner 1982) (Fig. 2.21), again in which identical chains of $\mathbf{R}_2^2(8)$ rings are formed in the two structures. The combination of space groups and molecules in the asymmetric unit ($P4_12_12$ ($Z = 4$) and $P1$ ($Z = 16$) respectively) in this pair of polymorphs is quite unusual for organic materials. The difference in the relationship among neighbouring layers is shown in Fig. 2.21(b).

2-VI   2-VII   2-VIII   2-IX

An additional example for a *cocrystal* of N,N′-bis(*para*-Br-phenyl)melamine **2-VIII** and barbital **2-IX** Zerkowski *et al.* (1997). There are clearly a number of potentially different hydrogen-bonding donor and acceptor sites on each molecule (in fact that was part of the rationale for studying these cocrystalline materials). The two polymorphic structures have identical tape hydrogen-bonding patterns, but the tertiary structure, as defined by the arrangement of the tapes within the crystal is quite different (Fig. 2.22).

3

Controlling the polymorphic form obtained

Crystal growth is a science and an art. The scientist's role in the crystal growth process is that of an assistant who helps molecules to crystallize. Most molecules, after all, are very good at growing crystals. The scientific challenge is to learn how to intervene in the process in order to improve the final product. (Etter 1991)

I am very sorry, that to the many... difficulties which you meet with, and must therefore surmount, in the serious and effectual prosecution of Experimental Philosophy, I must add one discouragement more, which will perhaps as much surprise you as dishearten you; and it is, that besides that you will find... many of the experiments published by Authors, or related to you by the persons you converse with, false or unsuccessful, ... you will meet with several Observations and Experiments, which though communicated for true by Candid Authors or undistrusted Eye-witnesses, or perhaps recommended to you by your own experience, may upon further tried disappoint your expectation, either not at all succeeding constantly, or at least varying much from what you expected.

This Advertisement may seem of so discouraging a nature that I should much scruple the giving it to you, but that I suppose the trouble at that unsuccessfulnesse which you may meet with Experiments, may be somewhat lessened, by your being forewarned of such contingencies. And that you should have the luck to make an Experiment once, without being able to make the same thing again, you might opt to look upon such disappointments as the effect of and unfriendliness in Nature or Fortune to your particular attempts, as proceeding from a secret contingency incident to some Experiments, by whomever they be tried. (Boyle 1661)

Control nevertheless <u>is</u> important in science—tremendously so—not as an end but as a component of proof. The ability to control is the strongest possible demonstration of true understanding. Many doubted whether Becquerel, Curie, Bohr, Oppenheimer and the rest really understood what causes what inside the atom. But after July 16, 1945, when the day dawned prematurely to the northwest of Alamagordo, at White Sands, New Mexico, no one could possibly doubt any more, for the atom bomb was plainer than the sun. With a demonstrated ability to control, the good scientist may sign off like the mathematician at the end of a proof: *Quod erat demonstrandum*. (Huber 1991)

3.1 General considerations

Crystallization is a process that has fascinated both scientists and casual observers throughout the ages. It is indeed remarkable that upwards of 10^{20} molecules or ions, distributed essentially randomly throughout some fluid medium (gas, liquid, or solution) coalesce, very often spontaneously, to form a regular solid with a well-defined structure, or in the case of polymorphs, with a limited number of well-defined structures. Those structures are invariant across a wide variety of conditions, in some cases almost under any conditions for which crystals form. Two of the principal questions to be asked for such a process is how it begins and how it proceeds, especially in the context of polymorphic systems. A great deal of work has been devoted to attempts to answer these questions, and in spite of considerable progress especially on experimental and empirical fronts, there is still much to be learnt in developing current models. Historical treatments of the classic notions of crystallization and recrystallization, including many important references, have been given by Tipson (1956) and van Hook (1961). A more recent thorough account may be found in Mullin's book (1993).

For any substance it is possible in principle to define experimentally the solvents, temperature range, rate of evaporation or cooling, and many of the other conditions under which it will crystallize. This collection of conditions has been called the *occurrence domain* (Sato and Boistelle 1984). That domain exists for any substance, but rarely, if ever, are its contents completely known. The contents of the occurrence domain for any material—in the present context, any polymorph—are not necessarily unique. In regions in which there is an intersection of domains, one may expect that two or more polymorphs would crystallize under essentially identical conditions. On the other hand, determining which regions of the domain are unique to a particular polymorph can be advantageous in determining crystallization strategy. This chapter deals with a number of the factors which should be considered in making such a determination, along with examples of the phenomena associated with competitive polymorphic crystallizations.

3.2 Aggregation and nucleation

The thermodynamics and kinetics outlined in Chapter 2 attempt to treat the question of crystallization on the macroscopic scale. On the microscopic scale we would like to be able to answer questions about the critical size and structure of a collection of molecules that will grow into the eventual crystal. In particular, how and when will polymorphs be obtained or be prevented from forming? Classically, the first stage of crystallization is viewed as nucleation, the spontaneous formation or introduction of a nucleus, or centre of crystallization, in the crystallization medium from which crystals may grow, although nuclei may also be destroyed before growing into larger crystals. The size of such nuclei has been a matter of considerable discussion. On the one hand, for instance, Ostwald (1902) claimed that particles containing between 10^8 and 10^{12} molecules are not sufficiently large to induce crystallization from supersaturated solutions, but later work indicated that much more modest numbers (e.g. $10-10^5$) may be considered a critical size to generate crystals (McIntosh 1919; Tamman and von Gronow 1931). This amounts to a cube of approximately 100 Å on an edge and a 'crystal' nucleus weighing as little as 10^{-18} g. Additional aspects of the question of the size of a crystal nucleus are discussed by Mullin (1993).

In an attempt to avoid some of the confusion extant in the current literature on the nature of nucleation, Mullin has provided a useful schematic classification for various terms in use:

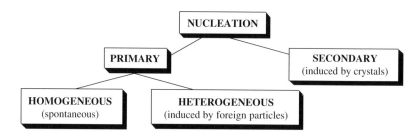

Primary nucleation refers to those systems that do not contain crystalline matter. When no foreign bodies are present (i.e. the crystallization results from the spontaneous formation of nuclei of the crystallizing material) the process is referred to as homogeneous. The presence (intentional or unintentional) of foreign particles can also induce nucleation, which is then termed heterogeneous.

Secondary nucleation deals with the situation in which nuclei are generated in the vicinity of crystals of the solute already present in a supersaturated solution. The solute crystals may have resulted from primary nucleation or may be deliberately added. This subject has also been covered by Mullin, as well as in a number of reviews (Strickland-Constable 1968; Botsaris 1976; DeJong 1979; Garside and Davey 1980; Garside 1985; Nyvlt *et al.* 1985).

Mullin has argued that the minimum number of molecules in a stable crystal nucleus can vary from about ten to several thousand. A model based on the simultaneous collision of this number of molecules with the degree of order required for it to be recognized by additional molecules as a crystal is highly unlikely. A more likely scenario is that the nucleus would be generated by a sequence of bimolecular additions in which the so-called critical cluster would be built up stepwise:

$$A + A \rightleftharpoons 2A$$

$$A + 2A \rightleftharpoons 3A$$

$$A + (n - 1)A \rightleftharpoons nA \text{ (critical cluster } A_1)$$

In Mullin's model, further molecular additions to the critical cluster results in nucleation.

A solution or melt can contain a variety of clusters $A_1, \ldots, A_n$, in a system of competing equilbria. Each cluster in turn is a potential critical cluster for the nucleus of one or more polymorphic crystal modifications. In the context of polymorphic structures, in particular those which crystallize under similar conditions, there must be a number of processes of this type, all involved in competing equilibria. This is the idea behind Etter's (1991) extension of this model, describing the clusters as aggregates,

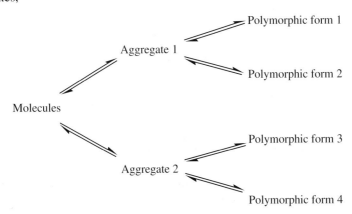

which must contain the structural essence of the eventual crystal structure(s), and are therefore likely to be dominated by the same intermolecular interactions. Because such a system involves multiple equilibria, once nucleation occurs for one of the polymorphic forms, the equilibrium will be displaced in favour of that form at the expense of other forms. On a qualitative basis, this demonstrates the competition between kinetic and thermodynamic factors. For instance, even if Polymorphic Form 1 were the thermodynamically most stable one, Polymorph 3 might be the only one obtained if Aggregate 2 nucleated crystal growth faster than Aggregate 1. When these factors are equal, or very nearly so, then two or more modifications may result from the same aggregate or from different aggregates, leading to concomitant polymorphs (see Section 3.5).

Twenty years before Etter's model for competing aggregate structures in the formation of polymorphs, Powers (1971) clearly stated the fundamental question regarding the challenge of understanding the nucleation process:

It would appear almost certain that the development of these at least transient aggregates are the precursors to the development of phase transitions to the ordered solid nucleus, in harmony with the local thermodynamic conditions. Yet though a vast amount of study—conferences, books, etc.—have been devoted to nucleation, it is still uncertain how this last transition takes place.

Some experimental evidence for the presence of different aggregates in solution leading to polymorphic structures has been presented recently by Näther *et al.* (1996*a*,*b*), and there have been attempts to relate nucleation rates with proposed structures in solution based on molecular modelling (Petit *et al.* 1994). The number of studies of this nature is sure to increase with increasing sensitivity and sophistication of both experimental and computational tools; those are the kinds of investigations that can provide answers to Powers' challenge.

From a practical point of view, control over nucleation, and in cases of polymorphic systems, control over the polymorph obtained as a result of nucleation, has been the concern of those industries for which crystallization is a crucial or final step in the production process: for example, sugars, amino acids, pharmaceuticals, and fatty acids. A number of examples of studies regarding polymorphic variation and preferences for nucleation may be found in the literature from those disciplines.

The initiation of crystal growth has been a problem for the sugar industry since its infancy. Aqueous solutions of sugars often tend to form syrups—indeed, that has become one form of marketing, although clearly not the preferred one. As noted above, Powers has reviewed the role of nucleation in the sugar industry, including much of the accumulated experience involving sucrose. As in many industries, successful techniques developed over decades or centuries and were considered trade secrets or even commonplace practice without being scientifically recognized or understood. Thus, traditionally, sucrose crystallizations, carried out from the huge copper vats in which solutions were concentrated, were initiated (i.e. nucleated) by the mechanical shock of hammering on the vat (Fig. 3.1). Yet it was only in 1912 that Young (1911) described mechanical shock as a factor in nucleating supersaturated solutions (Powers 1971). Two other curiosities relating to nucleation of sugars demonstrate some of the difficulties encountered. Turanose was long considered to be a liquid at room temperature (Powers 1971), until it spontaneously crystallized; following that event fresh batches of the material always crystallized. In another case more closely related to polymorphism, α-D-mannose had been prepared routinely until the appearance of β-D-mannose, following which the α form could not be induced to crystallize in the same laboratory (Levene 1935; Dunitz and Bernstein 1995). As Powers noted, both of these cases can be attributed to unintentional seeding, a topic treated in more detail in Section 3.6.

Black and Davey (1988) describe a number of the interrelationships and practical aspects of the control of nucleation, crystal growth, and polymorphic transformation of amino acids. The factors described and demonstrated for primary nucleation of L-glutamic acid include temperature, critical nucleus, relationship of interfacial tension to solubility, thermal history, induction time, agitation, and effect of additive.

Kitamura (1989) studied many of these nucleation factors in the competitive crystallization of the α and β forms of L-glutamic acid. He found that at 25 °C only the α modification nucleates and grows. In this system, at least, the effect of temperature on the relative nucleation rates of the two polymorphs is more 'remarkable' than the effect of the supersaturation ratio: as the temperature is increased with a constant supersaturation ratio, the amount of α decreases. He also reported that the β form tends to nucleate in stagnant solutions, while at 25 °C essentially only α nucleated homogeneously.

In an example from the pharmaceutical industry Sudo *et al.* (1991) studied the relative nucleation properties of forms A and B of cimetidine, which is reported to have four polymorphic non-solvated forms and three polymorphic monohydrates. Modification A is preferred for pharmaceutical formulations. The 'waiting time method' was used to study the primary nucleation process (Harano and Oota 1978), mainly for competitive crystallization of the A and B modifications. A is a thermodynamically metastable form and is more soluble than B in any solvent. At high supersaturation

CRYSTALLIZING PANS

Fig. 3.1   Detail from an 1850 drawing of the London sugar refineries of Messrs. Fairrie and Co., showing a copper crystallizing pan for sugar. The worker to the right of the pan is holding a mallet which was used to bang on the pan to induce nucleation of the crystallization process. (Reproduced from Fairrie 1925, with permission.)

$(S_A \geq 4.5)$ modification A is obtained, in the presence or absence of seeds. At $S_A \geq 3.6$, A was obtained regardless of the form of seed. At $S_A \leq 2.0$ the form of the seed determined the form obtained.

Stearic acid is often considered a prototype for the long chain acids used in many processes and applications. Sato and Boistelle (1984) studied the occurrence and crystallization behaviour of three of the polymorphic modifications (A, B, and C) by varying conditions such as temperature, supersaturation, and solvent from which they determined occurrence domains for the existence of the three forms. Polymorph A is thermodynamically unstable at all temperatures studied; below 30 °C form B is most stable, while form C is more stable above 30 °C. Forms A and C nucleate preferentially from non-polar solvents at high supersaturation, whereas polymorph B nucleates more readily at lower supersaturations. Nucleation of Form B is preferred at higher supersaturations from polar solvents. This solvent effect could be influenced by the rate of stirring. If stirring is sufficiently 'violent' it increases nucleation, which enhances the formation of Form B.

The mere existence of polymorphic structures can be used as a probe of the nucleation process. For instance in considering the aggregation process in supersaturated solutions of 2,6-dihydroxybenzoic acid, Davey *et al.* (2000) found a direct link between the relative occurrence of two polymorphic forms (from toluene and chloroform solutions) and the solvent-reduced self-assembly (aggregation) of the molecule.

Nucleation from the melt has been studied for palm oil, composed of triglycerides of palmitic and oleic acids, and exhibiting at least three polymorphs (van Putte and Bakker 1987). Nucleation curves (induction time τ vs temperature T) of palm oil and palm stearin show discontinuities at 297 and 306 °C respectively, indicating the onset of nucleation, and the demarcation of the occurrence of the polymorphs, as confirmed by isothermal Differential Scanning Calorimetry (DSC) studies (Ng 1990*a,b*).

3.3   Thermodynamic vs kinetic crystallization conditions

Physical organic chemists have long been accustomed to making the distinction between 'thermodynamic' and 'kinetic' conditions when referring to reactions and reaction mechanisms. In chemical parlance, thermodynamic conditions essentially means those conditions under which thermodynamic equilibrium is maintained or very nearly maintained. Kinetic conditions refer to situations that are far from equilibrium (van Hook 1961).

In terms of crystallization (Ward 1997), thermodynamic conditions might refer to a slow evaporation, a very slow cooling, a slow crystallization from the melt at a constant temperature only slightly below the melting point, a slow sublimation for which there is only a small difference between the temperature of the solid and that of the cold finger on which the sublimate is crystallizing, etc. On the other hand, kinetic conditions might refer to a high degree of supersaturation, rapid cooling of a solution or melt, rapid evaporation of solvent, large temperature difference between the sample and cold finger in a sublimation, etc. A number of examples will serve to demonstrate how these principles have been applied to the crystallization of different modifications in polymorphic systems.

3-I

p-chlorobenzylidene-N-*p*-chloroaniline **3-I** is dimorphic (Bernstein and Schmidt 1972; Bernstein and Izak 1976). The thermodynamically more-stable orthorhombic form may be obtained by slow evaporation of a solution in which ethanol or methylcy-clohexane is the solvent. Prismatic crystals usually grow in hours to days, depending on the initial concentration of the solution. The metastable triclinic form is obtained by dissolving the maximum amount of substance in a boiling ethanolic solution, which is then immediately placed in a desiccator freshly charged with calcium chloride. Needle-like crystals appear within minutes, and the desiccant aids in accelerating the rate of evaporation. The crystals are metastable and may begin to spontaneously transform to the orthorhombic modification in periods ranging from hours to days.

Berman *et al.* (1968) noted that 'mannitol is unusual among carbohydrates in that exists in several polymorphic forms', indicating that a number of these are often obtained simultaneously. They describe the preparation of a number of these mod-ifications. The α form is obtained by slow crystallization from 96 per cent ethanol, the α' form by evaporation from 100 per cent ethanol and the β form from aqueous ethanolic solutions, all apparently under thermodynamic conditions. On the other hand the γ form is obtained kinetically by rapid cooling of a 1:1 water–ethanol solu-tion. An additional κ form was obtained (unexpectedly) upon evaporation of a boric acid/methanol solution (Kim *et al.* 1968).

Bock has studied a number of systems in which different polymorphs were obtained under thermodynamic and kinetic conditions. (2-pyridyl)(2-pyrimidyl)amine **3-II** is dimorphic. Modification I is readily crystallized thermodynamically 'from any sol-vent' (toluene was actually used) while modification II is obtained kinetically by fast evaporation of an ethereal solution or by resolidification of the melt (Bock *et al.* 1997).

3-II   3-III

In 2,3,7,8-tetramethoxythianthrene **3-III**, the less stable (lower in both density and absolute value of lattice energy) monoclinic modification is obtained under kinetic conditions: rapid crystallization from polar diisopropyl ether, whereas the more stable (higher density and lattice energy) orthorhombic modification is thermodynamically obtained from a non-polar hydrocarbon solvent.

In pharmaceutical applications the choice of polymorphic modification for formu-lation depends very much on the robustness of the crystallization process as well as the properties and characteristics of the preferred modification. Hence, consider-able effort is expended in gaining control over the polymorphic form obtained under various conditions. As noted above, up to four polymorphic modifications and three monohydrates have been reported for cimetidine (SmithKline Beecham's Tagamet®) (Bavin *et al.* 1979; Prodic-Kojic *et al.* 1979; Shibata *et al.* 1983; Hegedus and Görög 1985). In experiments to selectively crystallize the A form in preference to the more stable B congener it was found that with isopropanol as a solvent, A crystallizes exclusively at high supersaturation, in the presence or absence of seeds (Sudo *et al.* 1991).

In another example, an antiarrhythmic under development (McCauley *et al.* 1993) was shown to exist in two anyhydrous polymorphs, two dihydrated enantiotropic polymorphs, a monohydrate, and the solvates of several organic solvents. Following characterization of all of these modifications it was desired to selectively obtain one of the dihydrates, termed modification A, which is thermodynamically less stable at room temperature than another dihydrate, D, in contact with aqueous solutions, but A is more stable over a wider range of relative humidities. The enantiotropic transition point between these two crystal modifications is 37 °C. Procedures were developed for obtaining A preferentially. Above the transition point a thermodynamic crystallization is carried out at 50 °C, using type A seeds as an added precaution to force the crystallization to type A. The desired type A can also be obtained under kinetic conditions by spontaneous crystallization below the transition point followed by rapid filtration and removal of excess water. The latter procedure prevents a transformation from the A state (metastable below the transition temperature) to the D form in the crystallization medium. Similar considerations were applied to develop procedures for the selective crystallization of the α and β modifications of glutamic acid (Kitamura 1989).

3.4 Monotropism, enantiotropism, and crystallization strategy

The examples cited in the previous section involved distinguishing between thermodynamic and kinetic conditions for a crystallization. For a practising chemist that distinction is often made instinctively rather than consciously. However, it can be related directly to the monotropic or enantiotropic relationship between two polymorphic forms, as expressed in the energy vs temperature diagrams (Sections 2.2.3 and 2.2.4). Practical means for determining the monotropic/enantiotropic relationship of two phases are given in Chapter 4. This information in turn can be used to design strategies for attempts to obtain desired crystalline forms at the expense of the less desired ones (Ceolin *et al.* 1996). Here we summarize the ramifications of the particular monotropic/enantiotropic relationship on the crystallization strategy once that relationship has been determined, preferably by generating the energy vs temperature diagram. For a dimorphic system there are four possibilities:

1. Obtaining the thermodynamically stable form in a monotropic system: no transformation can take place to another form, and no precautions need be taken to preserve the stable form or to prevent a transformation.
2. Obtaining the thermodynamically stable form in an enantiotropic system: precautions must be taken to maintain the thermodynamic conditions (temperature, pressure, relative humidity, etc.) at which the G curve for the desired polymorph is below that for the undesired one.
3. Obtaining the thermodynamically metastable form in a monotropic system: a kinetically controlled transformation may take place to the undesired thermodynamically stable form. To prevent such a transformation it may be necessary to employ drastic conditions to reduce kinetic effects (e.g. very low temperatures, very dry conditions, storage in the dark, etc.)

4. Obtaining the thermodynamically metastable form in an enantiotropic system: the information for obtaining and maintaining this form is essentially found in the energy–temperature diagram.

3.5 Concomitant polymorphs

In situations where there is overlap between the occurrence domains of two or more polymorphs the modifications may appear simultaneously or in overlapping stages so that a particular procedure or process yields more than one form under identical conditions. This phenomenon is termed *concomitant polymorphism* and has been treated in considerable detail in a recent review (Bernstein *et al.* 1999).

The situations in which polymorphs concomitantly crystallize are determined by the experimental conditions in relation to both the free energy–temperature relationships and the relative kinetic factors. These situations may arise because either specific thermodynamic conditions prevail or the competing kinetic processes have equivalent or very similar rates. In thermodynamic terms we have seen that polymorphs can coexist in true equilibrium only at the thermodynamic transition temperature (where the G curves cross). The chance of carrying out a crystallization precisely at such a temperature must be small with the inevitable conclusion if concomitant polymorphs are produced that kinetics play at least some role in the overall process of crystallization. The final consequence of this situation is that a system of concomitantly crystallizing polymorphs will be subject to change in the direction favouring the formation of the most stable structure. If the crystals have grown from and remain in contact with solution, then the most likely route for this transformation is via dissolution and recrystallization. This situation is commonly exploited in production processes as a slurrying procedure to produce the most stable polymorph (McCrone 1965; Cardew and Davey 1985). If the crystals have formed from the melt or vapour phase or have been isolated from their mother phase, a solid state transformation is possible (Cardew *et al.* 1984).

Concomitant crystallization offers the investigator the opportunity to maximize the data from a single crystallization experiment. Since concomitant polymorphs are energetically equivalent structures, or very nearly so, they provide excellent and demanding benchmarks for theoretical and computational models for predicting crystal growth and crystal structure (Reed *et al.* 2000). The programmes and force fields employed in these programmes must reproduce the near-equivalency of the crystal energetics, even if the absolute energy is not reproduced or even not known. These issues are treated in more detail in Section 5.8.

The phenomenon of concomitant polymorphs has been recognized since Wohler and Liebig (1832) discovered the first polymorphic organic substance, benzamide. Although Groth (1906*a*,*b*) collected a number of examples of the phenomenon in both organic and inorganic systems in his book published nearly 75 years later, he still characterized the phenomenon as a 'peculiarity', which is 'analogous to the . . . indifference towards direct transformation between polymorphic forms.'

In spite of Groth's guarded skepticism there are many examples of concomitant polymorphism crossing a wide range of crystallization techniques and diverse chemical systems. We cite some of them here, since they demonstrate that diversity as well as the interplay and often delicate balance of thermodynamic and kinetic factors governing the competition between different crystal modifications. Additional examples and more details can be found in the above-referenced review and in the original papers. The phenomenon is likely considerably more widespread than is generally appreciated, but is difficult to recognize or search for in the literature since it is rarely specifically abstracted or noted, and is usually buried in the experimental section of a paper. With increased awareness of the phenomenon and the information it can provide on polymorphic systems it would seem that more direct reference should be made to such systems. For the purpose of this chapter we have chosen to concentrate on examples of concomitant polymorphism that demonstrate how the fine tuning of crystallization conditions may be used to *control* the polymorphic form obtained. The more common *screening* techniques of varying conditions (solvent, temperature, rate of evaporation, etc.) are used to identify the occurrence domain of each of the crystal modifications. By definition, regions in which occurrence domains do not overlap are those in which control over the polymorphic result is maintained. Concomitant polymorphs occur in the regions of overlap between unique occurrence domains, and controlling the polymorph attained essentially involves defining the borders between the overlap regions and the unique ones.

3.5.1   *Crystallization methods and conditions*

The crystallization behaviour of the copper complex **3-IV** demonstrates the effect of concentration on the simultaneous appearance of polymorphs and the role of concentration in altering the preference for one or the other. Kelly *et al.* (1997, 2001) prepared blue crystals of **3-IV**, which precipitates from 'reasonably concentrated [acetonitrile] solutions' with square-planar coordination geometry around the copper. Filtering off the precipitate and treating the remaining mother liquor with diethyl ether, followed by cooling in a freezer, leads to concomitant crystallization of the blue crystals with green crystals, in which the molecular structure exhibits pseudo-tetrahedral coordination geometry around the copper centre.

3-IV

There is ample evidence that this also is a genuine case of polymorphism. The blue, square-planar form can be dissolved to obtain the green tetrahedral one; indeed the authors state that the two form intergrowths, which apparently grow coincidentally. The green form is obtained only from dilute solutions, and once it is obtained the blue form cannot be regenerated. The sequence of events and the experimental conditions are compatible with Ostwald's Rule of Stages (Section 2.3), the slight thermodynamic preference for the green form over the blue one (at least in dilute solutions—the melting points are essentially identical) and the phenomenon of disappearing polymorphs (Dunitz and Bernstein 1995) (Section 3.6). Once seeds of the (even slightly) preferred green form are present in the immediate environment the blue one can not be obtained.

The effect of temperature is demonstrated in the polymorphic behaviour of the 'diphenylcarbamide' **3-V**, one of the systems originally cited by Groth (1906a), and more recently studied by Etter *et al.* (1990a,b) and Huang *et al.* (1995).

3-V

The three reported forms are: α, yellow prismatic needles; β, white needles; and γ, yellow tablets. Upon crystallization from 95 per cent ethanol in the range of 30–75 °C the α and β forms always crystallize concomitantly, even in the presence of seeds of one of them. The relative amounts can be regulated by varying the temperature, α being preferred at higher temperatures. Upon evaporation of the mother liquor at 'the ordinary temperature (13 °C)' (*sic*) γ crystallizes out alone; warming the same mother liquor leads also to some crystals of β; at 40 °C γ no longer appears, but small amounts of α appear. The authors conclude that on the basis of the experiments with 95 per cent alcoholic solutions γ is the stable form at room temperature (considerably lower than room temperature today), and with increasing temperature β and then α are the stable forms. However, this behaviour may be modified by solvent. This is clearly a rich system with a very delicate balance between the relative stability of the three forms, as well as the kinetic factors that govern their appearance.

Concomitant crystallization is by no means limited to crystallization from solution, nor to preservation of constant molecular conformation. As noted in Section 2.2.5 the classic pressure vs temperature phase diagram for two solid phases (Fig. 2.6) of one material exhibits two lines corresponding to the solid/vapour equilibrium for each of two polymorphs. At any one temperature one would expect the two polymorphs to have different vapour pressures. This, in fact, is the basis for purification of solids by sublimation. Nevertheless there are examples where the two have nearly equal vapour pressures at a particular temperature and thus cosublime. This could be near the transition temperature or simply because the two curves are similar over a large range of temperatures or in close proximity at the temperature at which the sublimation is carried out. For instance, the compounds **3-VI** and **3-VII** both yield two phases upon vacuum sublimation (Cordes *et al.* 1992*a,b*), while **3-VIII** yields three differently coloured modifications (Griffiths and Monahan 1976).

3-VI   3-VII   3-VIII

For **3-VI** slow sublimation at 140 °C and 0.1 Torr led to a mixture of a 'few feathery needles' (mp 192–193 °C) among the main product of 'lustrous coppery blocks' (mp 220–223 °C), which could be separated manually for further characterization. **3-VII** was also purified/crystallized by vacuum sublimation (120 °C/10^{-2} Torr) yielding manually separable deep red needles (α phase, mp 157–160 °C) and blocks (β phase, mp 165–168 °C). The crystal structure determinations of the two phases of **3-VII** indicated significantly different molecular geometry, a cofacial dimer in the α phase, and a *trans* antarafacial dimer in the β phase.

Another example of cosublimed phases is **3-IX** for which it appears that the sublimation conditions were systematically varied in an attempt to obtain a second polymorph (Cordes *et al.* 1992*a,b*). The α form was initially prepared as a single phase of golden needles by slow sublimation over several weeks at 10^{-6} Torr during which the sample was heated to 180 °C and the cold finger was maintained at 100 °C. Raising the pressure to 10^{-1}–10^{-2} Torr and increasing the sample temperature to 220 °C, with a cold finger in the range 120–140 °C also led to the same α-phase needles, but accompanied by blocks of an additional β phase. Clearly, the authors moved along the solid/vapour line of the α phase towards the intersection with the solid/vapour line of the β phase (see Fig. 2.6).

Electrocrystallization, in particular of complexes of two components, is a technique that affords additional degrees of freedom of crystallization conditions compared to 'conventional' crystallization: voltage, current density, counter ions, supporting electrolyte, electrode materials, etc. For the current and voltage one can also obtain a variable range of conditions, from thermodynamic (with a minimum perturbation from equilibrium) to kinetic (with a large or even systematically varying perturbation). Electrocrystallizations are employed particularly in the field of organic conductors and magnets for the preparation of complexes and salts. Considerable effort is generally expended in growing crystals and fine tuning conditions for obtaining crystals of sufficient quality for further structural and physical characterization. Thus a wide variety of crystallizing conditions are attempted and many individual crystals from a single crystallization might be subjected to structural and physical characterization (Wang *et al.* 1989). These are favourable circumstances for the preparation and recognition of polymorphism and concomitant polymorphs.

3-IX   3-X

Perhaps most prominent of these materials are the organic conducting and super-conducting salts based on the so-called 'ET' compounds, in which BEDT-TTF **3-X** is the donor (cation in the salt), generally in a 2 : 1 ratio with the acceptor (anion in the salt) (Williams *et al.* 1991). One of the most widely studied of these salts is $(ET)_2I_3$, for which at least 14 different phases have been reported (Carlson *et al.* 1990; Williams *et al.* 1991), although the α and β phases tend to dominate (Carlson *et al.* 1990; Shibaeva *et al.* 1990). It has been shown that α is the kinetically favoured product (>90 per cent) under conditions of high current density and a small amount of water or oxidant added to the crystallization medium. Under more nearly equilibrium conditions (i.e. much lower current density) and dry solvent (tetrahydrofuran), pure β phase can be obtained, suggesting that it is the thermodynamically preferred form. Intermediate conditions apparently lead to concomitant crystallization of the two forms (Carlson *et al.* 1990; Williams *et al.* 1991).

When the crystallization is carried out as an oxidation of ET in tetrahydrofuran with a mixture of $(n\text{-}C_4H_9)_4NI_3$ and $(n\text{-}C_4H_9)_4NAuI_2$ (16 : 1 w/w) with an intermediate current to that described above, the α form is the main product, concomitantly crystallized with small amounts of θ and κ polymorphs (Kato *et al.* 1987). The authors note the difficulty in identifying all three forms on the basis of crystals shape alone, and the three were characterized by a combination of X-ray diffraction and other physical measurements.

Montgomery *et al.* (1988) have reported that in systems in which attempts were made to prepare alloys of $\beta\text{-}(ET)_2I_3$ with $\beta\text{-}(PT)_2I_3$ (PT = **3-XI**) they apparently obtained 'single crystals' of $(ET)_2I_3$ which were in fact mixtures of α and β forms. The phenomenon in this system was initially detected and then confirmed by ESR measurements, which were subsequently used to develop a quantitative procedure for the determination of the polymorphic composition of such 'mixed single crystals'. There are other scattered reports in the literature of the phenomenon (e.g. Freer and Kraut 1965), which has been termed 'composite crystals' and has been discussed in detail by Coppens *et al.* (1990) and Fryer (1997).

3-XI

Interdiffusion of saturated solutions (as opposed to electrocrystallization) is another method for obtaining crystals of the potentially conducting salts. In most of the preparations of TTF[Pd(dmit)$_2$]$_2$ (dmit: **3-XII**, X, Y, Z = S) by diffusion of (TTF)$_3$(BF$_4$)$_2$ and (n-Bu$_2$N)[Pd(dmit)$_2$] mainly black shiny needles of the α phase were obtained. However, some experiments yielded, in addition, the so-called α' phase (due to its structural similarity to the α phase, but different electrical behaviour), and occasionally a third δ phase of plate-shaped crystals could also be physically separated from the batch (Legros and Valade 1988; Cassoux *et al.* 1991).[1]

3-XII   X, Y, Z=S, Se   3-XIII

The technique of solvent diffusion has also led to concomitant polymorphs, which implies that the different modifications have very similar solubility in the same *mixture* of solvents. Such is the case for 1,5,9,13-tetrathiacyclohexadecane **3-XIII** (Blake *et al.* 1993). The material is reported to be trimorphic, all three structures crystallizing in polar space groups [needles, *Pbc*2$_1$ (α); plates, *P*2$_1$ (β); and twinned, apparently *Fdd*2 (γ)]. The α and β forms are obtained concomitantly at ambient temperature by diffusion of hexane into methylene chloride solutions of **3-XIII**. Lowering the temperature to 130 °C for the same diffusion process leads to the γ form exclusively. The forms that crystallize concomitantly have similar but unusual molecular conformations. The melting points for the three modifications ($\alpha = 59.5$–60.2 °C, $\beta = 57.8$–59.0°C, $\gamma = 60.0$–60.9 °C) are very similar, which is not surprising for concomitant polymorphs, and the authors used the technique of mixed melting points to verify the existence of the three polymorphs.

3.5.2   *Examples of different classes of compounds*

In the previous section the emphasis was on the variety of different crystallization techniques that have led to concomitant crystallization of polymorphs. In this section we survey the additional diversity of chemical entities which have exhibited this phenomenon, since such diversity can also provide guidelines to developing methods for controlling the polymorph obtained.

Among the sugars, mannitol was mentioned earlier. It was one of the first concomitant polymorphs noted by Groth (1906*a*), citing work by Zepharovich (1888). In fact

---

[1] Cassoux *et al.* (1991) note that 'The occurrence of several phases for this compound complicates its study'. While the separation of concomitantly crystallizing phases may have been an experimental complication in their particular study, we believe that the existence of polymorphs greatly *facilitates* the study of structure–property relations, since all chemical parameters are constant among polymorphs of a particular substance and differences in properties can be related directly to differences in structure. Chapter 6 is devoted to this subject.

because of the pharmaceutical uses of mannitol as a solid excipient (Section 7.2.2), its polymorphic behaviour has continued to be of considerable interest (Burger *et al.* 2000). In another rather intensively studied example from the pharmaceutical industry, sulphathiazole has been shown to crystallize in at least five polymorphs (Hughes *et al.* 1999) and over one hundred solvates (Bingham *et al.* 2001), often appearing together or sequentially, depending on the solvent and crystallizing conditions (Blagden *et al.* 1998a,b).

m-nitrophenol has had a long history (Bernstein and Davis 1999), although recent interest was aroused because of its significant second harmonic associated with a non-centrosymmetric crystal, in spite of the fact that it appeared to crystallize in a centrosymmetric space group (Pandarese *et al.* 1975). The dilemma was resolved when it was discovered that approximately 20 per cent of the crystals from a benzene-grown batch belonged to the non-centrosymmetric orthorhombic space group $P2_12_12_1$, while the remainder crystallized in centrosymmetric monoclinic $P2_1/c$ thus explaining the source of the second harmonic signal. A method for purifying *m*-nitrophenol and growing single crystals was developed by Wojcik and Marqueton in 1989 and the structure of the orthorhombic form was published by Hamzaoui *et al.* in 1996.

Two-component (i.e. two molecules) systems also exhibit concomitant polymorphism, implying a balance for the equilibrium situations governing the formation of the isomeric complexes as well as the kinetic and thermodynamic factors associated with the crystallization processes. The often serendipitous nature of the discovery of concomitant polymorphs is also illustrated by an example of a hydrogen-bonded two-component system, pyromellitic acid **3-XIV** and 2,4,6-trimethylpyridine **3-XV** (Biradha and Zaworotko 1998). The first polymorph (A) was obtained by reacting **3-XIV** with four equivalents of **3-XV** in a methanolic solution. Using **3-XV** as the solvent yielded a second polymorph (B) in 15 min. Modification A was found to have crystallized as well in the same reaction vessel after about 24 h. These two polymorphs are not readily distinguishable by their morphology. However, the authors point out that the experimental evidence indicates that Form B is the kinetically controlled one, while Form A is the thermodynamically preferred one.

3-XIV   3-XV   3-XVI

An example of a π-bonded complex is the remarkable cyanine : oxonol system, **3-XVI** : **3-XVII**, for which at least fourteen different polymorphs or solvates have been identified (Etter *et al.* 1984). Two of these, a gold and a red form (each containing a molecule of CHCl$_3$ solvent per 1 : 1 complex, and hence true polymorphs) crystallize concomitantly and have been structurally characterized (Etter *et al.* 1984). Three of these polymorphs are shown in Fig. 3.2. Despite the fact that both of these dye molecules are known to be individually *self* aggregating (Cash 1981) the two

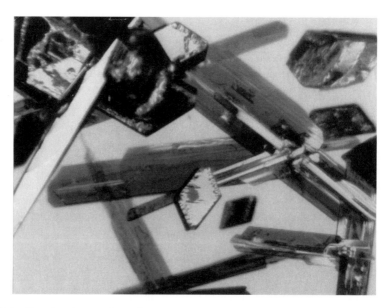

Fig. 3.2   (See also plate section.) Three of the concomitant polymorphs of the cyanine : oxonol dyes **3-XVI** and **3-XVII**. Gold (by reflection; otherwise red by transmission) and red forms mentioned in the text are easily distinguishable. The third form is a purple one, normally diamond shaped as on the middle right, but many of these crystals are undergoing conversion, as indicated by varying degrees of mottled surfaces. (From Bernstein *et al.* 1999, with permission.)

structures exhibit mixed stacks, with very significant differences in the relative orientation of neighbouring molecules along the stack (Fig. 3.3): in the case of the gold structure the molecular long axes are oriented nearly perpendicular to each other, while in the red polymorph the molecular long axes are very nearly parallel.

3-XVII   3-XVIII   3-XIX

Another π-complex system demonstrates concomitant polymorphism being manifested in differences in colour and in habit of different polymorphic modifications. **3-XVIII** is a relatively strong donor that was an early subject of study as a component of potential electrically conducting complexes (Kronick and Labes 1961; Kronick *et al.* 1964), and quite a few complexes of this donor were prepared (Matsunaga 1965, 1966; Koizumi and Matsunaga 1972; Inabe *et al.* 1996). The complex of **3-XVIII** and **3-XIX** was investigated in detail recently (Goto *et al.* 1996; Tamutsu *et al.* 1996)

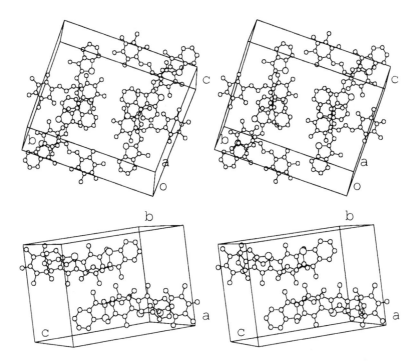

Fig. 3.3 Stereo views of the unit cells of the gold (upper) and red (lower) forms of the salt of **3-XVI** and **3-XVII**. In both cases the view is on the plane of the three central atoms of the oxonol molecule, with the bisector of the angle formed by the three oriented vertically for the oxonol molecule in the lower right hand corner of both figures. Chloroform molecules of solvation have been eliminated for clarity. Both figures indicate the relative orientation of cation and anion which is one distinguishing feature of the overall packing.

with results that indicate both the utility and some of the limitations of studies of polymorphic systems. The 1 : 1 complex is polymorphic, with a dark brown α form and a green β form. The two forms may be obtained simultaneously from benzene solution by both slow cooling and 'prolonged' slow evaporation. Similar results from these two equilibrium methods of crystallization indicate a similarity of the G vs T curves over the range of temperatures in the slow-cooling process.

A number of cases of concomitant polymorphs in which the variable of time or the equilibrium composition of the solution might be factors demonstrate the principles discussed in Section 1.2.

Ojala *et al.* (1998) reported on the crystallization of two *conformational polymorphs* (Bernstein 1987) (see Chapter 5) of acetone tosylhydrazone **3-XX**. A triclinic form and a monoclinic form are both obtained from anhydrous ethanol—sometimes together. If the crystallizing solution is allowed to evaporate completely, only the monoclinic form is obtained, suggesting that it is the thermodynamically preferred form at room temperature. This is consistent with Ostwald's Rule of Stages and McCrone's (1965) test for relative stability of polymorphs according to which the more stable polymorph will grow at the expense of the less stable one. The crystal structure determinations indicate that the conformations differ by ~15° about the S–C exocyclic bond. In this case then the solution has an equilibrium mixture of (at least) these two molecular conformations. Lattice energy calculations (Cerius2) are consistent with this observation, indicating that the triclinic polymorph is more stable than the monoclinic by about 1 kcal mol^{-1}, which suggests that the composition of conformers can vary with temperature.

3-XX

Another issue in the definition of polymorphs is that of *tautomerism*. An example of the crystallizing tautomeric structures of 2-amino-3-hydroxy-6-phenylazopyridine **3-XXI** has been reported by Desiraju (1983), **3-XXIa** being the 'low temperature' form as lustrous blue needles and **3-XXIb** being the 'high temperature' red form, both melting at 181–182 °C. They were obtained simultaneously from a recrystallization of the crude synthetic product from ethanol, but the relative amounts varied from batch to batch in subsequent crystallizations in which concentration and temperature of the crystallization were varied. The high temperature form was always obtained; the amount of low-temperature form varied with conditions. The tautomeric separation clearly takes place upon crystallization. To complete the picture and to make this arguably a case of polymorphism, it must be shown that the blue and red crystals dissolve to yield the same equilibrium mixture.

3-XXIa   3-XXIb

A case of concomitant crystallization which involves *configurational isomerism* of the benzophenone anil **3-XXII** has been reported by Matthews *et al.* (1991) was discussed in Section 1.2.2.

3-XXII(E)   3-XXII(Z)

One of the issues raised in the discussion definition of polymorphs (Section 1.2.1) deals with racemic mixtures vs enantiomerically pure crystals or conglomerates

(McCrone 1965; Dunitz 1995; Threlfall 1995). In principle, enantiomerically pure crystals are different from racemic ones, but if the enantiomers racemize quickly upon dissolution and/or racemates in solution spontaneously resolve upon crystallization it is still debatable whether the respective crystals are to be considered polymorphic substances. Masciocchi *et al.* (1997) characterized a concomitantly crystallized system that incorporates and illustrates many of these features.

The molecule under study was Pd[(dmpz)$_2$(Hdmpz)$_2$]$_2$, where Hdmpz = 3,5-dimethylpyrazole, **3-XXIII**. The material is trimorphic. The reaction mixture yields mostly (90 per cent) the monoclinic ($C2/c$, racemic) α phase, the remainder being the triclinic (P$\bar{1}$, racemic) γ phase. The latter can be removed by recrystallization from 1,2-dichloroethane, which suggests that it is the more soluble and hence, the less stable polymorph in that solvent. Masciocchi *et al.* (1997) found that mixtures of various amounts of α and γ polymorphs could be obtained by varying the solvent and precipitation temperature ($-70\,°C$ to $+50\,°C$), with α preferred at higher temperatures, consistent with the earlier observation of relative stability. Pure polymorph γ may be obtained by a different synthetic route which, when employing an excess of 3,5-dimethylpyrazole, leads to an approximately 50 : 50 mixture of polymorphs α and γ. This system thus also represents a case in which the polymorph obtained, or the polymorphic mixture obtained, depends on the synthetic route to the desired material. It is probably more correct to state that as usual, the polymorph or polymorphic mixture depends on the crystallization conditions, and these will clearly differ in the solvent/reagent/product compositions resulting from different synthetic conditions and routes.

3-XXIII

The tetragonal (*I*422, chiral) β 'polymorph' of **3-XXIII** is obtained quantitatively by a solid–liquid synthesis. The product is a conglomerate of enantiomeric crystals, which the authors claim does not transform into the α phase because of the impossibility of a solid/solid transformation. Dissolution of the β phase in 1,2-dichloroethane and subsequent evaporation quantitatively restores a mixture of α and γ forms. Despite different space groups the gross features of packing modes are very similar, the molecules being arranged about a pseudo (α, γ) or real (β) four fold axis. Such a view of the crystal structure is consistent with observations of Gavezzotti and Desiraju (1988) and Braga and Grepioni (1991) on the general similarities of packing features and coordination numbers of organic and organometallic compounds.

In view of the many experiments carried out in achieving the crystallization of proteins and subsequently refining those conditions to maximize crystal size and quality for X-ray structure determination, it is not surprising that examples of concomitant crystallization are found among proteins. We cite two here. Fu *et al.* (1994) reported the simultaneous crystallization of three polymorphs of an *m*-class glutathione

Fig. 3.4 Photograph of concomitantly crystallizing forms of rat liver glutathone S-transferase. The three forms are labeled. (From Fu *et al.* 1994, with permission.) (See also colour plate section.)

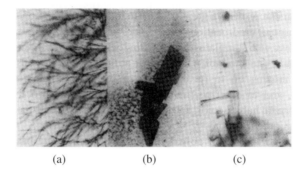

(a)   (b)   (c)

Fig. 3.5 Example of the stage-like growth in cytochrome c from *valid membrane faciens*. (a) Tree-like arrays of small triclinic crystals, (b) triclinic crystals, which are frequently seen to dissolve at the expense of the orthorhombic prism, shown in (c). (From Day and McPherson 1991, with permission.)

S-transferase from rat liver (Fig. 3.4). Day and McPherson (1991) reported crystallization of two crystalline forms in stages in accord with Ostwald's Rule for cytochrome c from *Valida membran-aefaciens* (Fig. 3.5). In every case of crystallization they obtained arrays of thin triclinic plates (Fig. 3.5a), some of which grew up to 0.5 mm in the largest physical dimension (Fig. 3.5b). In some cases, some of these dissolved (in accord with Ostwald's Rule of Stages (Section 2.3)) to give rectangular prisms (Fig. 3.5c), which turned out to be orthorhombic.

3.5.3   *The structural approach*

To this point this account of instances of concomitant polymorphs has been phenomenological. We have discussed the thermodynamic and kinetic crystallization of polymorphs. There is still the question if any insight concerning controlling the polymorph obtained can be gained from the study of the crystal structure of concomitantly crystallized polymorphs. A qualitative attempt was made to see if details of the crystal structures may provide clues to the reasons for concomitant crystallization, and the near energetic equivalence of the two forms which can be assumed from that concomitant crystallization. The squarylium dye, 3-**XXIV** crystallizes from methylene chloride in a triclinic violet form and a monoclinic green form (Bernstein and Goldstein 1988). The cell constants, Table 3.1, do not suggest any similarity of the structures. However, a projection on the molecular plane which includes two neighbouring molecules (Fig. 3.6) indicates that the stacking of the planar molecules is virtually identical in the two structures. In both cases the two molecules are related by a lattice translation: in the triclinic structure it is along the c axis and in the monoclinic structure along the b axis. Although the views appear the same to the eye the vertical separation between planes differs (3.40 Å vs 3.86 Å), which is a manifestation of the different axial lengths involved. The similarity of these diagrams strongly suggests that the stacking is the dominant influence in the crystal growth process for both—hence the concomitant crystallization. The crystal structures differ, of course, in that stacks are related by translation in the triclinic structure and by a screw axis in the monoclinic one, as shown in Fig. 3.7, but the fact that these crystallize concomitantly would be consistent with the assumption that energetically, at least, these interstack interactions contribute in a less significant way to the total lattice energy than those within the stack.

3-XXIV

It is of interest to note that another similar squarylium dye 3-**XXV** has also been shown to concomitantly crystallize in a green monoclinic phase and a purple triclinic phase (Ashwell *et al.* 1996). The structure has been reported for the former, but crystals of the latter were not of sufficient quality to determine the crystal structure. 3-**XXV**

Table 3.1 Crystallographic cell constants for the two polymorphic forms of 3-**XXIV**

	Triclinic form	Monoclinic form
$a(\text{Å})$	11.911	15.72
b	7.401	7.283
c	6.501	9.591
$\alpha\ (°)$	92.78	90
β	111.9	106.11
γ	98.08	90
Space group	$P\bar{1}$	$P2_1/c$

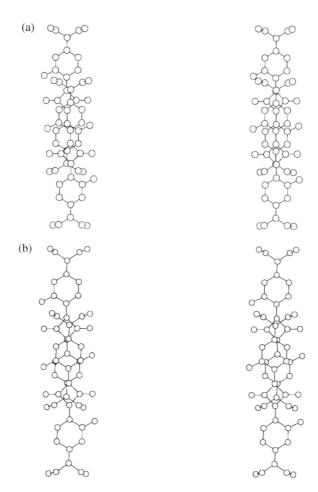

Fig. 3.6 Stereoviews of the overlap of translationally related molecules in the two structures of **3-XXIV**. In both cases the view is on the plane of the reference molecule: (a) triclinic structure, *c*-axis translation; (b) monoclinic structure, *b*-axis translation. (Adopted from Bernstein and Goldstein 1988, with permission.)

crystallizes in space group $P2_1/c$ with cell constants $a = 9.046$ Å, $b = 19.615$ Å, $c = 9.055$ Å, $b = 116.1$, which, lacking a short axis, essentially precludes any molecular overlap of the type seen in Fig. 3.7 for **3-XXIV**. The crystal structures of **3-XXIV** and **3-XXV** are significantly different so that the similarity in the colours of the polymorphs behavior of the two compounds appears to be entirely coincidental, albeit unusual.

These studies suggest the existence of similar aggregates leading to the different polymorphs, as a function of thermodynamic and kinetic conditions, as investigated recently by Wojtyk *et al.* (1999).

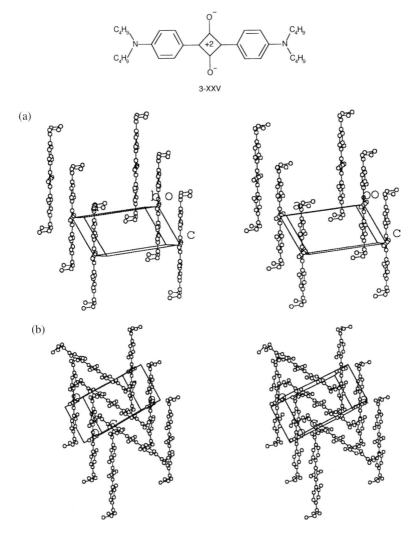

3-XXV

(a)

(b)

Fig. 3.7 Stereoviews showing the relationship between the stacks shown in Fig. 3.6 for **3-XXIV**. (a) Triclinic structure, translational relationship, (b) monoclinic structure, screw-axis relationship. (Adopted from Bernstein and Goldstein 1988, with permission.)

3.6 Disappearing polymorphs

Much of chemistry deals with controlling the reactions and interactions between molecules. Although this chapter deals with the principles behind controlling the polymorph obtained—that is in a sense controlling the interactions among molecules in the formation of crystals—in polymorphic systems that control is not always easy to achieve, or to maintain. The loss of control is indeed disturbing, and might even call into question the reproducibility of chemical processes. However, in spite of the fact that crystallization as a technique and a process is taken for granted by most practising chemists, it still has very much the nature of an art.

The appearance of a new polymorph in fact can change the environment in which the material is found, since there is now a competition between the new polymorph and the previously existing one(s). In fact, by Ostwald's Rule (Section 2.3), it is very likely that the new polymorph will be thermodynamically more stable than the others, and hence will be energetically preferred, but its effect depends as well on the overlap of its occurrence domain with that of earlier existing polymorphs. The resulting situation that can develop is that there are crystal forms that are observed over a period of time but are apparently displaced by a more stable crystal form. We have termed this phenomenon 'disappearing polymorphs' and although described in our review of this topic (Dunitz and Bernstein 1995), many chemists remain skeptical about a subject that calls into question the criterion of reproducibility as a condition for acceptance of a phenomenon as being worthy of scientific inquiry. Nevertheless, there are well-documented cases of such a phenomenon (e.g. ritonavir, Chemburkar *et al.* 2000; Bauer *et al.* 2001) and a number of them are presented in that same review.

Many other difficult questions arise when a situation described as a disappearing polymorph is encountered. Among them are: Why did the new polymorph appear at all (often after years of no hint of its existence)? Why does a previously robust process no longer yield the crystal form that had been obtained prior to the appearance of the new one? What crystallization parameters must be modified to obtain either the old or the new form exclusively and robustly?

As we have noted repeatedly in this chapter, there is a vast variety of conditions that can affect a crystallization, and for any polymorphic system it is indeed difficult to single out a particular factor that might dominate. However, it is certain that seeding can and often does play an important role in determining the fate of a crystallization, especially one in which competitive processes can lead to individual polymorphs, or a mixture of them.

Intentional seeding is a common practice among chemists who wish to coax the crystallization of a compound from solution or from the melt: small crystals or crystallites of the desired material (seeds) are added to the system (e.g. Pavia *et al.* 1988; Shriner *et al.* 1997). In this way, the rate-limiting nucleation step, which may be extremely slow, can be accelerated. For this method to be applied, it is of course necessary that a sample of the desired crystalline material is available; that is, the compound must have been already crystallized in a previous experiment. When polymorphic forms of a substance are known to occur, intentional seeding with one of the polymorphs is a useful and often the most successful way of preferentially producing it rather than the other form(s).

Unintentional seeding may also occur even if small amounts of the undesired polymorph are present as contaminants—in fact, in principle just one such seed is sufficient to act as a nucleation site.[2] Unintentional seeding is often invoked as an explanation of crystallization phenomena such as disappearing polymorphs which otherwise are difficult to interpret. We argue here in favour of such an explanation, although there is no consensus about the size and range of activity of such seeds, which have never been actually observed as such.[3]

Estimates of the size of a critical nucleus that can constitute a seed range from a few tens of molecules to a few million molecules (Mullin 1993). Even with a particle size of about 10^{-6} g (essentially the limit for visible detection) a compound of molecular weight 100 contains approximately 10^{16} molecules, sufficient to make 10^{10} nuclei of the size at the large end of the estimated range.

Thus, on the basis of quantity and size required to play a role, once a crystal form exists in a certain locale the presence of seeds is almost always possible, indeed often unavoidable. One can think of local seeding, where the unintentional source may come from e.g. the experimentalists' clothing, a portion of the room, an entire room, a building, or even, with increasing degrees of improbability, increasingly larger environments.[4]

However it is important to bear in mind that in principle only one seed is required to initiate a crystallization so that indeed small amounts of material can play a significant role in a crystallization process. Moreover, other bodies such as specks of dust, smoke particles, and other small foreign bodies can act as nucleating agents in promoting crystallization. Invoking the presence of seeds to account for crystallization phenomena often generates skepticism, even though the existence of unseen dust and smoke particles as solid particles, water vapour as liquid, and invisible scents due to individual molecules are commonly recognized and acknowledged.[5] Three earlier short compilations of examples of disappearing polymorphs have been given

[2] It is well known among practising chemists that it is often difficult to crystallize a newly synthesized compound. Subsequent crystallizations tend to be easier because of the presence of suitable seeds.

[3] The lack of direct observation of a proposed object or phenomenon is not contrary to the scientific method. The directly observable world constitutes but a small portion of the range of sizes believed to exist (see, for instance Morrison *et al.* 1982) but we consider many objects and phenomena outside of that range—modelled or understood on the basis of indirect evidence—as part of our well-established scientific body of knowledge. The existence and understanding of atoms was in such a category for almost two centuries, although modern techniques are enabling us to 'see' atoms, albeit with the aid of computer imaging. The atomic nucleus, being smaller by five orders of magnitude is certainly in the category of objects apparently quite well understood but never actually seen.

[4] On occasion, this has been expanded to make a claim for 'universal seeding' which taken literally is obviously absurd. The universe is estimated to contain about a millimole ($\sim 10^{20}$) of stars, so one seed per star, amounting to *one seed per solar system*, would require about 100 kg of a compound of nominal molecular weight of 100.

[5] A normal urban environment contains approximately one million airborne particles of 0.5 micron diameter or larger per cubic foot, the number being reduced by an order of magnitude in an uninhabited rural environment. A normal sitting individual generates roughly one million dust particles (≥ 0.3 micron diameter) per minute (for reference a visible particle is usually 10 microns or greater in diameter). Clean rooms for various purposes (e.g. surgical theaters, biological or pharmaceutical preparations, semiconductor fabrication) employ sophisticated technology to remove these particles and to prevent subsequent contamination (ThaiHVAC 2001). Therefore the possible presence of seeds of even a newly formed crystal form in a laboratory, a manufacturing facility, or any location having been exposed to that form cannot be casually dismissed; indeed its presence would be hard to avoid.

(Woodard and McCrone 1975; Webb and Anderson 1978; Jacewicz and Nayler 1979), although with varying points of view about the nature of the phenomenon, with the more recent detailed accounts noted above (Dunitz and Bernstein 1995).

The fact that a particular crystal form may disappear does not doom it to chemical history, and we believe that once a particular polymorph has been obtained it is *always* possible to obtain it again; it is only a matter of finding the right experimental conditions. Redesigning a strategy to find those conditions is by no means a simple matter and may often require unconventional measures (e.g. Davey *et al.* 1993; Ludlam-Brown and York 1993; Braga *et al.* 2000). However, as proven by a number of recent case studies, consideration of a number of experimental variables can assist in the design of such a strategy. Among these are careful observations using hot stage microscopy (Section 4.2), information on the relative thermodynamic stabilities (Section 4.3), the energy–temperature diagram (Section 2.2.3) and the enantiotropic or monotropic relationships among polymorphs (Sections 2.2.4 and 3.4), design of kinetic conditions to obtain the less stable form, consideration of the influence of seeding (both intentional and unintentional), judicious choice of solvent or tailor-made additives (Section 3.7) to inhibit a particular undesirable form, etc. A number of cases of the successful design and application of such strategies have been given recently (Bernstein and Henck 1998; Henck *et al.* 2001).

The important point is that the determination of the crystallization conditions for various polymorphic forms need not be a completely random process. The combination of keen thoughtful observation with consideration of all the available crystal structures and thermodynamic information can provide extremely useful guidelines, if not for success, then at the very least for further experiments. Crystallization is almost never a sure-fire procedure, especially when one is trying to selectively produce a particular polymorph, and one that has proven consistently or suddenly elusive.

3.7   Control of polymorphic crystallization by design

The generation of the occurrence domain for any polymorphic system and the subsequent definition conditions for controlling the polymorphic modification has generally been an empirical process, although the possibility of systematic exploration of the occurrence domain has recently been investigated (Crystallics 2000; Higginson 2000). As noted above, considerable insight over and above the empirical approach can be gained by preparation and use of the energy/temperature diagram for the identification of zones of stability of the various forms. Additional quantitative data that can be of great use are solubility curves in various solvents (Khoshkhoo and Anwar 1993; Threlfall 2000). Over the past two decades a number of groups have used the knowledge of the detailed crystal structure of the polymorphic forms to design experiments and procedures to rationally monitor and control the polymorph obtained (Bürgi *et al.* 1998).

The crystal structure analysis provides information on the detailed packing and intermolecular relationships within the crystal structure. In most of the rational design experiments the crystal habit of each of the polymorphic forms is also characterized, including the shape and identity (i.e. Miller indices) of the developed crystal faces. In other cases a model of the predicted crystal habit may be prepared using estimates of the relative energy of attachment to various crystal faces (Clydesdale *et al.* 1994*a,b*; 1996; 1997). By combining this information on the bulk structure with the projection of the structure onto the various crystal faces one may obtain a view of what a molecule in solution 'sees' when approaching the various crystal faces. This information is then used for the 'tailor-made' design of molecular additives that can selectively attach to particularly chosen faces. The basis of the design is such that the part of the additive molecule that attaches to the appropriate crystal face is essentially identical to the host crystal—which is what allows it to become attached. The part of the molecule intended to extend outside the crystal is sufficiently different from the molecules in the host crystal that the attachment of subsequent molecules to that face is inhibited. Crystal faces onto which additive molecules do not attach easily are less affected, so that the net result of such experiments as described is a modification of the crystal *habit*, but not necessarily of the polymorphic modification.

However, the same principles have been extended to design additives to inhibit the growth of particular polymorphic modifications, and thus allow or induce the selective growth of another modification (Weissbuch *et al.* 1991; Lahav and Leiserowitz 1993; Weissbuch *et al.* 1994). In a classic example of the application of this strategy, considering the details of both the crystal morphology and the packing arrangement, the polar form of glycine was obtained from 4 per cent aqueous acetic acid in preference to the stable centrosymmetric form (Weissbuch *et al.* 1994). Blagden *et al.* (1998*a*) have used a similar approach to develop solvent choice for the selective growth of polymorphs of sulphathiazole. In another approach Bonafede and Ward (1995) selectively grew the less-preferred polymorph of a cyanine–oxonol (**3-XVI–3-XVII**) dye salt rich in polymorphic structural variety (Etter *et al.* 1984) by utilizing the surface available on single crystal succinic acid crystals, interpreting the process as being due to a ledge-directed epitaxy. All of these approaches show a great deal of promise for directing and controlling the polymorph obtained in a crystallization procedure or process.

Analytical techniques for studying and characterizing polymorphs

You can observe a lot just by watching (Yogi Berra)

4.1 Introduction

Since polymorphs represent different crystal structures essentially every physical or
chemical property may vary among the polymorphic structures of a material. A consequence of this is the fact that virtually any technique that measures the properties
of a solid material may in principle be used to detect polymorphism and to characterize the similarities and difference among polymorphic structures. Some techniques
are more sensitive to the differences in *crystal structure* or molecular environment,
as opposed to *molecular structure*, and in many cases these are to be preferred in
detecting and characterizing polymorphs.

The intent of this chapter is to survey the most commonly employed of these analytical techniques with an eye to demonstrating how they are being used to discover
and characterize polymorphs. There is considerable literature associated with each of
these techniques and in the sense of being able to cover all the details and ramifications of each technique we do not intend to be comprehensive here. Rather, we will
limit ourselves to briefly and qualitatively describing the principles of the technique,
when necessary, to providing some leading references for further reading and entry
to the literature, and to giving some examples of the application of the technique
to the characterization and study of polymorphs. Because every technique provides
different information and some techniques will not distinguish among a particular set
of polymorphs, the importance of utilizing a wide variety of techniques for the identification and characterization of a polymorphic system cannot be overemphasized
(Threlfall 1995; Brittain 1999*a*; Bugay 2001).

4.2 Optical/hot stage microscopy

Hot stage microscopy provides a rapid method for screening substances for the
existence of polymorphism. In spite of the utility of this technique, it has not
enjoyed the widespread use it deserves. Microscopy was much ignored during the

middle decades of the twentieth century, due, in part at least, to the subjective nature of the observations and 'measurement' and to the lack of practitioners to pass the knowledge and insight on to their scientific progeny. However, its utility and importance are being increasingly recognized (McLafferty 1990; Streng 1997; Bernstein and Henck 1998).

Three of the traditional centres of activity in chemical microscopy (Cornell University, University of Innsbruck and McCrone Asssociates) have provided a number of excellent literature sources for becoming acquainted with the polarizing microscope equipped with a hot stage and it's use in studying polymorphism (Kofler and Kofler 1954; McCrone 1957; Kuhnert-Brandstätter 1971; Chamot and Mason 1973; McCrone et al. 1978; Kuhnert-Brandstätter 1982; Cooke 1998). The use of the microscope declined, in part at least, in an age of increasing quantification, due to the lack of quantitative measures that can be attached to visual observations. Photographic records of microscopic observations were more in the realm of geologists and biologists than that of chemists. However, the ready availability and relatively low cost of video recording with the capability of digitally capturing individual images or sequences of images bodes well for the future of chemical/thermal microscopy in the study of polymorphism.

The most general use of the microscope for the study of polymorphism is simply for observing the homogeneity or diversity of a crystalline sample. Variations in size, shape or colour may indicate the presence of polymorphism and the need for further examination (see, for instance Fig. 3.2). As noted in Section 2.4.1, differences in crystal habit are not necessarily indicative of polymorphism. Further physical characterization of the individual crystals may involve measurements such as optical constants (Hartshorne and Stuart 1960, 1964; Wahlstrom 1969; McCrone et al. 1978) or interfacial angles (Winchell 1943; Porter and Spiller 1951; Terpstra and Codd 1961).

By far the most useful accessory to polarized light microscopy is the hot stage originally invented by Lehmann (1877a,b, 1888, 1891, 1910) (Fig. 4.1) and now used in the versions developed by Kofler and Kofler (1954) and the Mettler Company in 1968 (Woodard 1970; Julian and McCrone 1971) (See also Kuhnert-Brandstätter 1971, 1982). Melting point determination on small samples is perhaps the most obvious application for hot stage microscopy. With practice, the true melting point (i.e. the temperature at which the solid and its melt are maintained at equilibrium) can be readily determined. Melting point is perhaps one of the longest used physical constants for characterizing solids, although as Borka (1991) has pointed out different polymorphs can have similar melting points. An outstanding example is that of trimorphic D-mannitol (Burger et al. 2000) for which the melting points are Form I 166.5 °C; Form II 166 °C and Form III 150–158 °C incongruently. A variety of other properties may be examined and studied with this versatile tool, which should be the first option for any characterization of a solid material, especially in the search for polymorphs.

The systematic use of the hot stage microscope employing mainly the Kofler hot stage was originally described in the book by Kofler and Kofler (1954) (the founders

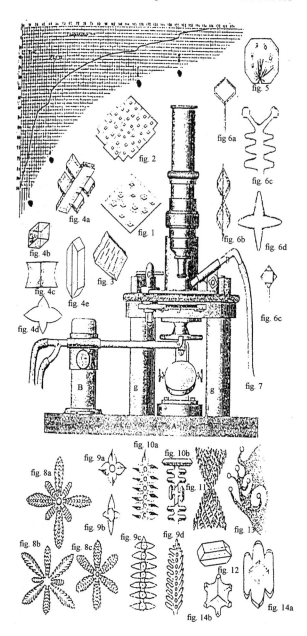

Fig. 4.1   An illustration from the original paper by Lehman (1877*b*) on 'Physical Isomerism' showing his hot stage microscope and various shapes of crystals observed. Another noteworthy feature of this figure is the time vs temperature cooling curve for ammonium nitrate in the upper left hand corner, showing four inflection points that indicate polymorphic transitions. (Reprinted by permission from *Zeitschrift fur Kristallographie*.)

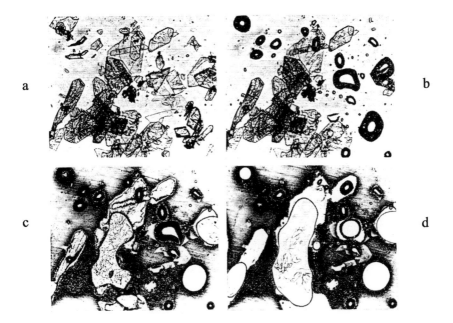

Fig. 4.2 Melting behaviour of sulphathiazole. (a) 175 °C, which is the melting point for commercially available Form II, (b) some Form II crystals melting at 175 °C, while others transform to the stable modification, (c) 200 °C, at which point some of the stable form begins to melt, (d) the equilibrium melting point (Kuhnert-Brandstätter 1971) for the higher melting form. (From Kuhnert-Brandstätter 1971, with permission.)

of the Innsbruck group).[1] The book of McCrone (1957) and it's 1965 revision follow a very similar approach, although he has pointed out that different experimental strategies guided the Innsbruck group and those in the US. In Innsbruck, most of the observations were made during heating, while McCrone's were made during cooling.

A typical example of a solid–solid transition resulting in the observation of two observed melting points is given in Fig. 4.2. The use of polarized light allows the ready detection of discontinuous changes in polarization colours during the heating process, the discontinuity in any property being symptomatic of a phase change. It is worth noting here that there may be optically observed phase changes that are barely detectable by other analytical techniques such as differential scanning calorimetry (DSC) (*vide infra*) (e.g. Burger *et al.* 1996).

The determination of the index of refraction of the melt (a procedure readily performed on the microscope (McCrone 1957; Kuhnert-Brandstätter 1971)) can serve as

---

[1] In addition to a through description of hot stage, the techniques this book contains a very useful chapter with a detailed description of experiments that may be done (including the specific substances that best illustrate the phenomenon) to characterize molecular crystals, including polymorphism. Unfortunately, the book is out of print and hard to obtain. At the time of this writing, a 1980 English translation was available through McCrone Associates, Chicago, IL.

a confirmation of the existence of polymorphs, since all polymorphic forms must melt to the same liquid. Such studies provide preliminary, but often key and detailed information on the temperature range over which polymorphs exist, the degree of stability of metastable forms with respect to such factors as temperature, time, mechanical or thermal shock, the presence of impurities, etc.

The determination of the enantiotropic or monotropic nature of a polymorphic transition (see Sections 2.2.4 and 4.3) can be made using the hot stage microscope, as demonstrated in Fig. 4.3. In a rarely used but potentially very powerful technique, transition temperatures and melting points can be determined by studying the behaviour of the corresponding modifications in a two component system and the measurement of their melting points (Müller 1914; Kofler and Kofler 1954; Kuhnert-Brandstätter 1971). A particular two component method involves the preparations of the unstable modifications in mixtures with a second component with which the material under investigation may or may not be isomorphous. The fact that the stability often increases with the presence of impurities allows the preparation of the melting point diagram from mixtures and the extrapolation of the melting point. As examples Kofler and Kofler (1954) cite of the work of Brandstätter (1947) on Form II of 1,3-dinitrobenzene (with 1,2,4-chlorodinitrobenzene) and those of Phillips and Mumford (1934) and Francis and Piper (1939) on the unstable forms of methyl esters of carboxylic acids.

An additional indirect method employs the determination of the eutectic temperature with an appropriate test substance. The method requires that the system does not form mixed crystals with the test substance, that it is monotropic, and that the equilibrium curves in the phase diagram be nearly parallel (Kofler and Kofler 1948). An elegant application of the eutectic method to determine the relative stabilities of a series of polymorphs of **4-I** was recently reported by Yu *et al.* (2000) (See Section 5.9). Hot stage microscopy can also provide a wealth of information on the nature of polymorphic transformations. For instance, it is possible to distinguish whether nucleation of a new phase takes place randomly throughout the crystal, at specific defects or crystal edges, and whether the phase change occurs throughout the crystal in a diffuse manner (Fig. 4.4) or at a front which moves through the crystal (Fig. 4.5).

4-I

Solvates may be readily detected and desolvation may be readily distinguished from a phase change using thermomicroscopy. The appearance of turbidity within the crystal upon heating is a sign of solvent being driven off, but a much more conclusive test involves covering the crystal with silicone gel or paraffin oil, which trap the bubbles of released solvent, as shown in Fig. 4.6.

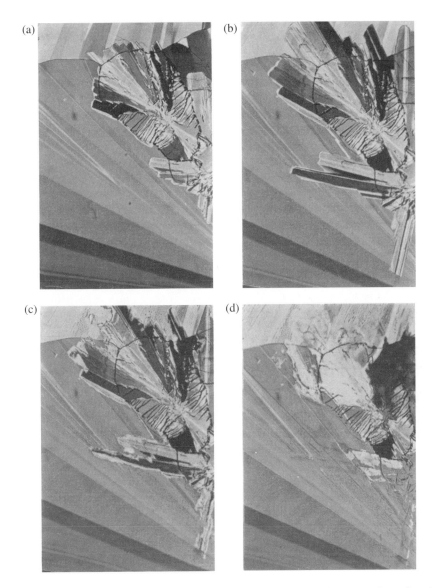

Fig. 4.3 (See also colour plate section.) An enantiotropic phase transformation in 2,4,6-trinitro-5-*tert*-butyl-*m*-xylene as observed on the hot stage microscope. Form II is stable at room temperature and the thermodynamic transition point is at 84 °C. (a) Room temperature stable Form II, the coarse crystals at upper right embedded in aggregate of Form I. (b) On heating Form II grows at the expense of Form I. (c) At 84 °C the transformation can be halted. (d) Above 84 °C Form I is stable and has grown at the expense of Form II. (From Kuhnert-Brandstätter 1971, with permission.)

Fig. 4.4 (See also colour plate section.) Phase transformation of β-naphthol, showing the diffuse nature of product phase formation. (From Kuhnert-Brandstätter 1982, with permission.)

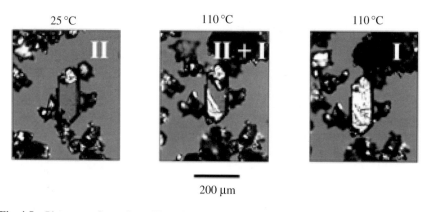

Fig. 4.5 Phase transformation of Form II orthorhombic paracetamol to the monoclinic Form I. At 25 °C (left photo), the crystal in the center of the field is extinguished under crossed polarizers. The conversion takes place at 110 °C along a front which moves from lower left to upper right. The conversion is approximately 50 per cent completed in the middle photo, and essentially complete in the right hand one. (From Nichols 1999, with permission; see also Nichols 1998.)

By thermomicroscopy the sublimation behaviour may be readily studied and conditions (e.g. temperature, amount of sample) may be varied to achieve different habits and even different polymorphic modifications (Section 2.4.1) (Fig. 4.7).

McCrone (1957) has described techniques for carrying out crystallization on a microscope slide. Another useful technique, also due to McCrone (1984), can be used to screen a variety of potential solvents for crystallizing on a very small (i.e. one crystal) sample. It is shown and described in Fig. 4.8.

Fig. 4.6 Bubbles of solvent evolved from a crystal heated in a silicone gel preparation. (From Kuhnert-Brandstätter 1971, with permission.)

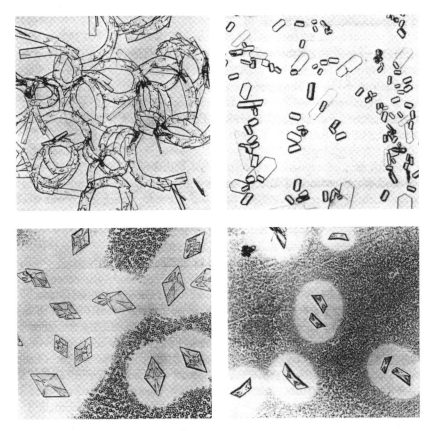

Fig. 4.7 Crystals of mephobarbital obtained under different sublimation conditions. All crystals are of the same polymorph but exhibit different habits. (From Kuhnert-Brandstätter 1982, with permission.)

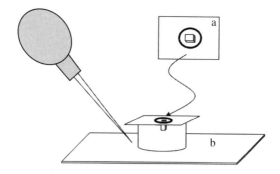

Fig. 4.8 Arrangement for surveying a variety of potential solvents for crystallization. A single *small* crystal is placed on a microscope cover slip. (It may be helpful to mark the location of the crystal by circling it with a felt tipped pen.) The cover slip is then inverted on a short (~4–6 mm) piece of Tygon tubing mounted on a microscope slide. A few drops of a trial solvent are placed on the side around the periphery of the tubing. They will seep into the space inside the tubing and evaporate, filling it with vapours that may or may not dissolve the crystal. Recrystallization of the sample may be followed on the microscope, and the same sample may be reused many times without loss of material. a: Cover slip with crystal in a marked circle. b: Microscope slide with a slice of tubing and the coverslip with the crystal (whose size has been exaggerated for clarity).

One of the tests described for the relative stability of polymorphs (Sections 3.5.1 and 4.12) involves the observation of competitive growth rates in a particular solvent (Fig. 1.4). Such experiments may be easily followed and recorded on the microscope (at different temperatures) as demonstrated in Fig. 3.5.

A great deal of information on the polymorphic behaviour of two component systems may be obtained using the contact preparation method, originally developed by Lehmann (1888), and apparently rediscovered independently by Kofler (1941) (Kuhnert-Brandstätter 1971; Kuhnert-Brandstätter 1982). The method involves melting the higher melting component between a microscope slide and cover slip so that about half of the intervening space is filled. After this melt is rapidly cooled, the lower melting component is introduced at the free edge of the cover slip. Upon melting, it flows under the slip until reaching the solid of the higher melting component. Preparations of this type allow one to determine the formation of one or more molecular compounds of a eutectic, a mixing gap of the liquid phases or mixed crystals (Fig. 4.9). The basic features of the melting diagram can be sketched out without

Fig. 4.9 (See also color plate section.) Photographs of the microscope slide Kofler preparations showing the various phases of the benzocaine : picric acid (**BC : PA**) binary system. *Top*: (a) the photomicrograph of the recrystallized contact preparation of **BC** and **PA** at 25 °C. The interference colours are due to the use of crossed polarizers. The pure compounds are at the extremities of the preparation, while in the region where the original compounds have merged a number of different areas may be observed, due to the formation of different crystalline species combining the two components. Heating this preparation on the hot stage microscope

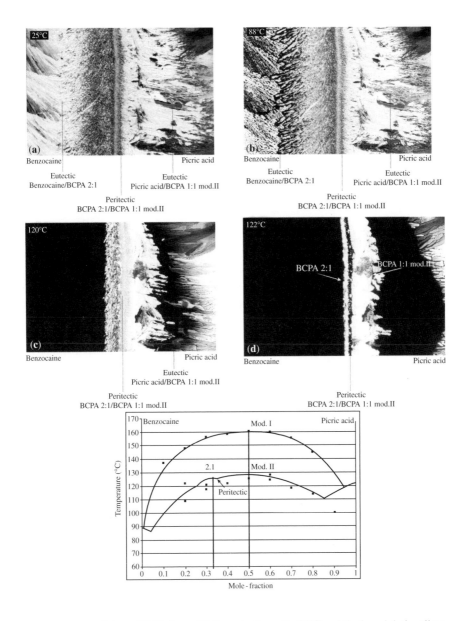

to a temperature of about 88 °C shows (b) the eutectic melt of **BC** and the broad dark yellow crystals of a new compound. Due to the crossed polarizers, the isotropic melt appears black. At about 120 °C (c) on the left side of the preparation **BC** is melted and the right side the eutectic between **PA** and the remaining crystals of the **BC** : **PA** 1 : 1 complex melts; (d) the situation which is observed at 122 °C. **PA** is almost melted and in the middle of the preparation an eutectic melt appears. Thus, two chemically different kinds of complexes between **BC** and **PA** have been formed. The one on the right side is the 1 : 1 complex while the small strip on the left side (the '**BC**' side) is a complex with the composition $(\mathbf{BC})_2$: **PA**. The former melts at 129 °C while the latter shows a melting point at 124 °C. (From Henck *et al.* 2001, with permission.) *Bottom*: The temperature–composition diagram derived in part from these hot stage observations. (From Henck *et al.* 2001, with permission.)

the necessity for weighing, or the determination of time consuming and material consuming cooling-curve determinations; quantitative aspects of the diagram can be obtained from the thermal methods described in the next section. McCrone (1957) estimated that a simple binary eutectic diagram can be determined microscopically in 2–4 h, while a ternary diagram might require two days, each additional polymorphic form doubling the time required. He also pointed out that Kofler's hot bench, or hot bar (Kuhnert-Brandstätter 1971, 1982) can serve as a useful auxiliary in determining composition diagrams (Kofler and Winkler 1950a,b).

4.3   Thermal methods

Whereas hot stage microscopy can be used to obtain qualitative information on polymorphic behaviour, thermal analysis provides quantitative information about the relative stability of polymorphic modifications, the energies involved in phase changes between them and the monotropic or enantiotropic nature of those transitions. The two techniques are best used in conjunction.

Thermal methods are based on the principle that a change in the physical state of a material is accompanied by the liberation or absorption of heat. The various techniques of thermal analysis are designed for the determination of the enthalpy accompanying the changes by measuring the difference in heat flow between the sample under study and an inert reference. These methods are all now commonly (and often mistakenly) referred to as DSC (differential scanning calorimetry), since there are a number of ways of carrying out these experiments, each yielding slightly different information (McNaughton and Mortimer 1975).

In the classical differential thermal analysis (DTA) system both sample and reference are heated by a single heat source. The two temperatures are measured by sensors embedded in the sample and reference. In the so-called Boersma system, the temperature sensors are attached to the sample pans. The data are recorded as the temperature difference between sample and reference as a function of time (or temperature). The object of these measurements is generally the determination of enthalpies of changes, and these in principle can be obtained from the area under a peak together with a knowledge of the heat capacity of the material, the total thermal resistance to heat flow of the sample and a number of other experimental factors. Many of these parameters are often difficult to determine; hence, DTA methods have some inherent limitations regarding the determination of precise calorimetric values.

On the other hand, in a genuine DSC instrument, sample and reference are each heated individually. A null balance principle is employed, whereby any change in the heat flow in the sample, (e.g. due to a phase change) is compensated for in the reference. The result is that the temperature of the sample is maintained at that of the reference by changing the heat flow. The signal which is recorded (dH/dt) (the heat flow as a function of time (temperature)), is actually proportional to the difference between the heat input into the two channels as a function of time (temperature), so that the integration under the area of the peak directly yields the enthalpy of the transition.

Thermogravimetric analysis (TGA) measures the change in mass of a sample as a function of temperature. It therefore provides information on the presence of volatile components, in the present context particularly solvents or water, which form the basis of solvates or hydrates respectively, as well as processes such as decomposition and sublimation.

Much of the literature on the thermal analysis of polymorphic systems up to 1995 has been cited by Giron (1995), along with many illustrative examples. Perrenot and Widmann (1994), Threlfall (1995), Kuhnert-Brandstätter (1996) and Vippagunta *et al.* (2001) have also provided excellent discussions of many of the practical aspects of the application of thermal analysis to polymorphic systems. The following discussion draws much from these contributions.

Many typical features of a DSC of a polymorphic system are demonstrated for sulphapyridine in Fig. 4.10. In the trace, the starting sample is in an amorphous state, obtained by rapidly cooling (shock cooling) the melt. This is one of the recommended procedures for detecting polymorphism in an uncharacterized sample. Initial heating starting at 40°C results in a second-order glass transition A to a supercooled liquid. A second-order transition is characterized by a change in the heat capacity with no heat absorbed or evolved, and is recorded as a lowering of the baseline. The resulting unstable supercooled liquid can crystallize spontaneously upon heating, yielding in the sharp exotherm B, corresponding to an unstable phase in accord with Ostwald's Rule (see Section 2.3). Further heating to *c.* 145 °C leads to an (monotropic) exothermic solid–solid transition, denoted by C, resulting in a metastable phase. At D, the latter melts at 175.30 °C and recrystallizes to the stable modification at E (at 177.30 °C, which in turn melts at 190.30 °C at F.

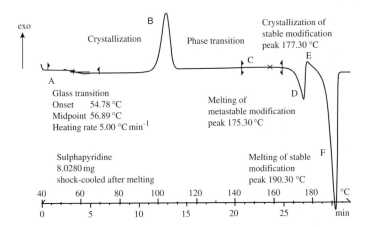

Fig. 4.10 Typical features of a DSC for the polymorphic system of sulphapyridine. (Reproduced with permission from Schwarz and de Buhr 1998.)

Some of these changes can be followed optically, as evidenced by the microscope images in Fig. 4.11. The quality of the thermal measurement, and consequently the amount of information that can be extracted from it depends on a number of experimental conditions. For instance, the heating rate is a crucial parameter, as demonstrated in Fig. 4.12. On the same material, vastly different traces are obtained at different heating rates. On the other hand, all of these contain information so that carrying out the measurement at a number of heating rates is always an advisable practice.

This may not always be a sufficient precaution, as evidenced in Fig. 4.13. In spite of the fact that the heating rate is identical to that in Fig. 4.10, this trace does not exhibit the melting of the metastable modification and subsequent crystallization into the stable form.

Other experimental factors that can influence the quality of the DSC measurement and the information that can be extracted from it are sample mass, particle size, the presence of impurities, the shape of the crystalline particles and the presence of nuclei or seeds of various polymorphs. For the investigation of solvates (pseudopolymorphism), the sample pan type also plays an important role (Giron 1995). Threlfall (1995) recommends routinely running both heating and cooling curves, while Perrenot and Widmann (1994) demonstrated the additional information that can be obtained by carrying out multiple heating runs on a particular sample.

In addition to providing information on the existence of polymorphs and the transformations among them, DSC measurements contain the quantitative information (complementary to optical microscopic observations) to aid in the preparation of the free-energy temperature diagram (Section 2.2.3). The principles behind this process (Burger 1982a,b; Giron 1995; Grunenberg et al. 1996) are outlined here.

The energy temperature diagram and characteristic DSC traces for the monotropic case are given in Fig. 4.14, while that for the enantiotropic case is given in Fig. 4.15. The connections between the diagram and the traces are described in the figure

Fig. 4.11 Photomicrographs of sulphapyridine at two temperatures. Left, ~ 110 °C; most of the field contains the metastable form that crystallized (peak B in Fig. 4.10). Right, ~ 180 °C (between peaks E and F in Fig. 4.11), most of the material has been converted to the stable form. (Reprinted from Schwarz and de Buhr 1998, with permission.)

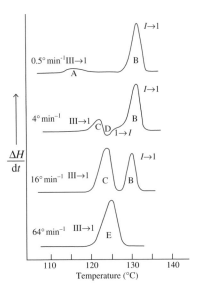

Fig. 4.12 A DSC measured at four different heating rates. At the slowest rate (0.5 °C min⁻¹) both a solid–solid phase transition (III → I) (A) and the melting of the more stable phase I (B) can be seen. At the fastest heating rate of 64 °C min⁻¹ modification III melts directly (E), but the heating rate is sufficiently fast to prevent the crystallization of Form I. At 4 °C min⁻¹ Form II melts (C) and recrystallizes to Form I (D) which subsequently melts. At the intermediate rate of 16 °C min⁻¹ the system does not reach equilibrium, so the recrystallization of Form I is masked by the direct melting of Form III. (From Burger 1975, with permission)

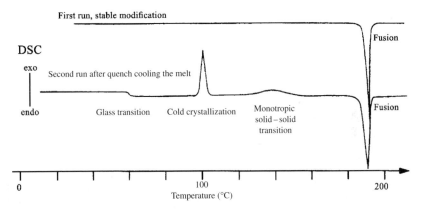

Fig. 4.13   DSC trace of sulphapyridine at 5 °C min⁻¹. (From Perrenot and Widmann 1994, with permission.)

captions. The principles are demonstrated for a system comprised of two polymorphic modifications, but of course, may be readily extended to more phases (e.g. Yu *et al.* 2000*a,b*). In practical terms the measured transition temperature for, say, a dimorphic enantiotropic system depends on the kinetic properties of the transition

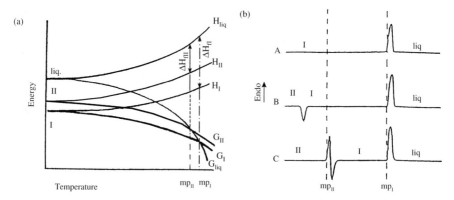

Fig. 4.14 Characteristic free-energy vs temperature diagram (a) and DSC traces (b) for the monotropic relationship between polymorphs. The G_I and G_{II} curves do not cross below their melting points mp_I, and mp_{II} indicated on the temperature axis. DSC trace A exhibits the melting of the thermodynamically high melting Form I from which the value of ΔH_{fI} may be extracted. In trace B, the low melting modification II transforms monotropically to modification I which subsequently melts. The $\Delta H_{II\rightarrow I}$ gives a measure of the gap between the H_I and H_{II} curves at that temperature. In trace C, modification II melts followed by the crystallization of modification I. The melting and crystallization give ΔH_{fI} and ΔH_{fII} for I and II respectively at the temperature of the process. (After Giron 1995, with permission.)

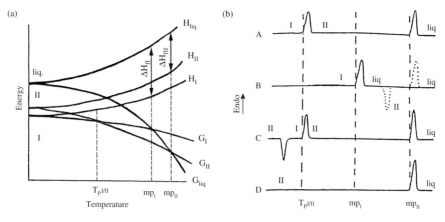

Fig. 4.15 Characteristic free-energy temperature diagram (a) and DSC traces (b) for the enantiotropic relationship between polymorphs. The G_I and G_{II} curves cross at the transition temperature $T_{I\rightarrow II}$ below their melting points mp_I, and mp_{II} all indicated on the temperature axis. DSC trace A: at the transition temperature modification I undergoes an endothermic transition to modification II, and the heat absorbed is $\Delta H_{I\rightarrow II}$ for that transition. Modification II then melts at mp_{II}, with the accompanying ΔH_{fII}. DSC trace B: Modification I melts at mp_I with ΔH_{fI} followed by crystallization of II with ΔH_{fII} at the intermediate temperature. Modification II then melts with details as above. DSC trace C: modification II, metastable at room temperature, transforms exothermically to modification I with $\Delta H_{II\rightarrow I}$ at that transition temperature. Continued heating leads to the events in trace A. DSC trace D: modification II exists at room temperature and no transition takes place prior to melting at mp_{II}, with the appropriate ΔH_{fII}. (After Giron 1995, with permission.)

under investigation and is a function of the experimental conditions. The thermodynamic transition point (see Section 2.2.4) can be estimated if the melting points and the enthalpy of fusion of the two polymorphs are experimentally accessible (Yu 1995).

The principles of TGA are demonstrated in Fig. 4.16 for glucose monohydrate, which is known to exist in an anhydrous form and as a monohydrate. The anhydrous form exhibits no weight loss on heating, while the monohydrate shows a weight loss of 7.1 per cent, slightly below the expected value of 9.1 per cent for a 1 : 1 molar ratio of glucose to water. Thermogravimetric analysis measurements are often accompanied by those of DSC, providing a great deal of information on the nature of the desolvation process, including heat of desolvation and subsequent transformations. For instance, the DSC measurement for glucose accompanying Fig. 4.16 is given in Fig. 4.17. While the anhydrous form exhibits only melting at about 161 °C, the monohydrate shows a broad endothermic peak accompanying the dehydration. The resulting anhydrate melts at a temperature below that of the pure anhydrate, due perhaps to incomplete crystallization or the presence of residual water. Of course, true polymorphism of hydrates and/or solvates may exist, and such cases will add complexity to the TGA and DSC traces. While other techniques are usually preferred for the quantitative analysis of mixtures of polymorphs there are some systems for which DSC has proven to be the preferred method (Carlton *et al.* 1996).

Mention should be made here of one of the recent developments in DSC technology, namely oscillating, alternating or modulated DSC (Barnes *et al.* 1993; Readings 1993; Dollimore and Phang 2000). An oscillating time/temperature function is applied to the sample with simultaneous heating at a constant rate. The oscillation allows the application of Fourier transform techniques to the signal and its separation into two components: the reversible component contains the specific heat C_p, while the irreversible component contains kinetic information. The technique permits separation of thermal events that overlap. The irreversible ones, such as desolvation and crystallization will appear on the kinetic curve, while fusion will appear on the reversible one.

Solution microcalorimetry is another thermal method for the determination of the difference in lattice energy of polymorphic solids. The difference in heat of solution of two polymorphs is also the difference in lattice energy (more precisely lattice enthalpy), provided of course, that both dissolution experiments are carried out in the same solvent (Guillory and Erb 1985; Lindenbaum and McGraw 1985; Giron 1995). The actual value for ΔH_{I-II} is independent of the solvent, as demonstrated in Table 4.1 for the two polymorphs of sodium sulphathiazole. Note also that the calculated heats of transition (ΔH_t) are virtually identical in spite of the fact that the heat of solution (ΔH_s) is endothermic in acetone and exothermic in dimethylformamide.

Since experimentally ΔH_{I-II} may be determined at a chosen temperature, the technique allows the determination of the difference in lattice energy at different temperatures. In addition the ΔH_{I-II} at one temperature (say T_1) may be used to

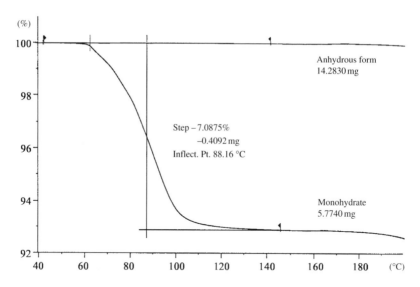

Fig. 4.16 TGA traces for the anhydrous and monohydrate forms of glucose. (Reprinted from Schwarz and de Buhr 1998, with permission.)

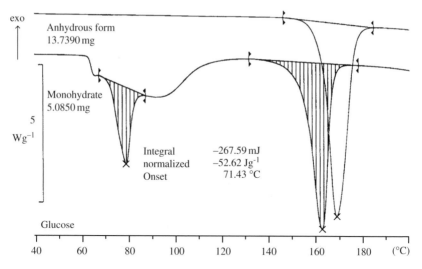

Fig. 4.17 DSC traces for the anhydrous and monohydrate forms of glucose. (Reprinted from Schwarz and de Buhr 1998, with permission.)

calculate the same quantity at a different temperature, T_2, from the difference in the heat capacities of the two polymorphs:

$$\Delta H_{\text{I–II}}(T_2) = \Delta H_{\text{I–II}}(T_2) + (T_1 - T_2)\,\Delta C_{p(\text{I,II})}.$$

Table 4.1 Heats of solution ΔH_s (kcal mol^{-1}) measured at 25 °C and calculated heats of transition ΔH_t for the two polymorphic forms of sodium sulphathiazole in two different solvents (adapted from Lindenbaum and McGraw 1985)

	Acetone	Dimethylformamide
$\Delta H_{s,I}$	2.853 ± 0.026	-1.113 ± 0.012
$\Delta H_{s,II}$	1.229 ± 0.016	-2.740 ± 0.023
ΔH_t (at 25 °C)	1.624 ± 0.042	1.627 ± 0.035

Some typical examples of these measurements are given by Lindenbaum *et al.* (1985). For quantitative analysis of mixtures, Botha *et al.* (1986) have demonstrated how the percentage composition of two polymorphs may be determined essentially over the entire range of composition by measurements of the heats of solution of the pure polymorphs. Additional details on many aspects of the applications of solution microcalorimetry to the study of polymorphs and solvates may be found in the book by Hemminger and Höhne (1984), the chapters by Grant and Higuchi (1990) and Brittain and Grant (1999) and the recent paper by Gu and Grant (2001).

4.4 X-ray crystallography

X-ray crystallographic methods, which reflect differences in crystal structure, in most cases can be definitive in the identification and characterization of polymorphs, and whenever possible should be included in the analytical methods utilized to define a polymorphic system.

In the application of X-ray diffraction methods to the study of molecular solids in general and polymorphic systems in particular a distinction is often made between powder methods and single crystal methods. Traditionally, the former have been used for the qualitative identification of individual polymorphic phases or mixtures of phases, while the latter have been employed for the determination of the detailed molecular and crystal structure. Fortunately, the gap between these two subdisciplines is being bridged, and much can be gained in the study of polymorphs by the cross-fertilization of these two techniques. For instance, single crystal structure solution techniques are being applied to powder data to solve larger and previously intractable crystal structures (Andreev *et al.* 1997; Shankland *et al.* 1997; Tremayne *et al.* 1997; David *et al.* 1998; Chan *et al.* 1999; Putz *et al.* 1999) with subsequent refinement by Rietveld methods (Young 1993), and this field is expected to grow rapidly with increasingly powerful X-ray sources and computing resources. The increasing application of synchrotron radiation to powder diffraction is also contributing significantly to closing this gap; in some cases revealing polymorphism not observed by laboratory scale X-ray sources (Sato 1999).

Recent and continuing developments in the field of electron diffraction of very small single crystals may also serve to bridge the gap between classical powder and single crystal diffraction techniques (Voigt-Martin *et al.* 1995; Dorset 1996; Dorset *et al.* 1998; Yu *et al.* 2000*b*).

The fundamentals of X-ray powder diffraction techniques are summarized in a number of texts and reviews (Azaroff and Burger 1958; Bish and Reynolds 1989; Jenkins and Snyder 1996). The X-ray diffraction pattern from a solid results from the satisfaction of the Bragg condition ($n\lambda = 2d \sin \theta$), where λ is the wavelength of the X-ray radiation, and d is the particular spacing between individual parallel planes. The condition can be satisfied when the angle θ between the incident radiation and that set of planes results in constructive interference. The X-ray powder diffraction pattern of a solid is thus a plot of the diffraction intensity as a function of 2θ values (or equivalently, d spacings) and may be considered to be a fingerprint of that solid. The values of the d spacings reflect the dimensions of the unit cell, while the intensities are due to the contents of the unit cell and the way the atoms and molecules are arranged therein. As polymorphs comprise different solids with different unit cells and different arrangements of molecules within the unit cell they have different fingerprints—most often as different as the X-ray powder patterns of two different compounds. Thus X-ray powder diffraction is probably the most definitive method for identifying polymorphs and distinguishing among them. Typical experimental powder patterns for two polymorphs are given in Fig. 4.18.[2]

Single crystal X-ray crystallographic techniques are employed to determine details of the molecular and crystal structure of the solid—bond lengths, bond angles, intermolecular interactions, etc., and are the source of some of the most precise metric data on these structural features (Dunitz 1979; Giacovazzo 1992; Glusker *et al.* 1994). The three-dimensional results are obtained from the collection of three-dimensional diffraction data. Those results may be used to simulate computationally the two-dimensional diffraction pattern to be expected from a powder of the same material. Such a calculated powder pattern may serve as a standard for the powder diffraction pattern, unencumbered by impurities, the presence of other polymorphs, or the experimental difficulties enumerated below. The other side of this coin is the attempt to obtain the full three-dimensional crystal structure from the two-dimensional powder diffraction pattern, as noted above. Some of the important features of powder diffraction are discussed here. Those emanating from single crystal studies are alluded to in many other places throughout this work.

There are some reported cases of two genuinely polymorphic structures exhibiting very similar powder diffraction patterns. One of these cases, the two polymorphs

[2] There have been a number of reports of polymorphic systems in which the reader might be led to expect some similarity in the powder pattern of two polymorphs because they crystallize in the same space group. There is no physical basis for this expectation. Except for the systematic absences of certain reflections due to space group symmetry polymorphic structures in the same space group and different cell constants will have different powder diffraction patterns. On the other hand, polymorphs with similar cell constants but different space groups may exhibit some similarity in X-ray powder diffraction patterns, but these cases are very rare (*vide infra*). See also Section 2.4.3.

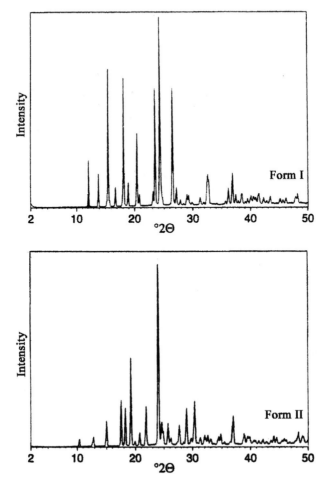

Fig. 4.18 X-ray powder diffraction patterns for the two polymorphs of paracetamol. (From Nichols and Frampton 1998, with permission.)

of terephthalic acid (Davey *et al.* 1994), is shown in Fig. 4.19. These two patterns are remarkably similar, save for the 2θ region just above the large peak at $\sim 30°$. In such cases, there must be a structural explanation for the similarities in powder patterns. The two structures crystallize in the triclinic space group $P\bar{1}$ with different cell constants as given in Table 4.2 (Bailey and Brown 1967, 1984; Herbstein 2001). As shown in Fig. 4.20 and discussed by Berkovitch-Yellin and Leiserowitz (1982) and Davey *et al.* (1994) both structures are characterized by the formation of infinite hydrogen-bonded chains. The chains are organized into two-dimensional sheets that differ to some extent between the two polymorphs in the relationship between neighbouring chains. The offset of neighbouring sheets is more significant, and comprises the principal difference between the two structures. However, since the major

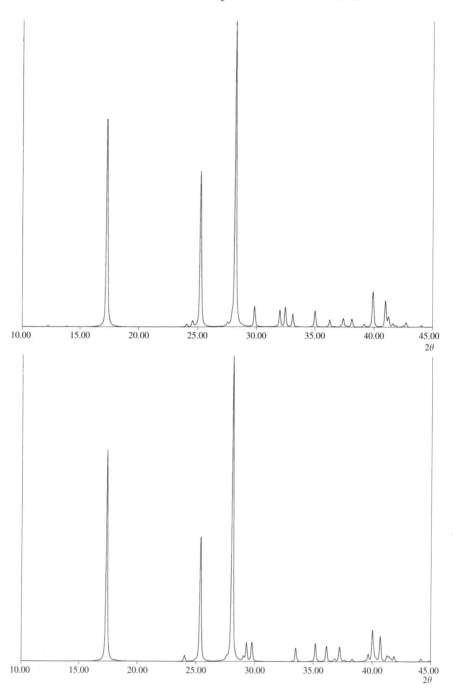

Fig. 4.19 X-ray powder diffraction patterns calculated from the crystal structures for the two forms of terephthalic acid. Upper, Form I; lower, Form II. (After Davey *et al.* 1994.)

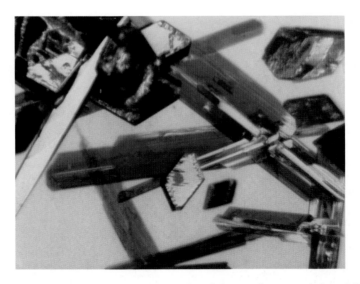

Fig. 3.2 Three of the concomitant polymorphs of the cyanine:oxonol dyes **3-XVI** and **3-XVII**. Gold (by reflection; otherwise red by transmission) and red forms mentioned in the text are easily distinguishable. The third form is a purple one, normally diamond shaped as on the middle right, but many of these crystals are undergoing conversion, as indicated by varying degrees of mottled surfaces. (From Bernstein *et al.* 1999, with permission.)

Fig. 3.4 Photograph of concomitantly crystallizing forms of rat liver glutathone S-transferase. The three forms are labelled. (From Fu *et al.* 1994, with permission.)

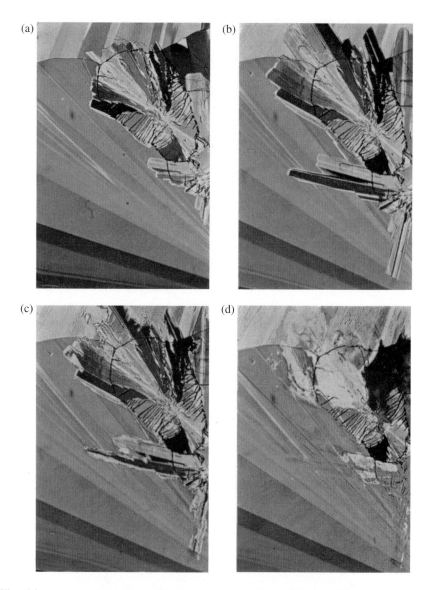

Fig. 4.3 An enantiotropic phase transformation in 2,4,6-trinitro-5-tertbutyl-*m*-xylene as observed on the hot stage microscope. Form II is stable at room temperature and the thermodynamic transition point is at 84 °C: (a) Room temperature stable Form II, the coarse crystals at upper right embedded in aggregate of Form I. (b) On heating Form II grows at the expense of Form I. (c) At 84 °C the transformation can be halted. (d) Above 84 °C Form I is stable and has grown at the expense of Form II. (From Kuhnert-Brandstatter 1971, with permission.)

Fig. 4.4 Phase transformation of β-naphthol, showing the diffuse nature of product phase formation. (From Kuhnert-Brandstätter 1982, with permission.)

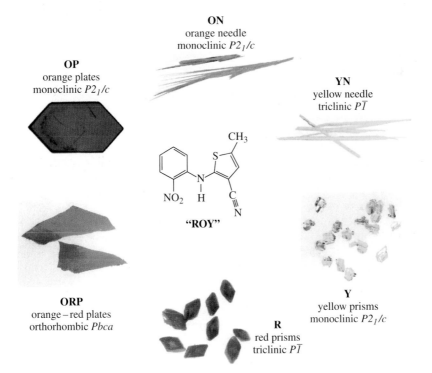

ON
orange needle
monoclinic $P2_1/c$

OP
orange plates
monoclinic $P2_1/c$

YN
yellow needle
triclinic $P\bar{1}$

CH_3

NO_2 H

"ROY"

ORP
orange – red plates
orthorhombic *Pbca*

Y
yellow prisms
monoclinic $P2_1/c$

R
red prisms
triclinic $P\bar{1}$

Fig. 5.9 Photomicrographs of the six polymorphs of ROY **5-XII**, showing the different colour (R = red, Y = yellow, O = orange) and morphology (P = plates, N = needles). From upper left, clockwise, OP, ON, YN, Y, R, ORP. (From Yu *et al.* 2000, with permission.)

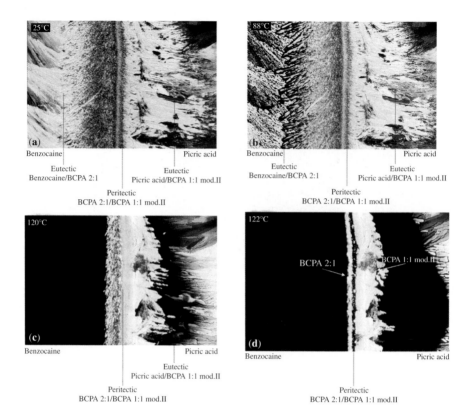

Eutectic
Benzocaine/BCPA 2:1

Eutectic
Picric acid/BCPA 1:1 mod.II

Peritectic
BCPA 2:1/BCPA 1:1 mod.II

Eutectic
Benzocaine/BCPA 2:1

Eutectic
Picric acid/BCPA 1:1 mod.II

Peritectic
BCPA 2:1/BCPA 1:1 mod.II

Eutectic
Picric acid/BCPA 1:1 mod.II

Peritectic
BCPA 2:1/BCPA 1:1 mod.II

Peritectic
BCPA 2:1/BCPA 1:1 mod.II

Fig. 4.9 Photographs of the microscope slide Kofler preparations showing the various phases of the benzocaine : picric acid (**BC** : **PA**) binary sysytem. Figure a shows the photomicrograph of the recrystallized contact preparation of **BC** and **PA** at 25 °C. The interference colours are due to the use of crossed polarizers. The pure compounds are at the extremities of the preparation, while in the region where the original compounds have merged a number of different areas may be observed, due to the formation of different crystalline species combining the two components. Heating this preparation on the hot stage microscope to a temperature of about 88 °C shows (Fig. b) the eutectic melt of **BC** and the broad dark yellow crystals of a new compound. Due to the crossed polarizers, the isotropic melt appears black. As seen in the Fig. c, at about 120 °C on the left side of the preparation **BC** is melted and the right side the eutectic between **PA** and the remaining crystals of the **BC** : **PA** 1 : 1 complex melts. Figure d shows the situation which is observed at 122 °C. **PA** is almost melted and in the middle of the preparation an eutectic melt appears. Thus, two chemically different kinds of complexes between **BC** and **PA** have been formed. The one on the right side is the 1 : 1 complex while the small strip on the left side (the '**BC**' side) is a complex with the composition (**BC**)$_2$: **PA**. The former melts at 129 °C while the latter shows a melting point at 124 °C. (From Henck *et al.* 2001, with permission.) (See also p. 103.)

Table 4.2 Crystallographic cell constants for the two polymorphs of terephthalic acid

	Reported cells[a]		Reduced cells[b]	
	Form I	Form II	Form I	Form II
a (Å)	7.73	9.54	3.76	5.027
b (Å)	6.443	5.34	6.439	5.36
c (Å)	3.749	5.02	7.412	6.991
α (°)	92.75	86.95	83.16	72.04
β (°)	109.15	134.6	80.87	76.03
γ (°)	95.95	104.9	88.53	87.09

[a] After Bailey and Brown (1967, 1987).
[b] Herbstein (2001). The cell constants for the reduced cell are based on those reported by Colapietro *et al.* (1984) and Domenicano *et al.* (1990) (for Form I) and by Fischer *et al.* (1986) (by neutron diffraction) for Form II. Since they follow the convention for reporting reduced cells (International Tables 1987) they appear in a different order from the original; also by convention the cell angles are defined as acute.

contribution to the X-ray scattering is from the sheets, the similarity in structures is manifested in the similarity in powder patterns.[3]

Some other examples of very similar X-ray powder diffraction patterns have been noted. One of these is D,L-leucine (Mnyukh *et al.* 1975) for which the spectral data are clearly distinguishable. Another is caffeine (Fig. 4.27) (Suzuki *et al.* 1985; Griesser 2000), which is discussed in the next section. Karfunkel *et al.* (1999) also quoted the 'surprisingly similar powder patterns' of some diketopyrrolopyrrole derivatives and have attempted to develop a model to describe the structural basis for similar powder patterns.

The preparation of samples for powder diffraction can lead to variations and inconsistencies among measurements on the same sample (e.g. Potts *et al.* 1994). Jenkins and Snyder (1996) have summarized the possible causes for compositional variations between the original sample and that prepared for the diffraction experiment; grinding of the sample (generally required to reduce preferred orientation) *vide infra* can lead to amorphism, strain in individual particles, decomposition, solid state reaction or

---

[3] A cautionary note is in order here. In principle there is considerable freedom in choosing the axial system in triclinic unit cells, and in some higher symmetry ones as well (International Tables 1987). The convention calls for choosing the so-called reduced cell (Mighell 1976; Burzlaff *et al.* 1983; Baur and Tillmanns 1986), and there are a number of computer programs available for doing carrying out this cell reduction (LePage and Donnay 1976; Macicek and Yordanov 1992). For example, both the unreduced and reduced cells are presented in Table 4.2. Most crystallographic databases now routinely check that the recorded cell is in fact the reduced one, and all data collection programs on automated diffractometers do the reduction. However, many literature values have not been checked for the presence or absence of the reduced cell. Cell reduction can also lead to the recognition of a cell of different symmetry (International Tables 1987). As a result, a comparison of unit cell constants which leads to the conclusion that two structures are polymorphs may be in error. One way to check for identity or difference in polymorphs is to calculate the powder pattern, which is invariant to the choice of cell, although different choices of cell will lead to different indexing of peaks for identical diffraction patterns.

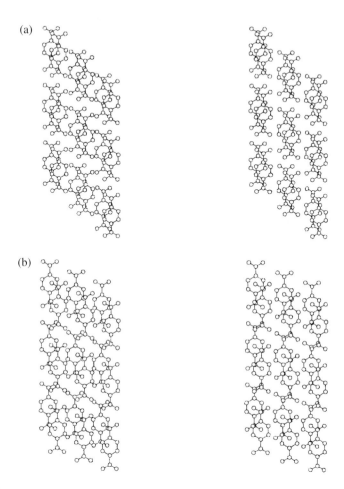

Fig. 4.20 Stereoviews of the crystal structures of the two polymorphs of terephthalic acid. In both cases, the view is on the molecular plane. (a) Form I; (b) Form II. Both polymorphs are triclinic, $P\bar{1}$, and are composed of very similar layers (which is essentially the plane of the paper) built up from linear chains of hydrogen-bonded molecules. The lateral offset of chains within a layer is also very similar in the two structures. The structures differ in the manner in which subsequent layers are offset from each other. (After Davey *et al.* 1994, with permission.)

contamination; the radiation used in the diffraction experiment can induce changes in the material, such as solid state reaction (e.g. polymerization), decomposition, or transformation to an amorphous state; the environment (humidity, temperature) can also effect the addition or loss of solvent, onset of reaction, decomposition, etc. All of these factors should be taken into account in determining and comparing powder patterns.

Perhaps the most pervasive problem influencing the intensities of powder diffraction lines is that of preferred orientation (Jenkins and Snyder 1996). Due to their non-spherical habit crystallites have a tendency to become oriented to efficiently occupy a minimum volume. For instance, plate-like crystals tend to lie flat on top of each other. Such orientational non-randomness will tend to diminish the intensities of those Bragg reflections that will not come into the diffracting geometry because of this preference for orientation of plates. A dramatic example of preferred orientation is shown for Form III of sulphathiazole in Fig. 4.21, in which a comparison with the powder diffraction pattern calculated from the crystal structure indicates that the intensities of almost all of the expected diffraction peaks have been severely suppressed. The effects of preferred orientation can be reduced by grinding the sample prior to mounting, or spinning the mounted sample. Jenkins and Snyder note that one of the most effective ways of eliminating preferred orientation is by spray-dried preparation of the crystalline sample. Unfortunately, there may be considerable loss of control over the polymorphic form obtained in using this process.

It is important to make the distinction between the determination of polymorphic identity and polymorphic purity. The former is essentially a qualitative determination, asking the question, 'Is a particular polymorph present in a given sample?' The latter is a question of quantitative analysis, and it is generally (though not always) assumed that the sample is *chemically* pure, so the analytical problem to be addressed is the determination of the relative amounts of different polymorphs in the sample. Recalling that different polymorphs are for all intents and purposes different solids, the determination of polymorphic purity is then no different in principle from quantitative determination of the composition of a mixture of solids. Such quantitative determinations comprise one of the traditional activities of analytical chemistry, especially when the materials are different chemical entities. In those cases, a variety of different analytical methods may be employed. In the case of polymorphic mixtures, or the determination of polymorphic purity, the choice of analytical method is considerably more restricted, and X-ray diffraction is one of the most definitive techniques (see e.g. Stowell 2001).

The techniques of quantitative X-ray powder diffractometry have recently been summarized in a book by the same name by Zevin and Kimmel (1995) and more briefly in a review by Stephenson et al. (2001). The technique is based on the assumption that the integrated intensity of a diffraction peak is proportional to the amount of the component (i.e. polymorph) present. Along with other factors that can influence the intensity noted above with regard to the determination of polymorphic identity, that intensity can also be severely affected by absorption of the incident radiation, for which appropriate corrections are available (Klug and Alexander 1974). Relative amounts of different polymorphs are determined by the relative intensities of a small

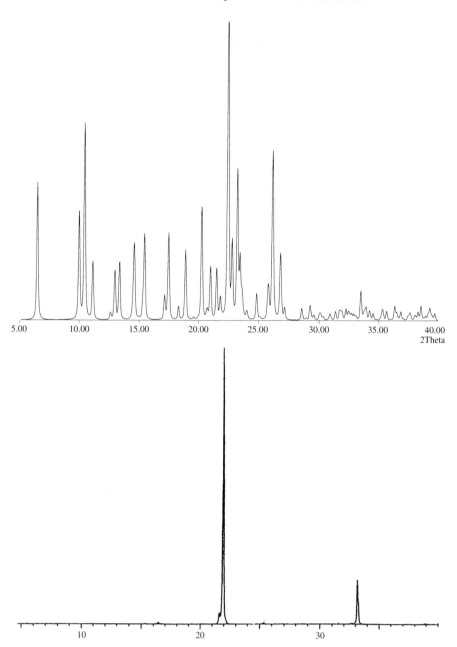

Fig. 4.21 The influence of preferred orientation on the experimental X-ray powder diffraction pattern of modification III of sulphathiazole. Upper, expected pattern calculated from the single crystal structure; lower, experimental powder pattern. (Adapted from Threlfall 1999, with permission.)

number of (ideally) distinctive and preferably relatively strong peaks in the diffraction patterns, which means that the overlap of neighbouring peaks is also a problem to be avoided or properly dealt with. Experimentally determined integrated intensities also are influenced by statistical errors, extinction effects and other systematic aberrations, and all of these will affect the precision of the quantitative analysis. Of course, the usual precautions for quantitative analysis of mixtures must also be observed: the calibration and validation samples must be homogeneous and truly representative of the concentration level.

Many of these difficulties can be monitored and overcome with the use of standards, either internal or external (Zevin and Kimmel 1995). For the internal standard method, a known quantity of standard material is added to an unknown mixture, and the ratio of the intensity of the standard component is compared to a previously determined calibration curve to determine the mass fraction of the unknown (e.g. one or more of the polymorphic components). In the external standard method, the entire composition of the unknown sample is determined simultaneously by comparing the measured intensities and respective calibration constants of reference intensity ratios (determined beforehand), which must all be with reference to the same reference standard.

As noted above, the traditional methods of quantitative analysis make use of one or a small number of non-overlapping peaks from the diffraction patterns of the different component (e.g. polymorphic) phases. With the advent of more powerful laboratory X-ray sources and synchrotron radiation, faster and more sensitive detectors, computer controlled diffractometers and the almost universal use of digitized data there is increasing use of the full diffraction pattern for quantitative X-ray diffraction analysis (Zevin and Kimmel 1995).

Qualitative and quantitative analytical applications of X-ray diffraction both require reference diffraction patterns to identify and quantify the different polymorphic modifications. Experimental powder patterns may be suspect for their use as standards as a result of experimentally induced errors or aberrations or the lack of polymorphic purity in the sample itself (which may even result from the sample preparation). The availability of full crystal structure determinations for any or all of the polymorphic modifications can considerably facilitate generation of standard powder patterns. A variety of public domain software is now available for calculating powder diffraction patterns from single crystal data (ICDD 2001, IUCr 2001).[4]

For example, the calculated powder patterns for four of the polymorphs of sulphapyridine are given in Fig. 4.22. The calculated pattern represents that of a pure sample. Using some of the more sophisticated programs to calculate the powder pattern, one can assume the absence of preferred orientation or alternatively some specified degree of preferred orientation. The line shape can be varied to match experimentally observed line shapes. If the crystal structures of all the polymorphic forms (and impurities) in the mixture are known, then the diffraction patterns

---

[4] Access to powder patterns generated from single crystal structures has been considerably facilitated by a cooperative arrangement between the Cambridge Crystallographic Data Centre (the repository of the experimental crystal structure data) and the International Centre for Diffraction Data (the repository for powder diffraction data). Either one of these organizations may be contacted for further details.

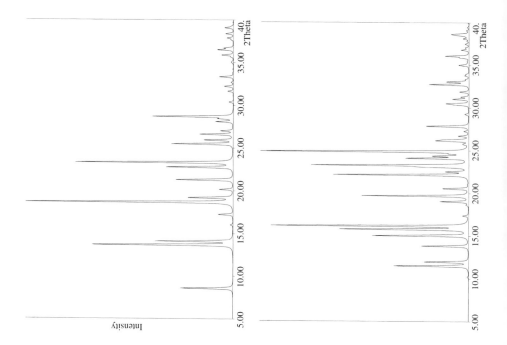

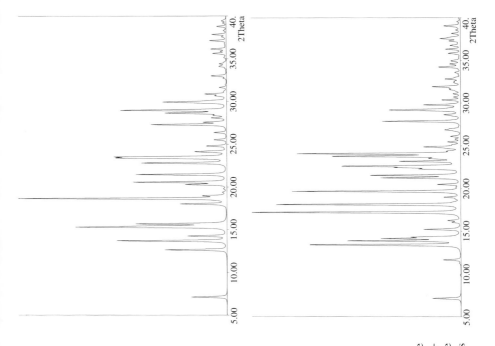

Fig. 4.22 X-ray powder diffraction patterns calculated from the crystal structure determinations of the four polymorphs of sulphapyridine (Bar and Bernstein 1985; Bernstein 1988). For ease of comparison such patterns should be stacked with identical scales on the abscissa. From top to bottom, Forms II, III, IV and V.

Table 4.3 Crystallographic constants for two polymorphs

	Form A	Form B
Crystal system	Orthorhombic	Monoclinic
Space group	$P2_12_12_1$	$P2_1/c$
a (Å)	10.260 (3)	14.5878 (3)
b (Å)	33.335 (3)	14.111 (3)
c (Å)	10.101 (3)	18.101 (3)
β (°)	—	111.85
Z	8	8

of synthetically generated mixtures may be calculated for calibration or for use as benchmarks for experimental mixtures.

Many of the features of the use of X-ray diffraction in the analysis of a mixture of two unsolvated polymorphs (A and B) are demonstrated in a recent study by Newman *et al.* (1999). Under ambient conditions, Form A has been shown to be thermodynamically more stable and the crystal structures of both forms have been determined. They crystallize in different space groups with very different cell constants (Table 4.3); nevertheless, the laboratory generated powder patterns (Fig. 4.23) are quite similar. According to Newman *et al.* the polymorphs are also not easily distinguishable by their DSC traces or IR, Raman and NMR spectra. However, there are some clearly distinct features between their powder patterns measured using synchrotron radiation (Fig. 4.24). Since manufactured lots appeared to contain both forms with Form B generally as a minor component, it was necessary to develop a quantitative method to determine the amounts of the two forms in any batch. In developing the analytical protocol the authors followed the guidelines of the International Commission on Harmonization (1996), taking into account the specificity, working range, accuracy, precision, minimum quantifiable limit and robustness of the procedure.

Because of the general similarities in the diffraction patterns, and the lack of clearly resolvable distinguishing peaks, they employed the Rietveld method (Young 1993). In the Rietveld method, the entire experimental diffraction pattern for each solid phase is used as a basis for comparison. For structure determination using powder diffraction, this comparison is made with a structural model used to generate a calculated pattern. In quantitative analysis of polymorphic phases, the known crystal structures are used to generate the standard diffraction patterns and these are then refined against the experimental powder pattern of the mixture to obtain the relative amounts of the polymorphs.

The development of the experimental procedure then involves the preparation of standard mixtures to prepare a calibration curve, with due care paid to corrections for particle size distribution, background, illuminated volume of sample and preferred orientation. A typical calibration run is shown in Fig. 4.25. Determinations on a series of similar 'spiked' mixtures leads to the calibration curve in Fig. 4.26. Analysis of the resulting data led to the determination of a minimum quantifiable limit of 5 per cent, a working range of 5–50 per cent Form B and an RDS of 16 per cent. The method

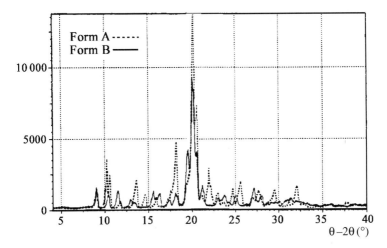

Fig. 4.23 Experimental X-ray powder diffraction patterns for Forms A and B of a polymorphic system. (From Newman *et al.* 1999, with permission.)

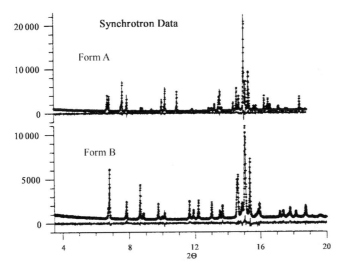

Fig. 4.24 Synchrotron X-ray powder diffraction patterns for Forms A and B of Figure 4.23. (From Newman *et al.* 1999, with permission.)

can be used to routinely monitor quantitatively the composition of mixtures of these two polymorphs for production lots, the control of processes, the stability of samples and the monitoring and manipulation of process parameters to prepare each form or mixtures of them with predetermined proportions.

It is important to point out here that each polymorphic system, indeed each polymorphic modification of polymorphic system must be considered unique and must be individually characterized. For instance, the 5 per cent minimum quantifiable limit for the above system is well above the technical feasibility. Depending on the nature

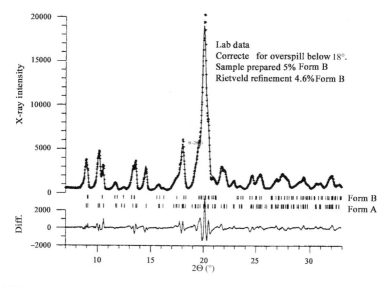

Fig. 4.25 A typical calibration run of a mixture of two polymorphs using the Rietveld analysis. The calibration sample was prepared using 5 per cent of Form B in a mixture of Forms A and B. The upper trace shows the laboratory data for this sample. The next two rows indicate the positions expected for the diffraction peaks of Forms B and A. The bottom trace shows the rms deviation resulting from the refinement of the combination of the full patterns for the two forms against the measured pattern. The best fit is obtained for a value of 4.6 per cent Form B. (From Newman *et al.* 1999, with permission.)

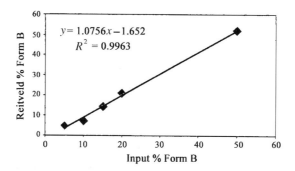

Fig. 4.26 Calibration curve for mixtures of polymorphs A and B using laboratory X-ray data and Rietveld analysis. The value for the 50 per cent mixture was obtained from synchrotron data. (From Newman *et al.* 1999, with permission.)

of the system values at least as low as 1 per cent are obtainable using quantitative X-ray analysis (Tanninem and Yliruusi 1992).

Artificial neural network (ANN) theory (Zupan and Gasteiger 1991), which has been applied to other areas of quantitative chemical analysis (Bos and Weber 1991) has recently been claimed to be useful in the quantitative analysis of mixtures of polymorphs using X-ray diffraction methods (Agatonovic-Kustrin *et al.* 1999, 2000). In studying the two polymorphs of ranitidine hydrochloride, and comparing the method to the more conventional method employing polynomial regression, the authors showed that the ANN methods yielded a smaller standard deviation and relative error, especially for the region of lower concentrations of Form 2, as low as 1–2 per cent. Depending on its general applicability, the ANN methodology may be a potential alternative to the more traditional methods of quantitative analysis of mixtures, and the authors suggest extending the use to entire patterns (rather than a few selected peaks as in their study) as has been done with partial least squares methods in the quantitative analysis of mixtures using IR methods (see the next section).

4.5 Infrared spectroscopy

Infrared spectroscopy is of course a standard technique for the characterization of compounds in the current context of solid materials. Because it is based on the measurement of the vibrational modes generally of bonded atoms, with absorptions usually in the range of 400–4000 cm^{-1} it is primarily a tool for investigating *molecular* properties rather than solid state properties.

Nevertheless, for half a century (Ebert and Gottlieb 1952), it has been one of the most widely used methods for investigating the propensity of materials to form polymorphs (Kuhnert-Brandstätter and Junger 1976), including thermodynamic details such as transition points and number of components (Gu 1993). Fourier transform infrared spectroscopy (FTIR) is clearly the current method of choice, having replaced traditional grating and prism spectrophotometers (Krishnan and Ferraro 1982; Brittain 1997). Since the characteristics of bonds or bonded atoms are monitored with this method, it is the perturbations of those vibrations due to variations in conformational or environmental factors among polymorphs that can lead to the differences in spectra. In general, many molecular features are constant from polymorph to polymorph and the effect of environment on particular bonds and their vibrational manifestation may not be sufficiently large to be evident as differences in the IR spectra among polymorphs (Fig. 4.27), e.g. caffein (Suzuki *et al.* 1985; Griesser 2000) and trovafloxacin mesylate (Norris *et al.* 1997), especially when presented in graphical form on a reduced scale. Hence, IR spectra of different polymorphs quite often exhibit many similar features with differences showing up in particular bands, which may, however, provide distinctive markers for polymorphic characterization and distinction (Fig. 4.28). The FTIR instrumentation and technique provides precise location of absorption bands, and that information, coupled with the IR assignments and comparisons for the various polymorphs, allows a means of characterization and comparison which is not available to the reader who is presented with a series of graphical spectra. Much information about the similarities and differences among polymorphs is often lost in such a graphical presentation. An example of the preferred mode of presentation is in Fig. 4.29 and Table 4.4 for two forms of lamivudine. While it is not always possible to make assignments of the bands to molecular bonding or

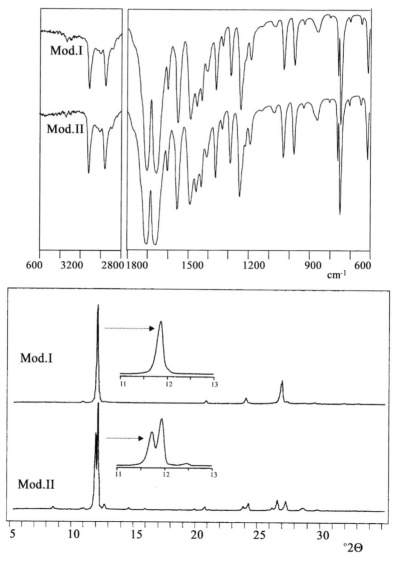

Fig. 4.27 Upper: FTIR spectra of the two anhydrous modifications of caffein, exhibiting very similar features. Lower: X-ray powder diffraction patterns of the same modifications, showing considerable similarity, but distinguishable with the use of an expanded scale in the region $11° < 2\theta < 13°$. Other differences can be detected upon close inspection. (From Griesser 2000, with permission.)

functional groups, the tabulation of distinguishing peaks among polymorphic systems should be encouraged as standard practice. Evaluation of the statistical significance of similarities and differences in the spectra is considerably facilitated by the inclusion of information on the experimental precision of peak location.

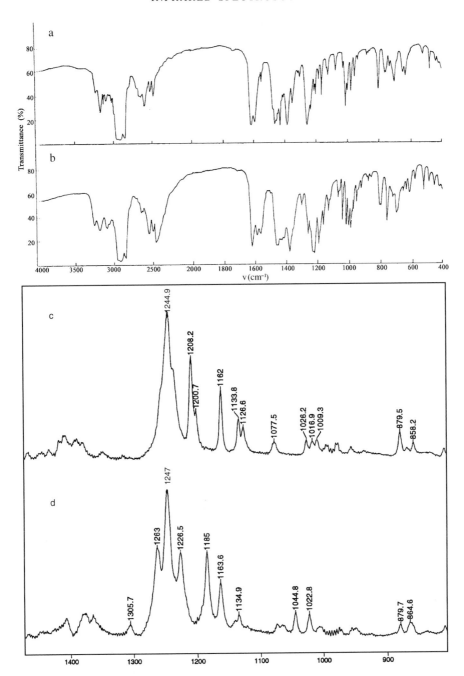

Fig. 4.28 (a, b) IR spectra of (a) Form 1 and (b) Form 2 of ranitidine hydrochloride. The spectra are very similar (though not identical) in many respects, with one particularly distinguishing band at 1045 cm^{-1} in the spectrum of Form 2. (From Cholerton *et al.* 1984, with permission.) Detail of the (c) Form 1 and (d) Form 2 FTIR spectra of the two polymorphs of ranitidine hydrochloride, including the instrument-determined location of individual peaks. (From Forster *et al.* 1998, with permission.)

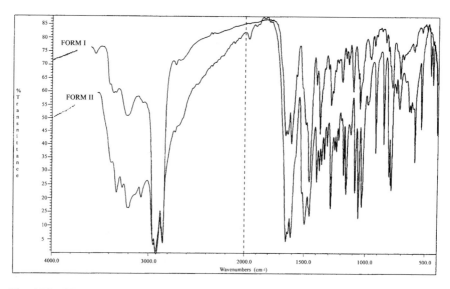

Fig. 4.29 FTIR spectra of the two forms of lamivudine. Note that the two spectra are pre-sented as nearly overlaid as possible to facilitate comparison. In addition, the vibrational origin of significant peaks has been identified and the peak positions for the two polymorphs are compared in Table 4.4 (Harris *et al.* 1997). (Figure from Lancaster 1999, with permission.)

Table 4.4 IR assignments and frequencies for the two polymorphs of lamivudine (From Harris *et al.* 1997, with permission.)

Assignment	Form I	Form II
ν_{OH}[a]	3545	—
ν_{NH}, ν_{OH}	3404, 3365, 3341, 3232	3376, 3328, 3270, 3201
ν_{CO}	1662	1652
ν_{CN}	1643	1636
ν_{CC}	1613	1613
ν_{CO}/δ_{OH}	1053	1060

[a] Of a hydration water molecule.

Since determination of the IR spectra of polymorphs is often carried out for the purpose of comparison, sample preparation becomes a particularly important factor in the experimental procedure (Bugay 2001). Threlfall (1995) has reviewed many of the factors that must be taken into consideration in this regard, including labelling of polymorphic form (obvious, but in view of the confusion in polymorph nomenclature (see Section 1.2.3) still a potential problem), sample purity (both chemical and poly-morphic purity), crystal size, crystal habit and orientation (Griesser and Burger 1993; Kobayashi *et al.* 1994), instability to pulverization and grinding (Farmer 1957; Hoard and Elkovich 1996; Threlfall 1999), solubility in the mulling medium or hydration (Kuhnert-Brandstätter and Riedmann 1989), desolvation of a solvate (Burger and Ramburger 1979*a,b*) or instrumental variables (Free and Miller 1994).

A number of experimental alternatives to traditional IR transmission spectroscopy are suitable for overcoming some of these complicating experimental factors. In the technique of diffuse reflectance infrared Fourier transform spectroscopy (DRIFTS) (Hartauer *et al.* 1992; Neville *et al.* 1992) the sample is dispersed in a matrix of powdered alkali halide, a procedure which is less likely to lead to polymorphic trans-formations or loss of solvent than the more aggressive grinding necessary for mull preparation or pressure required to make a pellet (Roston *et al.* 1993). For these rea-sons, Threlfall (1995) suggests that DRIFTS should be the method of choice for the initial IR examination of polymorphs. He has also discussed the possible use of atten-uated total reflection (ATR) methods in the examination of polymorphs and provided a comparison and discussion of the results obtained on sulphathiazole polymorphs from spectra run on KBr disks, Nujol mulls and ATR.

In spite of many of the potential experimental pitfalls and difficulties (which should be viewed here as caveats rather than as deterrents), IR spectroscopy is still one of the simplest and most widely and routinely employed analytical tools in the study and characterization of polymorphs. Some other modifications, developments and 'hyphenated techniques' are worthy of note here, since they often considerably enhance the potential of the technique while reducing the drawbacks. Perhaps the most obvious of these is the combination of microscopy with FTIR spectroscopy for visual examination and spectral characterization of small areas in heterogeneous samples or identity and analysis of the spatial distribution of components of mixtures (e.g. pharmaceutical formulations) (Messerschmidt and Harthcock 1988).

Photoacoustic spectroscopy is based on the absorption of modulated, rather than direct, radiation; hence, it is independent of the linear absorption, thus eliminating attenuation due to excessive sample thickness or high extinction coefficients. Spectra can be measured on neat samples without dilution (Vidine 1982; Graham *et al.* 1985; McClelland *et al.* 1992). This avoids the polymorphic changes that might be brought about by grinding, but particle size plays a role in such measurements (Rockley *et al.* 1984). An example of a study of a polymorphic system by photoacoustic spec-troscopy on polymorphic forms sensitive to mechanical perturbations has been given by Ashizawa (1989).

Just as polarized light enhances the utility of optical microscopy in the study of poly-morphic systems, so polarization can be used in conjunction with IR spectroscopy. As we shall show later in greater detail (see Section 6.3.2) polarized spectroscopic methods provide detailed information on the *directional* properties which distinguish the spectral features of polymorphs. Thus, for instance, the directional properties of a polymorphic transformation of fatty acids (Kaneko *et al.* 1994a–c) and inferences about the differences in packing modes (Yano 1993) have been investigated with polarized IR methods.

As with other analytical techniques previously widely used for polymorph charac-terization (i.e. polymorph identity), IR spectroscopy is being increasingly employed as a technique for quantitative analysis (e.g. polymorphic purity) (Aldridge *et al.* 1996; Blanco *et al.* 2000; Bugay 2001; Stephenson *et al.* 2001; Patel *et al.* 2001).

A typical example has recently been given by Bugay (1999) for cefepime dihydrochloride (Bugay et al. 1996). The question of polymorphic purity (more correctly in this case purity of crystal modification) involved in determining the amount of a dihydrate in a marketed monohydrate form. As seen in Fig. 4.30(a), the two modifications have quite distinct IR spectra in the $3000–4000\,cm^{-1}$ range, and a mixture of 5 per cent dihydrate in the monohydrate clearly shows how the peak of the former just below $3600\,cm^{-1}$ may be used to quantify its presence. To extract quantitative information, a calibration curve of known w/w mixtures was prepared in the working range of 1.0–8.0 per cent, using diffuse IR reflectance. The statistical analysis led to the graph shown in Fig. 4.31, with a limit of detection of 0.3 per cent, a minimum quantifiable level of 1.00 per cent, and a good cross-validation with an independently developed assay based on X-ray powder diffraction (see Section 4.4).

For the cases of the quantitative analysis of mixtures of crystal modifications in which the distinction between two spectra is not as favourable or where unambiguous identification of many of the characteristic absorption peaks is desired (Zenith 1994), the method of partial least squares (Haaland and Thomas 1988) may be required. This technique originally developed for the use in the near IR spectral region uses the entire (digitized) spectrum, rather than an individual peak or small number of peaks. Some applications to the quantitative analysis of mixtures of polymorphs have been given by Geladi and Kowalski (1986), Hartauer et al. (1992), Tudor et al. (1993) and Jaslovsky et al. (1995) (the latter two with Raman spectroscopy), and it appears that this is a method that will see increasingly widespread use.[5]

An alternative procedure for the recognition of a minor polymorphic component in a polymorphic mixture has been reported by Aldridge et al. (1996). This method employs pattern recognition of the near-IR spectra of the polymorphs (Ciurczak 1987) to distinguish between them. Near-IR spectroscopy may be advantageous over mid-IR since sample preparation procedures have less chance of inducing polymorphic conversion and there may be spectral differences in the former which are not apparent in the latter (Bugay 2001). In this study the authors compared the near-IR reflectance spectra in the range $1100–2500\,cm^{-1}$ of the desired polymorph, and undesired polymorph, a hydrate, a solvate and the free acid of the material using two pattern recognition algorithms, the Mahalanobis distance method and the SIMCA residual variance method (Gemperline and Boyer 1995) to determine the preferred method for detecting unwanted small quantities of undesired crystal forms. The former algorithm gave better results, indicating that levels as low as 2% of the undesired polymorph may be detected, although the number of samples studied was too low to determine proper limits of detection.

---

[5] The desirability of using the entire spectrum (and hence every absorption peak) for both identification and quantification is particularly relevant in light of a recent court decision which required the plaintiff in a patent infringement litigation involving a particular polymorphic form to show that all the peaks in the patent were present in the defendant's accused infringing product. In that litigation the peaks were X-ray powder diffraction maxima, but the same principles would apply to peaks in IR spectra (see Section 10.2).

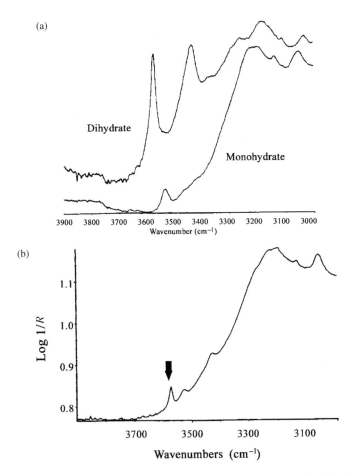

Fig. 4.30 (a) IR spectra for the monohydrate and dihydrate of cefepime dihydrochloride; (b) spectrum of a 5/95 w/w mixture of the dihydrate in the monohydrate. The arrow indicates the distinct dihydrate absorption used in the quantitative analysis. (From Bugay 1999, with permission.)

4.6 Raman spectroscopy

Infrared and Raman spectroscopy are often grouped together, since both techniques provide information on the vibrational modes of a compound. However, since the two spectroscopic techniques are based on different physical principles the selection rules are different. Infrared spectroscopy is an absorption phenomenon, while the Raman spectroscopy is based on a scattering phenomenon (Raman and Krishnan 1928). In general, infrared energy is absorbed by polar groups, while radiation is more effectively scattered in the Raman effect by symmetric vibrations and nonpolar groups (Colthup et al. 1990; Ferraro and Nakamoto 1994). For most molecules other

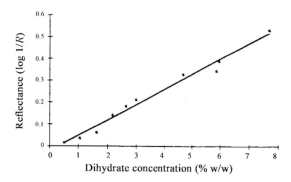

Fig. 4.31 Calibration curve for the presence of cefepime dihydrochloride dihydrate in cefepime dihydrochloride monohydrate as determined by diffuse reflectance IR spectroscopy. (From Bugay 1999, with permission.)

than the simplest most symmetrical ones, the selection rules are not strictly obeyed, so both Raman and IR bands are likely be active for virtually all the bonds, but their relative intensities will differ, the more symmetrical ones giving higher Raman intensities, while the less symmetrical ones exhibit higher IR intensities (Lin-Vien *et al.* 1991). As with IR spectroscopy, Fourier transform methods have revolutionized the field and led to increasingly sensitive and sophisticated techniques. Cases where both the IR and Raman spectra of polymorphs have been measured and compared have been given by Tudor *et al.* (1993), Terol *et al.* (1994) and Grunenberg *et al.* (1996). In view of the similarity and overlap of the two techniques, the principles demonstrated and examples given in the previous section will not be repeated here. Rather, we will cite some relevant references, which can provide additional details as well as an entry into the specific literature.

A typical example of the characterization of a polymorphic system by FT Raman spectroscopy has been given by Gu and Jiang (1995) while an application of the technique with near infrared excitation to the polymorphic cimetidine system has been described by Tudor *et al.* (1991). The FT Raman technique has been compared to infrared diffuse reflection spectroscopy in the study of the polymorphs of spironolactone (Neville *et al.* 1992), and the pseudopolymorphic transition of caffeine hydrate (i.e. loss of solvent) has been monitored using the technique (de Matas *et al.* 1996).

Some of the previous references contain descriptions of the use of Raman spectroscopy for quantitative analysis of mixtures of polymorphs. Additional examples may be found in Deeley *et al.* (1991), Petty *et al.* (1996), Langkilde *et al.* (1997), Findlay and Bugay (1998) and Bugay (2001).

One advantage of Raman spectroscopy over IR methods is that in general little or no sample preparation is required. This considerably facilitates the examination of samples *in situ* without undertaking experimental procedures for instance, grinding or pressing, which might induce polymorphic changes (e.g. Bell *et al.* 2000). One manifestation of this situation is the development of instrumentation that combines optical microscopy with Raman spectroscopy (e.g. Williams *et al.* 1994*a*; Webster *et al.* 1998). The initial applications of this technology were in quality control in industrial environments (Williams *et al.* 1994*b*; Hayward *et al.* 1994), but clearly the combination of microscopy and Raman spectroscopy can readily be applied to the analysis of polymorphic systems in a variety of environments, by facilitating the simultaneous optical and spectroscopic examination of multicomponent and/or heterogeneous samples. Typical systems would be the optical location and subsequent spectroscopic identification of individual polymorphs in a mixture polymorphic forms, or the determination of the polymorphic form of the active ingredient embedded in the excipients of a formulated drug substance (Bugay 1999).

4.7   Solid state nuclear magnetic resonance spectroscopy

In terms of the structural features that are probed with various analytical methods, solid state nuclear magnetic resonance (SSNMR) may be looked upon as representing a middle ground between IR spectroscopy and X-ray powder diffraction methods. The former provides a measure of essentially molecular parameters, mainly the strengths of bonds as represented by characteristic frequencies, while the latter reflect the periodic nature of the structure of the solid. For polymorphs differences in molecular environment and/or molecular conformation may be reflected in changes in the IR spectrum. The differences in crystal structure that define a polymorphic system are clearly reflected in changes in the X-ray powder diffraction. Details on changes in molecular conformation or in molecular environment can only be determined from full crystal structure analyses as discussed in Section 4.4.

Solid state nuclear magnetic resonance spectroscopy provides information on the environment of individual atoms. In essence, the change in environment of any atom can arise from two factors, which usually are not separable in the interpretation of the SSNMR spectra, but are conceptually independent. Since different polymorphs are different crystal structures, it is expected that the crystal environment of at least some atoms will differ from polymorph to polymorph (Section 2.4.2). In addition, since the molecular conformation may also vary among polymorphs (Section 5.6), the change in the environment of an atom due to conformational differences will also be reflected in the SSNMR (Levy *et al.* 1980; Bugay 2001; Strohmeier *et al.* 2001).

The theoretical basis and practical considerations for the application of SSNMR to the study of polymorphism may be found in a number of references, which themselves contain additional primary sources (for instance Yannoni 1982; Fyfe 1983; Komorski 1986; Bugay 1993; Harris 1993; Brittain 1997; Byrn *et al.* 1999). As with many of the analytical techniques described in this chapter, SSNMR is a rapidly developing field with great potential in the investigation of polymorphic systems. It is not limited to a single nucleus (although most studies to date have concentrated on the ^{13}C nucleus) and it is being adapted for quantitative analysis of polymorphic mixtures and other multicomponent systems.

The original development of the basis for SSNMR by Schaefer and Stejskal (1976) through the combination of high power proton decoupling with magic angle spinning and cross-polarization SSNMR techniques for ^{13}C is demonstrated in Fig. 4.32.

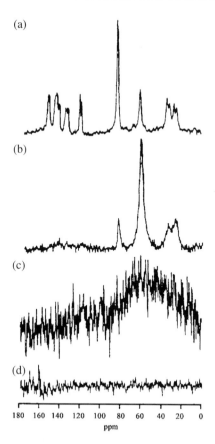

Fig. 4.32 Demonstration of the cumulative effects of techniques employed in ^{13}C CP/MAS SSNMR spectroscopy on a sample of dideoxyinosine. (a) High-power proton decoupling combined with magic-angle spinning (MAS) and cross polarization (CP); (b) high-power proton decoupling and MAS at 5 kHz; (c) high-power proton decoupling only and (d) conventional solution phase techniques. (From Bugay 1993, as modified by Byrn *et al.* 1999.)

A powdered crystalline sample measured by conventional solution-phase pulse techniques gives an essentially featureless spectrum. The use of high power proton decoupling, magic angle spinning (MAS) and cross-polarization leads to a spectrum with resolution similar to that obtained from solution, and the potential for extracting a great deal of chemical and structural information.

Perhaps the first application of this technique directly to polymorphic systems was by Ripmeester (1980), and Threlfall (1995) has more recently reviewed the subject in addition to references cited above. For the study of polymorphic systems SSNMR has a number of advantages. The signal is not influenced by particle size which may eliminate the complications of possible polymorphic transformations due to the grinding required in, say IR and X-ray powder diffraction techniques. The intensity of the signal is directly proportional to the number of nuclei producing it, so that the (qualitative) presence of mixtures of crystal modifications may be recognized (by noting the presences of overlapping spectra) or the quantitative composition of polymorphic mixtures (or, say, an active ingredient in a pharmaceutical preparation) may be determined. The lack of requirements for sample preparation means that investigations for the presence of polymorphism or polymorphic transformations may be easily carried out during any stage of the development or processing of a solid material. However, there is at least one caveat; the high spinning rates required for the MAS technique generate considerable mechanical stress and local heating which may be sufficient to lead to polymorphic transformations during the measurement.

In general, the similarity between the chemical shift in solution and solid state ^{13}C CP/MAS spectra allow for the assignment of the peaks in the latter. Byrn *et al.* (1988) have compared the range of chemical shifts for individual atoms in five crystalline forms of prednisolone with those obtained from solution. A similar study has been carried out on the five polymorphs of sulfathiazole (Apperley *et al.* 1999). The variation in the chemical shift from polymorph to polymorph enables one to recognize significant differences in the atomic environment among polymorphs. While the SSNMR spectra of two or more polymorphs may be considered fingerprint identifiers (e.g. Fig. 4.33), identification of the individual peaks and presentation of their chemical shifts in tabular form (e.g. Table 4.5) provides considerably more information and is to be encouraged.

In the characterization of polymorphs, SSNMR can provide important crystallographic information, even in the absence of single crystal samples for full structure determination. Since the technique is a probe of crystal environment, differences among polymorphs in the number of molecules in the asymmetric unit are expected to be manifested in the solid state spectra. Multiple molecules in the unit cell in principle lead to splittings for individual atomic peaks, since chemically equivalent atoms in crystallographically inequivalent molecules can have different surroundings

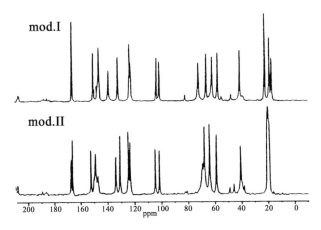

Fig. 4.33 Solid state ^{13}C CP/MAS NMR spectra of the two modifications of nimodipine. (From Grunenberg *et al.* 1995, with permission.)

Table 4.5   Atomic labelling and chemical shifts of nimodipine in CDCl$_3$ solution and in two polymorphic modifications. (From Grunenberg *et al.* 1995, with permission.)

	CDCl$_3$ (ppm)	Modification I (ppm)	Modification II (ppm)
C-1	148.15	150.43	148.86
C-2	123.28	123.39	124.54
C-3	150.09	150.43	152.31
C-4	134.72	139.06	133.82
C-5	128.61	132.13	130.67
C-6	121.28	122.63	123.10
C-7	40.10	47.70	40.76
C-8	103.00	101.26	104.51
C-9	145.38	148.26	147.76
C-10	144.60	146.20	147.01
C-11	103.81	103.31	101.35
C-12	166.62	166.44	166.03
C-13	62.99	62.09	63.95
C-14	19.39	17.46	20.28
C-15	19.54	19.12	20.28
C-16	167.13	166.44	167.29
C-17	67.37	66.45	67.86
C-18	22.12	22.49	20.28
C-19	21.80	22.49	20.28
C-20	70.54	72.36	69.30
C-21	58.83	57.89	58.78

and hence potentially different chemical shifts. (Similar environments or accidental redundancies can still lead to overlapping peaks.) A rather dramatic example of such a difference is presented in Fig. 4.34 for lamuvidine. Form 2 crystallizes in a tetragonal structure with one molecule in the asymmetric unit ($Z' = 1$), yielding a rather straightforward spectrum. On the other hand, Form 1 crystallizes in an orthorhombic structure with five molecules in the asymmetric unit. Some peaks are merely broadened in going from Form 2 to Form 1 (due to near overlap), while others are split into multiple peaks. Care must be exercised in interpreting such spectra in the absence of other confirming evidence for the polymorphic purity of a sample. For instance, the spectrum of a sample containing two polymorphs of a substance will

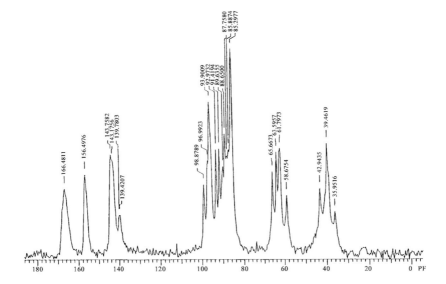

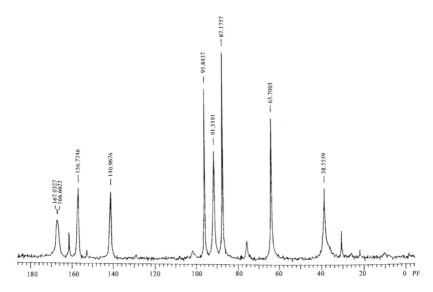

Fig. 4.34 Solid state ^{13}C CPMAS NMR spectra of the two modifications of lamuvidine. Lower, Form 2, tetragonal, $P4_32_12$, $Z = 8$, $Z' = 1$; upper, Form 1, orthorhombic, $P2_12_12_1$, $Z = 20$, $Z' = 5$. (From Lancaster 1999, with permission.)

appear as the sum of the two (Fig. 4.35) (see also Harper and Grant 2000), which might otherwise be interpreted as another form with multiple molecules in the unit cell.

In a crystal structure in which a molecule occupies a crystallographic special position, some chemically equivalent atoms can become crystallographically equivalent,

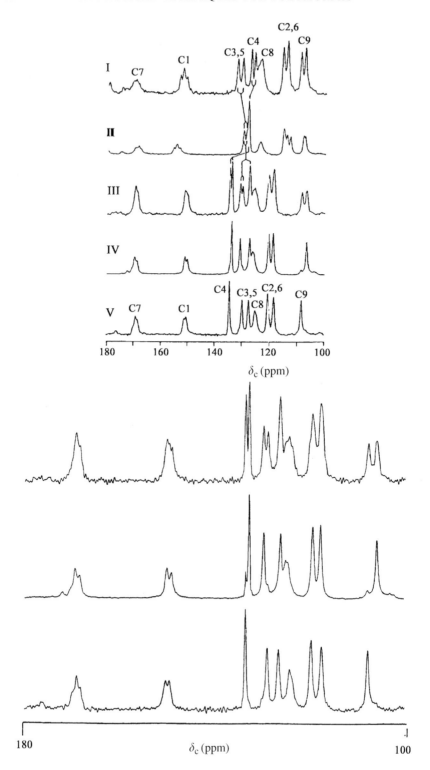

meaning that they have identical environments in the crystal. Such a situation will reduce the number of lines observed in the SSNMR spectrum. For example in Fig. 4.36, for Form I of 5, 5'-diethylbarbituric acid **2-IX** ($Z' = 1$), peaks for all eight carbon atoms can be identified; in Form II the molecule lies on a two fold axis that passes through carbon atoms 2 and 5, so that there are only five crystallographically unique carbon atoms and five unique peaks in the NMR spectrum.

As noted above, the characterization of polymorphs by SSNMR is by no means limited to ^{13}C spectra. For instance, Bauer *et al.* (1998), showed that ^{15}N CPMAS could easily distinguish between the two polymorphs of Irbesartan **4-I** and could be used to study the difference of the tautomeric behaviour in the tetrazole ring of the two polymorphic forms. Variable temperature spectra indicate that the ring in Form B is involved in an exchange process that does not occur in Form A (Fig. 4.37).

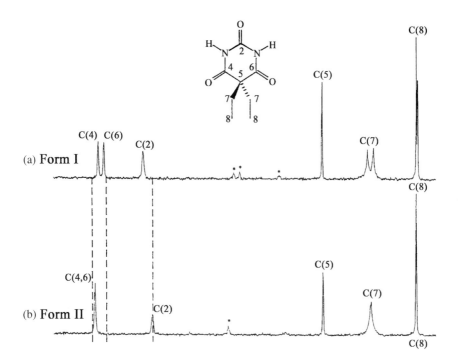

Fig. 4.36 ^{13}C CPMAS NMR of two of the polymorphs of 5,5'-diethylbarbituric acid. The atomic numbering is shown corresponding to the peak with the same number. (a) Form I; (b) Form II. Peaks denoted by asterisks are spinning side bonds. (From Navon *et al.* 2001, with permission.)

Fig. 4.35 ^{13}C CPMAS NMR spectra of sulphathiazole. Top: Spectra for a genuine sample of each of the five polymorphs. (From Apperley *et al.* 1999, with permission.) Bottom: effect of a mixture of polymorphs. The lower two spectra are of pure polymorphic forms, while the upper one is a mixture of the two lower ones. (From Threlfall 1999, with permission.)

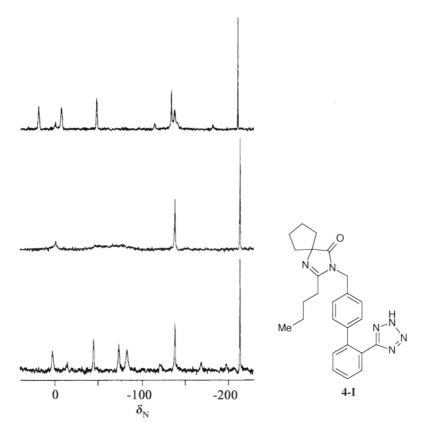

Fig. 4.37 ¹⁵N CPMAS spectra of Irbesartan **4-I**. Top, Form A at 295 K; Middle, Form B at 295 K; Bottom, Form B at 253 K. (From Bauer *et al.* 1998, with permission.)

In addition to the identification of crystal modifications, SSNMR has been used to monitor reactivity and phase changes in different polymorphic forms. For instance, Harris and Thomas (1991) followed the photochemical conversion of formyl-*trans*-cinnamic acid with ¹³C SSNMR (see also Section 6.4). Variable temperature techniques have been used to study the interconversion of four polymorphic modifications of sulphanilamide (*p*-amino-benzenesulphonamide), including interpretation of at least some of the molecular motions during the course of the transformation (Frydman *et al.* 1990). A similar combination was augmented with colourimetric techniques to study the coexistence of two phases in the course of a phase transition (Schmidt *et al.* 1999). Of course, differences between unsolvated and solvated or hydrated crystal modifications may also be readily characterized by the SSNMR technique, as was done with the anhydrous and monohydrate of oxyphenbutazone (Stoltz *et al.* 1991). Due to the availability of the crystal structures for both modifications the SSNMR results could be interpreted directly in terms of the different atomic environments, especially for the differences in hydrogen bonding in the presence and absence of the water of hydration. A recent application combined SSNMR with XRPD, to monitor changes in crystal modification of the monohydrate of neotame [N-(3,3-dimethylbutyl)-L-phenylalanine methyl ester] (Padden *et al.* 1999), which is the most stable form of the compound under ambient conditions. Other modifications can be generated under vacuum, and the original monohydrate can be regenerated by exposure to moisture. X-ray powder diffraction monitoring of these processes indicated that no structural changes had occurred, but the changes in the ^{13}CSSNMR spectra suggested the presence of many forms during the reconversion process. Considerably more complex substances, such as starches, often distinguished as 'A' and 'B' forms, can yield useful structural information from SSNMR studies (Veregin *et al.* 1986).

The SSNMR technique has also been applied to characterize polymorphic systems of organometallic complexes. In one early application Lockhart and Manders (1986) studied three polycrystalline methyltin(IV) compounds. They were able to isolate one of the pure polymorphic modifications and to determine the presence of other modifications by SSNMR techniques.

A comparison of the environment of atoms in different crystal modifications requires the assignment of peaks to individual carbon atoms. As noted above, this assignment may be complicated in the solid state by the variation of chemical shifts from their solution values and by the presence of crystallographically inequivalent molecules in the asymmetric unit. A number of methods have been used by Zell *et al.* (1999, and references 34–41 therein) to overcome these difficulties. These authors used two-dimensional SSNMR with high spinning speed and high ^{1}H decoupling power on uniformly labelled (Szeverenyi *et al.* 1982) samples of two of the three crystal modifications of aspartame. For the first time, they were able to follow the connectivity between resonances of the crystallographically inequivalent molecules in a polymorphic modification of a compound. The further development of such techniques should considerably aid in providing tools for the detailed comparison of the molecular environment of molecules in different polymorphic forms in the absence of full crystal structure analyses. Since that environment includes intramolecular interactions, such information can also be used to build conformational models which are useful starting points for computational methods used for developing models for crystal structures and crystal structure solution (see Sections 5.8 and 5.10). In this particular study, Zell *et al.* assigned the peaks for three crystallographically independent molecules in the asymmetric unit of one of the aspartame crystal modifications.

The qualitative and quantitative analysis of mixtures of polymorphic forms has also been carried out by SSNMR methods. Mannitol is one of the most abundant natural sugars and has wide applications as an additive in the food industry and as an excipient in the pharmaceutical industry (Burger *et al.* 2000). Trovão *et al.* (1998) crystallized mannitol from aqueous solutions and from acetone/water mixtures in the presence and absence of 'Keggin anions' [$PW_{12}O_{40}$]$^{3-}$ or [$SiW_{12}O_{40}$]$^{4-}$, and monitored the results with a number of techniques, including SSNMR (Fig. 4.38). In the absence of anions, they obtained the β form. In the presence of [$PW_{12}O_{40}$]$^{3-}$, they obtained a mixture of the κ and α forms, while in the presence of [$SiW_{12}O_{40}$]$^{4-}$ the α form was obtained.

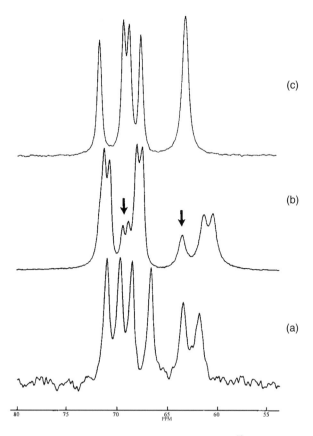

Fig. 4.38   Qualitative recognition of mixtures of polymorphs by ^{13}C CP/MAS NMR spectra of
D-mannitol obtained under different crystallization conditions: (a) absence of Keggin anions,
β form; (b) presence of $[PW_{12}O_{40}]^{3-}$, a mixture of κ and α forms (the latter indicated by
arrows); (c) presence of $[SiW_{12}O_{40}]^{4-}$, α form. (From Trovão *et al.* 1998, with permission.)

There is also a growing use of SSNMR in the quantitative analysis of polymorphic mixtures (e.g. Harris 1985; Suryanarayanan and Wiedmann 1990; Bugay 1993;
Stockton *et al.* 1998; Stephenson *et al.* 2001). A recent example has been given by
Bugay (1999) for Forms A and B of a polymorphic developmental drug substance.
For both developmental and regulatory purposes, it was required to quantitatively
monitor the presence of Form B in A. The SSNMR spectra of the two forms are given
in Fig. 4.39. In the region of 10–20 ppm each polymorph exhibits a large, essentially
single peak offset from the other polymorph. This region was therefore used for the
calibration of known mixtures and assay of the unknown mixtures (Fig. 4.39). The
assay had a 5 per cent w/w detection limit for Form B in A, and could be carried out on
100 mg and greater potency tablets, an important advantage of the technique being that
bulk samples could be assayed without having to consider issues of sampling protocol.

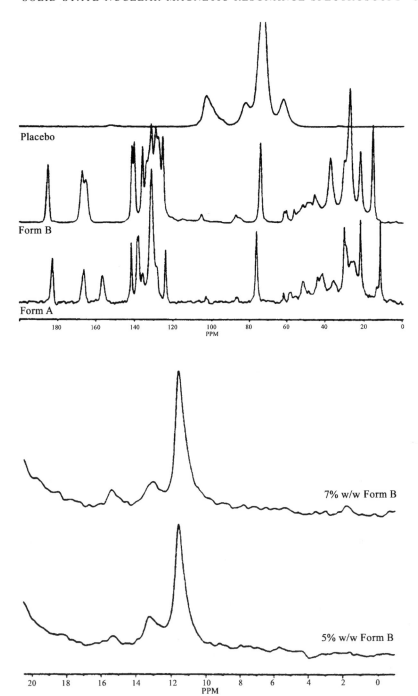

Fig. 4.39   Upper, ^{13}C CP/MAS NMR spectra of two forms of developmental pharmaceutical compound discussed in the text. Lower, the spectra of two mixtures of Form A spiked with known quantities of Form B. The peak at *ca.* 15 ppm is indicative of Form B. (From Bugay 1999, with permission.)

4.8   Scanning electron microscopy

Scanning electron microscopy (SEM) provides greater magnification than optical microscopy. In the study of polymorphs, it can be very useful in characterizing and understanding differences in the properties of polymorphs. For instance, comparison of the optical and SEM images of the two forms of paracetamol shown in Fig. 4.40 demonstrates their similarity. The differences in habit (monoclinic and orthorhombic (Section 2.4.1)) which is clear from both techniques, may not be obvious in the optical case for samples of small grain size. Scanning electron microscopy can be particularly useful for the investigation of the properties of surfaces.

The differences in morphology between the two polymorphs of ranitidine hydrochloride (Fig. 4.41) were used to account for the differences in filtering and drying behaviour. Form 1 crystallizes in poorly shaped plate-like particles that tend to agglomerate and adhere to each other thus blocking the filter medium and reducing the effectiveness of suction for filtering and drying. On the other hand, Form 2 tends to crystallize as more needle-like crystals which form a much more granular porous layer on the filter medium, considerably increasing the filtering efficiency and drying characteristics. These differences in filtering and drying were the principal advantages for granting of the quite lucrative patent on Form 2 (Crookes 1985) (Section 10.2).

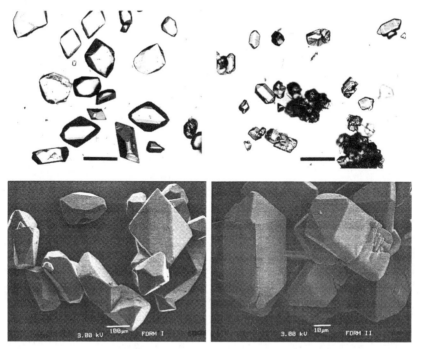

Fig. 4.40   (See also colour plates.) Comparison of optical and SEM images of polymorphs of paracetamol grown from IMS. Upper, optical photomicrographs of Form I (left, scale bar = 250 μm) and Form II (right, scale bar = 100 μm). (From Nichols 1998, with permission.) Lower, SEM images of Form I (left), and Form II (right). (From Nichols and Frampton 1998, with permission.)

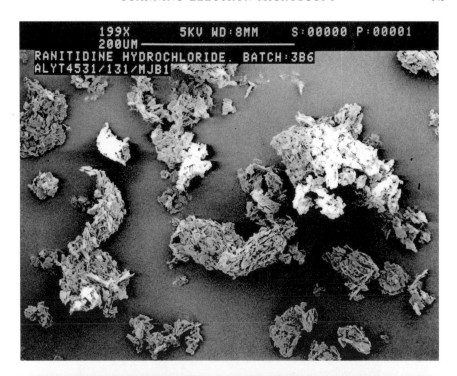

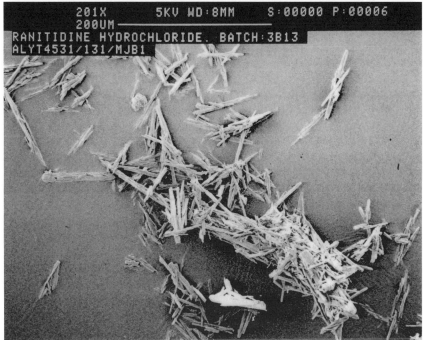

Fig. 4.41 SEM images of the two forms of ranitidine hydrochloride. Upper, Form 1; lower, Form 2. (Courtesy of Glaxo SmithKline.)

Fig. 4.42   SEM image of the growth of the fibrous needles of the dihydrate of carbamazepine on the larger faces of the monoclinic anhydrous modification. (From Rodríguez-Hornedo and Murphy 1999, with permission.)

In addition to the characterization of habit and surface features, the structural symbiosis between two crystalline modifications can be readily studied by SEM. Figure 4.42 shows the growth of the dihydrate modification of carbamazepine on the surface of the anhydrous form. Such information can considerably aid in the understanding of the process of transformations among crystal modifications and the development of robust procedures for the selective preparation of a desired crystal modification.

4.9   Atomic force microscopy and scanning tunnelling microscopy

As demonstrated above, surfaces play a crucial role in determining the polymorphic outcome of a crystallization. Hence, the study of the structure of surfaces and their interaction with a crystallizing material can provide detailed information on the nature of the process. That information may be used to design experiments and processes for the study of nucleation and the selective control of the growth of a particular crystal modification (Palmore *et al.* 1998) (Section 3.7). These AFM and STM techniques (and rapidly developing spin-off modifications of them) provide that information (Frommer 1992; Ward 1997). For instance, tapping mode atomic force microscopy

(TMAFM) has recently been employed to characterize the surface morphologies of the polymorphs of an organic radical, *p*-nitrophenyl nitronyl nitroxide (Fraxedas *et al.* 1999*a,b*).

4.10   Density measurements

The 'density rule' proposed by Burger and Ramberger (1979*a,b*) (See Section 2.10) is useful for determining the relative stability of polymorphic forms. Hence, the measurement of density or changes in density can provide the experimental information for making such determinations. A number of the techniques for carrying out these determinations are rather time-consuming and fraught with possibilities for error, but they should be included in the potential armory of techniques for studying polymorphs.

The density may be measured by flotation, pycnometry, or by volumenometry (Bauer and Lewin 1972; Andreev and Hartmanova 1989; Richards and Lindley 1999). Differences in density between polymorphs normally do not amount to more than 1–2 per cent of the density of one of the forms, so that both high precision and high accuracy are required for the measurements to be meaningful. In addition, virtually all the errors that can be encountered in the experimental procedures of the density tend to lower the experimentally determined value. A reliable computational alternative is the calculation of the density from the experimental crystallographic unit cell constants, routinely obtained from single crystal structure determinations (e.g. Glusker *et al.* 1995), but also available from the indexing of powder patterns (e.g. Jenkins and Snyder 1996). An added advantage of these 'indirect' methods is that the unit cell constants are usually accompanied by estimated standard deviations, so that the statistical significance of differences and similarities in the densities may be properly evaluated. As a *caveat*, Herbstein (2000) has recently shown that the standard deviations of the unit cell constants are often underestimated, suggesting a higher precision than the experiment warrants. With the increasing automation of both single crystal and powder diffractometry, and the accompanying sophistication of the appropriate software, comparative studies of the temperature dependence of the density of polymorphs may provide useful information on the phase relationships.

Dilatometry is one of the older classic methods for the determination of transition points between solids (Drucker 1925). The dilatometer usually consists of a large bulb connected to a capillary and filled with an inert liquid. Volume changes as a function of temperature or resulting from a solid–solid transition may be determined by changes in the volume of the inert liquid. The recent advances in the miniaturization of chemical instrumentation (e.g. Jakeway *et al.* 2000; Krishnan *et al.* 2001), and the high precision associated with that miniaturization may lead to a renaissance of the use of some of these classic techniques and their applications to the study of polymorphism.

4.11   New technologies and 'hyphenated techniques'

The potential application of chemical and physical techniques for the study of polymorphs is virtually limitless. The combination of techniques into a single apparatus in order to extract the maximum information from a single experiment and a single sample is particularly applicable to polymorphic systems. The drawback in combining techniques is that the capabilities and versatility of each individual technique may be compromised. The joining of visual methods with spectroscopic and calorimetric measurements has already been mentioned. These can be expanded further, as witnessed, for instance by a triple combination of microscopy, calorimetry and microphotometry (Cammenga and Hemminger 1990; Lin 1992; Rustichelli *et al.* 2000) on SMATCH spectroscopy (Simultaneous Mass and Temperature-Change FTIR) (Timken *et al.* 1990). The combination of TGA with FTIR allows simultaneous quantitative analysis of weight changes during thermal processes with the IR identification of the decomposition products (e.g. solvent) resulting from those processes (Materazzi 1997), and the amalgamation of thermal methods with a variety of other techniques (Han *et al.* 1998; Cheng *et al.* 2000) is very promising in research on polymorphic materials. One important technique where progress has been somewhat slower is the combination of X-ray powder diffraction with other techniques, although some promising beginnings have been reported (Klein *et al.* 1998). Earlier reports on the potentially very useful combination of X-ray powder diffraction and DSC (Fawcett *et al.* 1989) have not yet seen commercialization, in spite of some successful results in characterizing polymorphic systems (Fawcett *et al.* 1986; Fawcett 1987; Landes *et al.* 1990; Loisel *et al.* 1998). It is hoped that the increasing sensitivity and speed of detectors, combined with more powerful and convenient laboratory X-ray sources that powder diffraction will be conveniently combined with optical, thermal and spectroscopic in single unit instruments capable of more comprehensive studies of polymorphic systems, especially the metastable phases of those systems. The capability of measuring such (often fleeting) phases *in situ* is clearly a distinct advantage in the efforts to achieve the complete characterization of a polymorphic system.

4.12 Are two samples polymorphs of the same compound?

Historically, most polymorphs have been discovered by serendipity, rather than as the result of a systematic search (e.g. Tutton 1922; Cholerton *et al.* 1984; Chemburkar *et al.* 2000; Lommerse *et al* 2000). On the one hand, this testifies to the general unpredictability of polymorphism, but on the other hand, it also testifies to the intellectual curiosity and powers of observation and analysis of the discoverers. What generally characterizes polymorphs of a compound is that some or all of their properties will differ. Thus, in principle, an investigator armed with the knowledge that a particular compound is chemically pure, can employ almost any sufficiently sensitive physical measurement to determine if two crystalline samples of that compound may be polymorphs or not. As noted earlier, however, in practice the full characterization of the polymorphic behaviour should involve as many techniques as possible (e.g. Chiang *et al.* 1995). Previous sections outline many of these physical measurements and provide examples of the distinctions between polymorphs. In addition to those, however, the answer of McCrone (1965) to the question posed here is worthy of repetition, since it demonstrates the principles and the relatively simple and straightforward techniques that are involved.

First, it is necessary to eliminate tautomerism or dynamic isomerism by determining that the two materials give identical melts. X-ray diffraction can determine that crystal strain (which can be mistaken for polymorphism) is not a factor. Then, McCrone

(1957, 1965) suggests the following additional tests for polymorphism, using the polarizing microscope equipped with a hot stage. The two samples are polymorphs if

1. They have different crystal properties (axial ratios, X-ray diffraction, indices of refraction, melting point, etc.) and they can be converted into each other through a solution or solid phase transformation.
2. They differ in all crystal properties (as determined say, by the methods outlined throughout this chapter), but both melt to a liquid with the same refractive index and same temperature coefficient of the refractive index. In addition, if the two solids are mixed and melted to the melting point of the lower melting materials, and the temperature is maintained at the melting point, in the case of polymorphism the two will both melt (if the melting points are very similar) or will crystallize as the high melting form (whose melting point would be above that temperature). The persistence of a mixture of melt and solid at the melting temperature of the low melting form indicates the presence of two different compounds rather than polymorphs.
3. The crystal properties are different, yet the mixed fusion between them is identical.
4. Seeding of a fused sample of one of them with crystal of each of the two, and observation under cross-polarizers on a polarizing microscope, indicates that one of the phases is growing through the other phase. If there are fronts between different areas of the growth of a solid that do not undergo change with time, then the two samples are likely the same material.
5. Mixing crystals of the two on a dry microscope slide (beneath a cover slip) and allowing a saturated solution of one of them to run under the cover slip leads to a phase transformation.
6. Heating crystals of both in proximity on a hot stage leads to a solid–solid transformation of one, followed by melting of both. Also, if on continued heating, one form melts and then can be induced to solidify by seeding with the other form, and then melted upon further heating, then the two are polymorphs.
7. During heating of a mixture there is a vapour phase transition.

4.13 Concluding remarks

Since polymorphism is a structural manifestation of kinetic and thermodynamic factors in crystallization effects and leads to a variation of physical and chemical properties, its characterization and investigation require a multidisciplinary approach. Different analytical techniques and methods provide a variety of structural and thermodynamic information, often complementing each other and adding to overall understanding of the relationships among the various phases. For the investigator, the more information available the greater the control over the polymorph obtained and over the resulting properties; the message here is that the broadest range of the analytical techniques discussed in this chapter, many of which are summarized in tabular form in Table 4.6 (Yu *et al.* 1998) should be employed in the study of polymorphic systems (Bauer *et al.* 2001).

Table 4.6 Summary of information obtained from different physical techniques for various types of crystal modifications (From Yu et al. 1998, with permission.)

Type of crystal modification	Single crystal X-ray crystallography	X-ray powder diffraction	IR/Raman spectroscopy	Solid-state NMR spectroscopy	Thermal methods	Microscopy
True polymorphs	Same chemical composition. Unique unit cell parameters molecular conformation and packing.	Unique Diffraction peaks. Useful for determination of phase purity and % crystallinity.	Characteristic spectra. Sensitive to H bonding.	Unique Chemical shifts. Useful for determining phase purity, molecular mobility.	Unique mp, heat capacity, heats of fusion/transition, solubility. Useful for determining relative stability of forms.	Characteristic indices of refraction, birefringence, dispersion colour, crystal habit.
Solvates	Same as true polymorphs	Same as true polymorphs	Unique solvent bands. Shifted molecular bands. Sensitive to H bonding.	Unique solvent resonances. Shifted molecular resonances. Solvent mobility can be determined.	Low-temperature transitions due to desolvation (TGA loss)	Same as true polymorphs. Desolvation observable by hot stage microscopy.
Isomorphic desolvates	NA	Diffraction pattern only slightly changed from parents solvates.	Solvent bands disappear. Molecular bands shifted.	Solvent resonances disappear. Molecular resonances shift.	Low-temperature desolvation absent. Events due to crystallization or lattice relaxation	Birefrigent microcrystalline domains, with cracks and fissures.
Amorphous solids	NA	No diffraction peaks	Broadened spectra	Broadened spectra	Glass transition seen. Often followed by crystallization and melting. 'Fragility' related to width of T_g.	No birefringence, irregular particle shape.
Polymorphic mixtures	NA	Composite pattern of crystalline components	Composite spectrum of all components	Nuclei-specific composite spectrum of all components	Thermal behaviour indicative of phase diagram (e.g. mp-depression, eutectic melting, dissolution).	Composites of distinct crystalline and amorphous particles

5

Conformational polymorphism: intra- and intermolecular energetics

Can crystal structures be predicted? No! (Gavezzotti 1994*a*)

5.1 Introduction

To a great extent, the attraction and appeal of X-ray crystallography as an analytical tool for molecular structure determination is due to the precision with which that determination can be made. Bond lengths and bond angles with e.s.d.'s of a few thousandths of Ångstroms or tenths of a degree respectively are difficult to obtain as routinely by other methods (Dunitz 1979; Glusker *et al.* 1994). Moreover, the accumulated experience and data of approximately seventy years of such determinations have provided evidence for the limited variability of such parameters, so that characteristic values may be determined and used as benchmarks for additional studies (Allen *et al.* 1987; Orpen *et al.* 1989). But molecular structure is defined as well by torsion angles, and these tend to exhibit considerably greater ranges of values, for both electronic and stereochemical reasons. For instance, in a particular grouping of the four atoms that define a torsion angle, the variability can be manifested in different angles as that group appears in different molecular settings. In the context of the subject of polymorphism we are particularly concerned with the cases in which a molecular conformation as defined essentially by the torsion angles can and does vary from one polymorph to another. Such cases are of intrinsic interest for determining the characteristic value or ranges of values of a torsion angle, but more important, they provide particularly useful cases for investigating the interplay of intra- and intermolecular energetics,[1] which is the subject of this chapter.

---

[1] Kitaigorodskii (1970) suggested four strategies for investigating the general problem of the influence of crystal forces on molecular conformation: (1) comparison of compounds in the gaseous and crystalline states; (2) comparison of the geometry of crystallographically independent molecules in the same crystal; (3) analysis of the structure of a molecule whose symmetry in a crystal is lower than that of the free molecule; (4) comparison of molecules in different polymorphic modifications. The advantages and disadvantages of these various approaches have been considered elsewhere [Bernstein (1984, 1992)].

5.2   Molecular shape and energetics

As noted above, molecular shape is defined by three different types of molecular parameters: bond lengths, bond angles, and torsion angles. Variations in the geometry of a molecule are then simply defined as changes in these parameters: bond stretching or compression, bond bending or deformation, and bond twisting or torsion, and these distortions are characterized by certain energy domains. Typical force constants for bond stretching (in 10^5 dyne cm^{-1}) range from 4.5 for the single bond in ethane to 15.7 for the triple bond in acetylene (Brand and Speakman 1960). Bond angle bending is less expensive. Mislow (1966) has shown that for many carbon bond angles the empirical relationship for the potential energy (1) holds, where θ is the bond angle. Thus, an angular distortion of 10 degrees involves the same amount of energy as the distortion of a single bond by about 0.05 Å.

(1) $V_\theta \sim 0.01(\Delta\theta)^2 \, \text{kcal}^{-1} \, \text{deg}^{-2}$

   Torsional changes involve the rotation about the bond axis; the barrier to rotation about the single bond in ethane is about 2.8 kcal mol^{-1}, which is approximately the difference between *trans* and the *gauche* conformations. We previously gave estimates of the energy 'cost' of these distortions, based on the principles of molecular mechanics (Bernstein 1987). These have recently been compared with values derived from *ab initio* calculations at the MP2/6-31G* level by Hargittai and Levy (1999), as summarized in Table 5.1. Thus, as a rule of thumb, bond stretching is roughly one to two orders of magnitude more expensive energetically than rotations about single bonds, with bond angle deformations falling in the intermediate range.

5.3   Intermolecular interactions and energetics

On the intermolecular level, in molecular crystals one is dealing potentially with a variety of interactions, most of them quite weak compared to those involved in chemical bonding. A crystal structure corresponds to a free energy minimum that is not necessarily the global minimum. The existence of a number of energetically closely spaced local minima is the thermodynamic rationale for the possible appearance of

Table 5.1 Comparison of the estimates of energy 'costs' resulting from distortions of molecular geometric parameters[a] (From Hargittai and Levy (1999), with permission.)

	Estimates from molecular mechanics	Calculated using *ab initio* methods
Bond length, C–C	0.1 Å, 14	0.02 Å, 0.6
Bond angle, C–C–C	10°, 10.3	2°, 0.4
Torsion angle, C–C–C–C	10°, 0.9	5°, 0.2

[a] For each entry the first number is the distortion from the nominal equilibrium value, and the second number is the (average) estimated energy involved in that distortion, in kJ mol^{-1}.

polymorphism, assuming that these minima are kinetically accessible. While intuitively it is convenient to associate a local minimum with a situation in which attractive interactions are maximized, Dunitz and Gavezzotti (1999) have pointed out that

... molecules in the bulk of a crystal are held together by mutual attraction, but are clamped in their places—in contrast to molecules in a liquid—mainly by resistance to compression, in other words, by repulsions that oppose the disentanglement of interlocking molecules and thus hinder any displacement from their equilibrium positions and orientations.

At any rate, a minimum does represent a situation in which attractions and repulsions are balanced (Brehmer *et al.* 2000). The nomenclature of these intermolecular interactions is quite variegated and the terms are not always clearly defined or distinguished from one another. Some in common usage include van der Waals interactions, London forces, dipole–dipole interactions (and higher terms), dispersion forces, steric repulsion, hydrogen bonds, charge-transfer interactions (also called donor–acceptor interactions), electrostatic interactions, exchange repulsion forces, etc.

There has been some convergence of thought about the use of these terms, at least among those who deal with 'crystal forces', especially from the computational point of view. Generally, the intermolecular interactions fall into three classes: (a) non-bonded, non-electrostatic (van der Waals, London, etc.); (b) electrostatic (coulombic); (c) hydrogen bonding. The lines of distinction between these general classes are not always particularly sharp so, for instance, hydrogen bonding has been treated by a combination of the first two general types of interactions (Hagler and Lifson 1974; Hagler *et al.* 1974). Another important distinction is that the non-bonded and electrostatic interactions are generally treated as isotropic, although it has long been recognized (Starr and Williams 1977*a*,*b*; Pertsin and Kitaigorodsky 1987) that a more realistic physical representation requires the inclusion of anisotropicity into that treatment (Price 2000). Hydrogen bonds, by their very nature, are directional and anisotropic.

The non-bonded (van der Waals or London) forces are generally weak interactions between uncharged atoms or molecules. The separation of a molecule from its crystal environment requires overcoming all the attractive forces acting on it. The energy involved in this process may be viewed as the sublimation enthalpy of the crystal. Thus the magnitude of the sum of forces acting on the molecule or the energies involved in the interactions of individual atoms of the molecule with atoms of surrounding molecules may be estimated from the sublimation energy of those molecular crystals in which other interactions are essentially absent. For most molecular crystals the experimental sublimation enthalpy is roughly in the range $10\text{--}25 \, \text{kcal} \, \text{mol}^{-1}$ (Chickos 1987). For a molecule with an intermolecular 'coordination number' of 8–12 (Kitaigorodskii 1961; Braga *et al.* 1990, 1992), the interaction energy thus is about $1\text{--}2 \, \text{kcal} \, \text{mol}^{-1}$ per molecular neighbour. A sub-class of these interactions, which are generally more anisotropic in nature, is the charge-transfer (i.e. $\pi\text{--}\pi$ or $\sigma\text{--}\pi$) type, for which the energies involved rarely exceed $5 \, \text{kcal} \, \text{mol}^{-1}$ (Foster 1969). The electrostatic interactions can vary over a much wider range, depending on the distance and on the degree of polarization of the molecule or parts thereof, which computationally is manifested in the assignment of partial atomic charges to the various atoms. Intuitively, electrostatic interactions for, say, hydrocarbons might seem to be negligible; however, for many organic crystals they have been shown to comprise a significant part of the total lattice energy (Williams 1974; Cox and Williams 1981; Popelier *et al.* 1989). Hydrogen bond strengths are generally estimated to be in the range 1–10 kcal mol^{-1} (Jeffrey and Saenger 1991; Desiraju and Steiner 1999).

These intermolecular interactions all fall on the low end of the scale of energies required to bring about distortions of molecular geometry. This already suggests that if the crystal environment has any influence on the molecular geometric parameters, then those parameters most likely to be affected will be the torsion angles around the single bonds, rather than distortions in bond angles or bond lengths which require substantially larger energies to bring about significant changes. Hence we will concentrate here on the changes in torsional parameters which may be due to the influence of the crystal environment. In spite of these arguments, the possibility of perturbations in bond lengths and bond angles brought about by intermolecular interactions should not be totally ruled out. Such situations are well documented and have been discussed in considerable detail (Colapietro *et al.* 1984; Colapietro *et al.* 1984*a,b*; Domenicano and Hargittai 1993; Wolff 1996; Wagner and Englert 1997).

As Dunitz (1979) has pointed out, a crystal structure yields information about the 'preferred conformations' (*sic* plural) of a molecule and that any arrangement of atoms or conformation 'cannot be very far' energetically from the equilibrium structure of the molecule. Thus a number of conformations of a molecule may be energetically equivalent, or nearly so (e.g. Dobler 1984). An important consequence of this fact is that different conformations may appear in different crystal structures, or for that matter in the same crystal structure when the number of molecules in the asymmetric unit exceeds one. Certain crystal packing motifs are more favourable than others (Kitaigorodskii 1961; Desiraju 1989; Wolff 1996), which may lead to a predominance of one conformation in a crystal structure, while in solution a number of different conformations may be present, including, most likely, the one(s) in the crystal structure(s). It is important to emphasize here that the energetic situation representing the combination of the crystal and molecular structures is a minimum which is not necessarily unique, and at energies not far from the global minimum there may exist a number of possible molecular geometries of very nearly the same energy for both the molecular conformation and the crystal structure. It is this proximity of molecular conformational energies which makes possible the existence of different molecular conformations in a single crystal structure when there is more than one molecule in the asymmetric unit, or of the same molecule in different crystal structures. The energetic justification for the existence of different conformations in different polymorphic structures comes from the fact that the differences in lattice energy among different polymorphic forms can be expected to be in the range 1–2 kcal mol (Kuhnert-Brandstätter 1971; Kitaigorodskii 1970), although computed values for known structures exceeding that range have been reported (for example Buttar *et al.* 1998; Stockton *et al.* 1998; Starbuck *et al.* 1999). Gavezzotti (1991) has estimated that the differences in *total* energy (including molecular and lattice terms) will be of the order of 1 kcal mol^{-1}. Allen *et al.* (1996) have shown through a statistical analysis of the CSD that torsion angles associated with an intramolecular strain energy exceeding c.1 kcal mol^{-1} are relatively rare occurrences in crystal structures, including, of course, polymorphic structures. From the estimates of the magnitudes of intermolecular interactions, this is clearly in the range of energy required to bring about changes in molecular torsional parameters about single bonds, but it is generally not sufficient to significantly perturb bond lengths and bond angles. Therefore, for those molecules that do possess torsional degrees of freedom, various polymorphs may exhibit significantly different molecular conformations. This, then is the energetic rationale for the phenomenon of *conformational polymorphism*, a term apparently initially coined by Corradini (1973), and adopted more widely by Panagiotoupoulis *et al.* (1974) and others (Bernstein and Hagler 1978). Corradini's definitions of a number of situations for the arrangement in a crystal of molecules of the same chemical composition but different conformations, are illustrated in Fig. 5.1.

Conformational polymorphism is the existence of different conformers of the same molecule in different polymorphic modifications, as represented in Fig. 5.1(a),(b). Additionally, Corradini has specified that the conformers are nearly isoenergetic, and also includes as an example, racemic and optically active crystals for the case of chiral molecules. While both the general definition of polymorphism and chemists' general understanding of conformation are fraught with nuances and special complicating circumstances, we prefer the general definition of conformational polymorphism here, without Corradini's additional qualifications. Specific questionable or unusual cases can be dealt with as encountered.

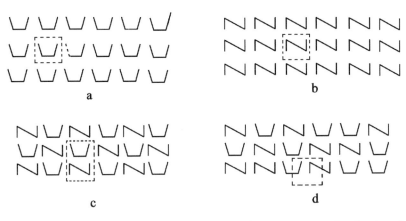

Fig. 5.1 Schematic illustration of three possibilities for arrangement of molecules of the same chemical composition but different conformations in the crystal. For each case two conformations represent symbolically cisoid and transoid dispositions around a single bond. The rectangle defined by a broken line represents one possible choice of the unit cell. (a) and (b) conformational polymorphs; (c) conformational isomorphism; (d) conformational synmorphism. (From Bernstein 1987, with permission, as adopted from Corradini 1974, with permission.)

Corradini also defined *conformational isomorphism* as the existence of different conformers of a molecule in the same crystal structure, a situation which can arise when there is more than one molecule in the asymmetric unit ($Z' > 1$). In such a case, each crystallographic site in the unit cell is always occupied by the same conformer and the ratio of conformers is defined by whole numbers as in Fig. 5.1(c).

Conformational synmorphism describes the situation in which different conformers of a molecule are distributed randomly throughout the crystal lattice. Such a situation usually exists when two or more conformers have similar overall molecular shapes. Thus at any particular molecular site a number of conformations may be adopted, the relative population being determined by the relative intermolecular and intramolecular energies involved, as in Fig. 5.1(d).

5.4 The search for examples of conformational polymorphism

How common is the phenomenon of conformational polymorphism? As is often the case in chemistry, the answer depends on the conditions specified in the definition of the search. For literature examples, the obvious place to begin such a search is the CSD (Section 1.3.3, with the caveats described therein). From the analytical point of view, one may ask what actually constitutes conformational polymorphs? How different must be the conformations of two or more molecules in different crystal structures in order for them to be called conformational polymorphs? There is no hard and fast rule for such a determination. Chemists will generally recognize conformational differences whose significance is relevant to the problem at hand, and different problems may require different quantitative criteria. Nevertheless, it is possible to apply quantitative criteria to evaluate such systems; the question is one of defining different conformations. For molecules with a small number (say, up to 3–4) of conformational degrees of freedom, differences in conformation will usually arise from one or more significant differences in the torsion angles. These will be easily recognizable by a one-to-one comparison of torsion angles between the molecules in two or more structures.[2] The same may be true for molecules with more conformational degrees of freedom, but the overall conformations of two molecules may also differ considerably due to the accumulation of a large number of small differences in torsion angles. Such cases of conformational polymorphism would not be recognized by a one-to-one comparison of torsion angles, but might be recognized, for instance, by a comparison of a matrix of intramolecular distances or simply of end-to-end distances. Such techniques for comparison of molecular conformation are widely employed among protein crystallographers (Liebman 1982).

---

[2] Such a one-to-one comparison is not always straightforward. Often the crystal structure determinations of the polymorphs of a particular compound were reported by different authors, with different atom labelling systems. Since the CSD entry uses the atomic labelling reported by the authors, there may be no correspondence between the atomic labelling of the chemically identical molecules in the different structures. It is possible to develop algorithms for the automatic generation of such correspondence tables (Kedem and Bernstein 2001), but the existence of this problem should serve as a warning to those seeking an automated method for searching the CSD for conformational polymorphs.

5.5   Presenting and comparing conformational polymorphs

The preceding section indicates the importance of developing some quantitative tabular methods for the comparison and presentation of conformational polymorphs. For the overall understanding of molecular and crystal structures and the relationship between them, the qualitative visualization of similarities and differences often plays an important role and can lead to considerable insight and understanding, for instance regarding the nature of particular interactions, or the geometric nature of the 'reaction coordinate' for the transition between phases (Bürgi and Dunitz 1994). To facilitate such comparisons, all the polymorphic crystal structures to be compared should be plotted on the same *molecular* reference plane in the *same molecular orientation*. Such a plane might be a phenyl ring or other suitably rigid structural unit and should be chosen so that significantly different torsion angles emanate directly from that unit or within one bond from it. This will serve to highlight the similarities and differences in molecular conformation and facilitate visual comparison. The crystal structure with which these molecular conformations are associated can then be readily compared by generating the molecules (usually 8–12) surrounding the reference molecule. It may be difficult to produce an understandable packing diagram that includes molecules directly above or below the reference molecule. In that case, an additional view rotated by 90 degrees about the original in-plane horizontal or vertical axis may be required. Also, the initial view may be improved (for instance by elimination of some overlap) by *slight* (e.g. up to 5 degree) rotations from the initial reference orientation. Many of the currently available software packages for plotting molecular and crystal structures have the facilities for generating such figures. For instance, the PLUTO plotting package of the CSD software is very conveniently designed for this purpose, and has the added advantage of being readily accessible from the CSD search software routines.

5.6   Some examples of conformational polymorphism

The number of examples of conformational polymorphism has increased significantly since the first formal recognition of the phenomenon (Corradini 1973; Panagiotoupoulis *et al.* 1974; Bernstein and Hagler 1978) and a more recent review (Bernstein 1987). In this section, we will give only a brief overview of the current situation, in light of the diversity of the systems studied, to highlight some of the aspects of structures exhibiting conformational polymorphism. Many of the earlier examples are still classic ones, and will be included here as well. The increasing ease and proliferation of crystal structure analysis, along with the growing awareness and importance of polymorphism has prompted the study of many polymorphic systems, providing a fascinating variety of examples of conformational polymorphism. As we will try to demonstrate here with a necessarily limited number of examples, any class of compounds with conformational degrees of freedom may exhibit conformational polymorphism, and those seeking examples for particular compounds, functional groups, structural motifs, etc. should be able to readily identify them by making use of the extensive data and software available on the CSD.

Very often, other physical methods (e.g. Chapter 4) may be used to predict the existence of conformational polymorphs, in the absence of full crystal structure analyses, for example, n-propyl acetate (Ogawa and Tasumi 1979), 1-bromopentane (Ogawa and Tasumi 1978), 2-(4-morpholinothio)benzothiazole (Guzman and Largo-Cabrezio 1978) and dibenzo-24-crown-8 (Stott et al. 1979; Weber et al. 1998). The prediction of conformational polymorphism based on the IR spectra (Tomita et al. 1965) was realized in the trimorphic iminodiacetic acid 5-I system, the details of which were determined from the full crystal structures (Boman et al. 1974; Bernstein 1979). Of the four independent molecules in the three structures, two are nearly identical; however, there are torsional differences of up to 30° about C–N bonds. In two cases the hydroxyl hydrogen of one carboxyl group is *trans* to the nitrogen, while in two cases it is eclipsed.

$$HO_2C \overset{H_2}{\underset{}{C}} \overset{}{N} \overset{H_2}{\underset{}{C}} CO_2H$$

5-I

Because of its sensitivity to differences in molecular conformation and molecular environment, solid state NMR can be a particularly useful technique to recognize conformational polymorphs. Often, detail can be extracted to structurally characterize the polymorphs which may be useful in the absence of crystal structure analyses (Smith et al. 1998; McGeorge et al. 1996).

Information on space group and crystallographic site symmetry may also be sufficient to provide evidence for conformational polymorphism. For instance, in one of the forms of 2,2',4,4',6,6'-hexanitroazobenzene 5-II (with space group $P2_1/a$ and $Z = 2$) the molecule must lie on a crystallographic inversion centre, requiring the two phenyl rings to be in parallel planes, while in the second form the molecule lies on a crystallographic general position ($P2_1/a$ and $Z = 4$), making no restrictions on the molecular conformation in the crystal. These implications of molecular and crystallographic symmetry are borne out in the conformations observed in the crystal structures (Graeber and Morosin 1974) as seen in Fig. 5.2. This is an example in which the identity of space group for the two forms clearly is not manifested in any similarity in molecular or crystal structure, which should be considered the general case, as noted earlier.

The existence of conformational polymorphism can be an important factor both in drug design and in attempting to understand the molecular basis of drug action. Since much of our knowledge and understanding of precise molecular structure is derived from crystallographic studies, it is well to be aware of the fact that different polymorphs may exhibit different conformations of the same molecule. A number of examples will serve to demonstrate this point. In both the antiviral agent virazole 5-III

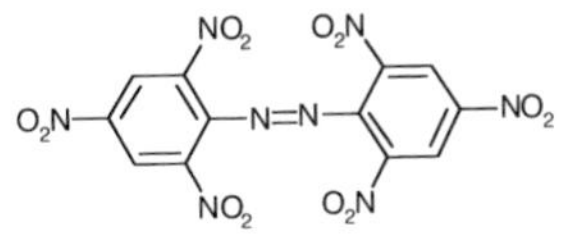

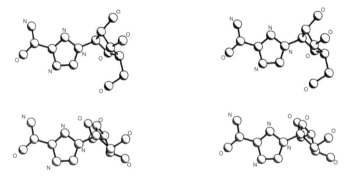

5-II

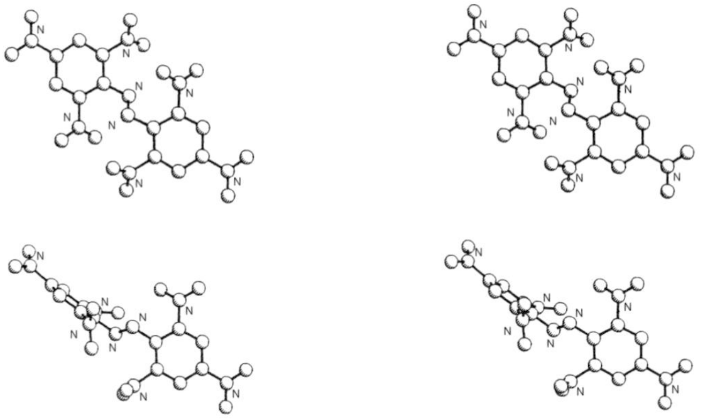

Fig. 5.2 Stereoviews of the molecular conformation observed in the two forms of **5-II**. Upper, $P2_1/a$ and $Z = 2$ structure; lower $P2_1/a$ and $Z = 4$ structure. In both cases the molecule is plotted on the plane of the phenyl ring on the right hand side of the molecule. For clarity, carbon and oxygen atoms have been left unlabelled. (From Bernstein 1987, with permission.)

Fig. 5.3 Stereoviews of the two forms of virazole **5-III**. In both cases the view is on the best plane of the triazole ring. For clarity, the carbon atoms have been left unlabelled. (From Bernstein 1987, with permission.)

(Prusiner and Sundaralingam 1976) (Fig. 5.3) and the important nucleotide adenosine-5′-monophosphate **5-IV** (Neidle *et al.* 1976; Kraut and Jensen 1963) (Fig. 5.4), there are significant differences in the torsion angles about the exocyclic single bonds leading to significant conformational differences. In the latter, the torsion angle about

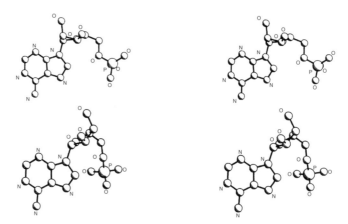

Fig. 5.4 Stereoviews of the two forms of adenosine-5′-monophosphate **5-IV**. Upper, mono-clinic; lower, orthorhombic. In both cases the view is on the best plane of the six-membered ring of the base. For clarity, carbon atoms have been left unlabelled. (From Bernstein 1987, with permission.)

the sugar-base bond differs by 46.8° and the sugar pucker differs, being C(2)′-*endo* in the orthorhombic form and C(3)′-*endo* in the monoclinic form.

5-III   5-IV

Similarly significant differences in conformation have been recognized, for instance in the anxiolytic drug Abecarnil, and the cholesterol-lowering drug Probucol. In the former, the structures of three polymorphs have been reported (Bock *et al.* 1994); the molecular conformations are essentially identical in two of the modifications, but differ in the third. Probucol is reported to be dimorphic with conformational differences arising mainly from a differences of ∼ 85° about an S–C–S–C torsion angle (Gerber *et al.* 1993).

In general, the very extensively studied family of steroids (Duax and Norton 1975; Duax *et al.* 1982b) have exhibited a remarkable degree of conformational consistency (Duax *et al.* 1982a). Quite a few of them do exhibit polymorphism, however (Kuhnert-Brandstätter 1971) and as has been pointed out elsewhere (Duax *et al.* 1982a), any variety observed in the conformations of these molecules can provide useful informa-tion on the range of conformations accessible, and even on the mode of physiological action of a particular molecule or family of molecules. For instance, cortisone acetate **5-V** is known to crystallize in at least three unsolvated forms (Kanters *et al.* 1985)

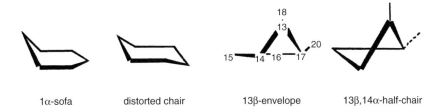

1α-sofa   distorted chair   13β-envelope   13β,14α-half-chair

Fig. 5.5 Ring conformations of steroids described in text. (From Bernstein 1987, with permission.)

Fig. 5.6 Stereoviews of two superimposed molecules of cortisone acetate **5-V** found in two different crystal structures. The atoms of the A, B, and C rings have been fitted to each other by least squares to emphasize their conformational similarity and the conformational differences in other regions of the molecule (see text). (From Bernstein 1987, with permission.)

for which the structure of Form I and Form II (DeClercq *et al.* 1972) have been reported. In both structures, ring A is a distorted 1α-sofa conformation while rings B and C are distorted chair. Differences arise in ring D, however; in Form I it exhibits a 13β-envelope and in Form II a 13β, 14α-half chair conformation (Fig. 5.5). The conformations of the molecules in both structures are compared in Fig. 5.6. This figure deserves some comment since we believe that it is a particularly vivid method for comparing molecular conformations and should be employed with greater frequency. The atoms of the A, B, and C rings of the second structure were fitted to those of the first by a least squares procedure described by Nyburg (1974) which produces coordinates for all remaining atoms in the framework of the first. These may be input into an appropriate molecular stereo plotting programme. In this case, it is possible to see easily the differences in the conformations of the side chains of the two molecules. The major differences are about the C(17)–C(20) and C(20)–C(21) bonds and amount to 16° and 14° respectively. Individually these differences are rather small, but the cumulative effect noted earlier in this chapter is dramatically demonstrated in this example. Conformational differences of a similar magnitude are observed in 6-bromo-testosterone acetate (Duax *et al.* 1981), and are manifested in different IR spectra.

5-V

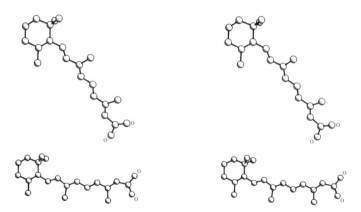

Fig. 5.7 Stereoviews of the two forms of vitamin A acid **5-VI**. Upper, triclinic form; lower, monoclinic form. In both cases the view is on the plane of the three lower atoms of the cyclohexenyl ring. For clarity only oxygen atoms are labelled (From Bernstein 1987, with permission.)

In contrast to the cumulative effect of a number of small differences along a chain, a large conformational difference can be generated by a single rotation, for instance in a dithiahexyl anthracene derivative (Reed *et al.* 2000) and in the triclinic and monoclinic modifications of vitamin A acid **5-VI**, a member of the family of visual pigments. The cyclohexenyl ring has essentially the same conformation in both forms (Stam and MacGillavry 1963; Stam 1972). However, as seen in Fig. 5.7 the triclinic form exhibits an s-*cis* conformation about the exocyclic bond and clearly a non-planar molecular conformation. The monoclinic form is s-*trans* about the same bond and the stereoview reveals a much more planar structure.

5-VI   5-VII

The very rich structural chemistry of organometallic compounds also provides many examples of conformational polymorphism. For instance, cyclic ligands often have low barriers to rotation so that conformational variations between polymorphs might not be unexpected. Riley and Davis (1976) reported the structures of the dimorphic sandwich compound **5-VII**, a system in which the molecular geometry is determined by the crystallographic site symmetry for both polymorphs (Fig. 5.8). In the triclinic form, the molecular site symmetry is C_i, requiring the '*trans*' orientation

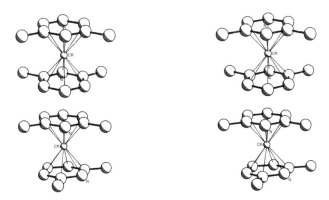

Fig. 5.8   Stereoviews of the two conformers of bis(2,6-dimethylpyridine) chromium **5-VII**. Upper, triclinic; lower, orthorhombic. Carbon atom labels have been deleted for clarity. Note that in this case a view on one ring as the reference plane would not have been a good choice for comparison. Instead, we have chosen to view structures nearly along the N· · · *p*-carbon axis of the upper pyridyl ring, with that axis tilted slightly downward, and the two methyl groups pointing toward the viewer. (From Bernstein 1987, with permission.)

of the two pyridyl rings, while in the orthorhombic form, the molecule lies on a 2-fold axis (site symmetry C_2) leading to the '*gauche*' orientation.

Some other notable examples of conformational polymorphism in organometallic complexes have been given by Foxman *et al.* (1981) (discussed in some detail in Bernstein 1987), Braga *et al.* (1992) and Wagner and Englert (1997), all of which exhibit some conformational variations in the orientation or conformation of the ligands.

Variation of the coordination geometry about the metal atom is also exhibited in polymorphic systems. **5-VIII** crystallizes in four modifications (von Stackelbar 1947; Lingafelter *et al.* 1961; Clark *et al.* 1977), two of which are green and two of which are brown. In the brown γ form the Cu atom is 0.5 Å from the plane of the naphthalene part of the ligand and there are dimers leading to a 5-coordinate (4 + 1 tetragonal pyramid) arrangement of ligands about the metal with an oxygen atom from the centrosymmetric dimer pair providing the fifth coordination site. In a second brown modification (β) (Clark *et al.* 1975) the corresponding distance from the plane is 0.63 Å and the metal is 6-coordinate (4 + 2, pseudo-octahedral). One of the two green forms (δ) (Martin and Waters 1973) exhibits a 4-coordinate square-planar geometry with the Cu being only 0.09 Å from the ligand plane.

5-VIII

In a more recent example, Kelly *et al.* (1997, 1999) showed that **5-IX** exists in two forms. Crystallization from hot acetonitrile yields blue crystals with square-planar coordination about the metal. Slow evaporation from methylene chloride/petroleum ether yields green needles with pseudo-tetragonal coordination about the metal.

5-IX   5-X

As in the steroids and the organometallic complexes, the differences in molecular structure in conformational polymorphs are not always strictly related to simple rotations about single bonds. Cyclic systems that appear to be quite rigid also exhibit conformational differences. The stable triclinic form of lepidopterene **5-X** (Gaultier *et al.* 1976) contains two crystallographically independent half molecules in the asymmetric unit while the metastable orthorhombic form (Becker *et al.* 1984) contains three full molecules in the asymmetric unit. In violation of the 'density rule' (see Section 2.2.10) the calculated density of the orthorhombic phase ($1.28\,\mathrm{g\,cm^{-3}}$) is slightly higher that of the stable triclinic phase ($1.26\,\mathrm{g\,cm^{-3}}$), which translates to a difference of $8\,\text{Å}^3$ per molecule. As measured by the angles between the planes of rings A and B or that between C and D the molecules in the orthorhombic phase are more splayed than in the triclinic structure. The increase in density may be responsible for these differences in geometry, the latter in turn being responsible for differences in the excimer spectra of the two polymorphs (Section 6.3.3).

5.7 What are conformational polymorphs good for?

Advances in chemical understanding are often based on making generalizations about systems, identifying the exceptions to those generalizations and studying the reasons for those exceptions. In arriving at the initial generalizations, one would like to minimize the number of variables considered, so that relationships can be clearly defined with cause and effect unambiguously assigned. Conformational polymorphs provide almost ideal systems for such an approach to the study of structure/property relationships, since the number of chemical variables is reduced to zero, the role of substituent being eliminated by the very choice of polymorphic systems. Thus changes in properties may be directly related to changes in structure. The subject of utilizing polymorphs in general and conformational polymorphs in particular to study structure/property relationships is treated in detail in Chapter 6. In conformational polymorphs in particular, differences in molecular structure *must* be due to differences in crystal environment; hence, these are ideal systems for the study of the relationship between the two. Furthermore, the identity of the molecular component among polymorphs, coupled with the differences in conformation and the relatively small, but important energy differences among them noted in Sections 5.2 and 5.3 make these systems excellent, but often demanding test cases for the development and use of computational methods and parameters used in the calculation of lattice energies. These aspects of conformational polymorphism will be treated in subsequent sections of this chapter.

5.8 Computational studies of the energetics of polymorphic systems

In studying a system exhibiting polymorphism in general and conformational polymorphism in particular, one is seeking the answers to a number of questions:

1. What are the structural differences among the polymorphs?
2. What are the differences in energy, if any, in the molecular conformations observed in the various crystal forms?
3. How does the energetic environment of the molecule vary from one crystal form to another?
4. Can the differences in energetic environment be understood on the basis of particular intermolecular interactions?

To answer these questions, a typical study might proceed according to the following scenario:

1. Determination of the existence of polymorphism in the system under study.
2. Determination of the existence of conformational polymorphism by the appropriate physical measurements.
3. Determination of the crystal structures to obtain the geometrical information—molecular geometries and packing motif—of the various polymorphs.
4. Determination of the differences in molecular energetics, by appropriate computational techniques.
5. Determination of differences in lattice energy and the energetic environment of the molecule by appropriate computational methods.

The first three of these are covered in Chapters 1, 3, and 4. Computational methods for the study of molecular energetics, from semi-empirical approaches to sophisticated *ab initio* methods, although constantly under development and improvement, are now standard techniques and tools for most chemists (Burkert and Allinger 1982; Pople 1999).

As discussed in Chapter 2, at a given temperature and pressure, the energies of polymorphs correspond to different Gibbs free energies. The relative stabilities of those polymorphs are then the differences in those energies (ΔG). These are expressed as:

$$\Delta G_{T,P} = \Delta U_{T,P} + P\Delta V - T\Delta S \qquad (5.1)$$

For the computational evaluation of these Gibbs free energy differences, in principle it is necessary to calculate all three terms on the right hand side of eq (5.1). The third term is often neglected, since differences in lattice-vibrational energies are usually very small (Gavezzotti and Filippini 1995). The $P\Delta V$ term can be calculated, but may be neglected for normal pressures due to the small value of ΔV. Hence $\Delta U_{T,P}$ is usually taken as an estimate for $\Delta G_{T,P}$. The main error in this approximation is the assumption of small entropy differences, which may not be correct in all cases of polymorphic phase transformations. As Dunitz has pointed out, as the temperature is increased through a transition point, the polymorph stable at $0\,K$ may become metastable and transform to another polymorph. For the new, more stable polymorph, the entropy increase with temperature will always be faster than for the less stable polymorph (Dunitz 1996).

Therefore most computational methods are aimed at calculating ΔU. The semi-empirical methods were originally developed by Kitaigorodskii (1973a,b) and much of the basic theory and developments through 1985 have been summarized by Pertsin and Kitaigorodskii (1987). The basic model is built on the atom–atom potential method, and assumes that the lattice energy can be obtained by a sum of pair-wise van der Waals interactions ($V(r_{ij})$) calculated using a Lennard-Jones (5.2) or Buckingham (5.3) potential. A coulombic term, $q_i q_j / r_{ij}$, is now normally included, especially since Williams (1974) has shown that even for hydrocarbon crystals such a contribution may approach 30 per cent of the total energy.

$$V(r_{ij}) = \frac{B}{r_{ij}^n} - \frac{A}{r_{ij}^6} \quad n = 9, 12 \tag{5.2}$$

$$V(r_{ij}) = B' \exp(-C'r_{ij}) - \frac{A'}{r_{ij}^6} \tag{5.3}$$

The individual partial charges on atoms, q_i, q_j, may be estimated from bond dipole moments (Hagler et al. 1974), quadrupole moments (Hirshfeld and Mirsky 1979) or charge densities obtained from molecular orbital calculations, and are assumed to be isotropic. The assignment of a partial charge to a particular atom from calculations remains an approximation and the derivation of that single charge is still a subject of investigation (Cornell et al. 1995; Masamura 2000). However, both experimental observations (Sarma and Desiraju 1985) and theoretical evidence (Williams and Hsu 1985; Price and Stone 1992) indicate that anisotropic considerations may be required for such calculations (Price 2000), and as computing power increases these will become more widely used.

The limitations of these methods should be understood. Their application requires the determination of the empirical parameters in eqns (5.2) and (5.3), as well as the partial atomic charges in the coulombic term. The former are usually parameterized from experimental solid state data such as vibrational frequencies or sublimation enthalpies, which in themselves contain some experimental uncertainties and vari-ability from system to system. The partial atomic charges can and do vary with the choice of basis set for the calculations from which they are derived. The function cho-sen and the complete set of parameters are often collectively termed a 'force field'. Ideally, one would like to develop a 'universal' force field, but given the diversity of chemical systems there are doubts as to whether this is attainable.[3] Furthermore, the computations generally do not take temperature into account, so the temperature for a particular calculation is often undefined. If the parametrization is based on *ab initio* calculations, the computed energies correspond to 0 K and neglect zero-point vibrational effects.

In spite of these caveats, there is intense activity in the application of these methods to polymorphic systems and considerable progress has been made. Two general approaches to the use of these methods in the study of polymorphism may be distinguished. In the first, the methods are utilized to compute the energies of the known crystal structures of polymorphs to evaluate lattice energies and determine the relative stabilities of different modifications. By comparison with experimental thermodynamic data, this approach can be used to evaluate the methods and force fields employed. The other principal application has been in the generation of possible crystal structures for a substance whose crystal structure is not known, or which for experimental reasons has resisted determination. Such a process implies a certain ability to 'predict' the crystal structure of a system. However, the intrinsically approximate energies of different polymorphs, the nature of force fields, and the inherent imprecision and inaccuracy of the computational method still limit the efficacy of such an approach (Lommerse *et al.* 2000). Nevertheless, in combination with other physical data, in particular the experimental X-ray powder diffraction pattern, these computational methods provide a potentially powerful approach to structure determination. The first approach is the one applicable to the study of conformational polymorphs. The second is discussed in more detail at the end of this chapter.

It has been found that in general for molecular crystals an interaction radius of 12–20 Å for the pairwise atom···atom interactions is sufficient for convergence of the energy (Hagler and Lifson 1974). The sublimation energy (i.e. the crystal 'binding energy' of a single molecule) is then minimized by altering the position and orientation of the reference molecule, as well as the unit cell parameters. If the starting model is a known crystal structure, usually only small perturbations in the structure result from the minimization procedure, and any crystallographic symmetry is maintained, even if not initially accounted for explicitly in the computational procedure. The resulting computed sublimation energy may then be compared to the experimental value to test the result and the reliability of the potential function as well as the parameters employed for each atom in the calculation. More detailed information on the nature of the intermolecular interactions may be obtained by partitioning the total lattice energy into non-bonded and electrostatic contributions. The fact that the calculation is based on the use of atom···atom potentials in principle allows an additional partitioning of

---

[3] There is a question whether such an approach is even practical or realistic. If a force field is viewed as a tool for calculating the energetics of real or possible crystal structures, then one might imagine using different tools for different systems. Just as a carpenter's toolbox contains a variety of tools for a variety of tasks, the computational chemist should probably have (and use) different force fields for different chemical systems. For instance, one would not expect a force field developed and parameterized on peptides to be effective or reliable on aromatic hydrocarbons, or vice versa, at the precision demanded for polymorphic systems.

the total energy or components thereof into individual atomic contributions. This latter method of partitioning is particularly useful in the investigation of the relationship between crystal forces and molecular conformation.[4]

It is important to note here some distinguishing characteristics of the computational approach to conformational polymorphism. With regard to the molecular energetics considerable computing effort is saved by noting that for the purposes of these investigations our interest lies really in those conformations that are observed in the crystal structures studied. Hence, rather than attempting to explore many regions on the multidimensional potential energy surface, it is often sufficient to compute the energies only for those few points actually observed in the different polymorphs. Moreover, our immediate interest in these studies is the differences in energies between various observed conformations rather than the total molecular energy, or even whether any particular observed molecular conformation is at the global minimum. Starbuck *et al.* (1999) have pointed out the drawbacks and methodical biases in such an approach, in which the energetics of only experimentally determined structures are studied. Their suggested strategies to overcome these biases include optimization of hydrogen atom positions (which apparently ameliorates many of the inconsistencies), optimization of all degrees of freedom save the conformational parameter in question, and full optimization to the nearest minimum. In any case, comparison should be made between energetic points obtained by identical computational approaches (see also Section 5.9). Clearly, in many cases studied, the size of the system taxes the available software or computing power, especially for *ab initio* calculations with higher level basis sets. However, the conformational energetics of large molecules may be obtained from molecular mechanics or *ab initio* calculations on model molecules representing the conformational parameters in question. Alternatively, semi-empirical molecular orbital methods often give good estimates of energy differences among conformations, even if other properties may not always be estimated very well. The considerations on choice of computational method for estimating conformational energy differences are rapidly changing with developments in computer power and availability.

For the lattice energies, a good test of the computational methods and parameters is a comparison with the sublimation energy, which may be measured fairly readily when sufficient quantities of material with suitably high vapour pressures are available (Daniels *et al.* 1970; Chickos 1987). Alternatively (although with considerably less precision), they may be estimated from the sublimation energies of analogous model compounds and group contributions (Bondi 1963; Cardozo 1991; Arnautova *et al.* 1996). These latter empirical methods generally cannot be used to estimate the differences in energy for polymorphic modifications, although some promising attempts have been made in that direction (Gavezzotti and Filippini 1995). Unfortunately, but perhaps not surprisingly, the literature contains very few experimental determinations of the sublimation energies of more than one member of a polymorphic system, and filling that gap remains a challenge.

---

[4] Unfortunately, the algorithms of some of the programs used for lattice energy calculation and minimization do not readily enable the extraction of this partitioned energy.

5.9   Some exemplary studies of conformational polymorphism

One of the first reports of a study of conformational polymorphism employing many of the aspects of this strategy appeared over twenty years ago (Bernstein and Hagler 1978). The subject was a dichloro-N-benzylideneaniline **5-XI**, which exhibited a number of features that made it particularly suitable for such a study: it is a relatively small molecule with only two conformational parameters; the two known polymorphs (triclinic, $P\bar{1}$ and orthorhombic $Pccn$) exhibit very different molecular conformations, and two very different, but readily describable crystal structures; the molecular energetics were studied using model compounds (although such an approximation would not be necessary today); atom and group types for the atom$\cdots$atom potentials are quite standard and hence, suitable for computation of lattice energies.

5-XI

The details of the studies on these and related molecules have been presented elsewhere (Bernstein and Hagler 1978; Bar and Bernstein 1987; Bernstein 1987); however, a number of relevant general points are worthy of note. In order to avoid any computational biases, in almost all cases, a number of different potential functions were employed in the lattice energy calculations, and some of the potentially problematic atom potentials were systematically varied. While these measures do result in different computed absolute energies, the *differences* in lattice energies could always account for the fact that the triclinic lattice could provide sufficient energy to stabilize the higher energy planar molecular conformation found therein, compared to the lower energy non-planar conformation found in the orthorhombic structure. Until considerably more experience has been gained from using these methods in a wide variety of systems, this variational testing of potential forms and potential parameters is highly recommended. It also contributes to setting limits of confidence in the methods and the results of such calculations.

Also worthy of note are some observations from analysis of the total energy partitioned into individual atomic contributions. Although qualitatively (i.e. from distance considerations) it appeared that chlorine $\cdots$ chlorine interactions may have dominated the structures and contributed in a major way to the relative stability of the two polymorphs, the partitioning of the energy into individual atomic contributions indicated that the difference was due to small contributions from many atoms, rather than a limited number of dominant interactions dominated by chlorine. These ideas are compatible with the notions recently expressed by Dunitz and Gavezzotti (1999), but should continue to be tested with similar partitioning analyses following computational generation of overall lattice energies.

In the intervening years, there has been increasing sophistication and ease of use in virtually every aspect of the techniques employed in a study of conformational polymorphism. Different groups have naturally taken slightly modified approaches with different emphases, but an increasing number of broad studies attests to the general applicability of this strategy of studying conformational polymorphs to probe crystal packing phenomena in general, and the relationship between crystal forces and molecular conformation in particular (Näther *et al.* 1996*b*; Buttar *et al.* 1998; Stockton *et al.* 1998; Starbuck *et al.* 1999; Yu *et al.* 2000). A review of the highlights from some of these studies will serve to put the field into current perspective.

One of the common features of the growing number of studies on conformational polymorphs is the utilization of a number of analytical and computational techniques to characterize the systems (e.g. Bauer *et al.* 2001). The emphasis differs from study to study, as demonstrated below, but these multidisciplinary approaches are to be encouraged and hopefully expanded. Another common feature of these investigations is the nature of the molecules investigated—relatively small molecules, with a limited number of conformational parameters. This allows a more direct comparison of conformational differences and the energies associated with those differences. As our understanding of the phenomenon increases and the computational capabilities to deal with larger systems improve, we can expect that more and more complex systems will be studied.

5-XII

The study of **5-XII**, (termed ROY by the investigators (Yu *et al.* 2000) for the red, orange, and yellow colours of the various polymorphs) demonstrates quite a few of these aspects. Six polymorphs, all existing at room temperature, have been prepared and characterized. They are readily distinguished by their colour and morphology, as shown in Fig. 5.9. All six can be obtained from methanolic solutions, occasionally crystallizing as concomitant polymorphs (see Section 3.5) from a supercooled melt. In spite of considerable effort to determine conditions for the selective crystallization of individual polymorphs by solution methods, the apparent similarity in thermodynamic stability often led to concomitant crystallization or rapid conversion. Due to thermal instability of two of the forms [YN (yellow needles) and ORP (orange red plates)] it was possible to determine the melting points of only four of the forms, which fall in the range 106.2–114.8 °C. Nevertheless, the authors succeeded in determining the crystal structures of all six forms.

The energetic relationships among the polymorphs were established by combining a number of techniques, providing a prototypical example of how this can be done. The pure melting endotherms of four forms (R, Y, OP and ON) could be determined by DSC (at 10 °C min^{-1} Fig. 5.10). Each one exhibits homogeneous melting, without

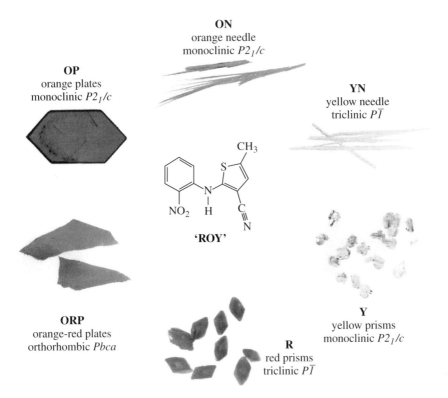

ON
orange needle
monoclinic $P2_1/c$

OP
orange plates
monoclinic $P2_1/c$

YN
yellow needle
triclinic $P\bar{1}$

'ROY'

ORP
orange-red plates
orthorhombic *Pbca*

Y
yellow prisms
monoclinic $P2_1/c$

R
red prisms
triclinic $P\bar{1}$

Fig. 5.9 (See also color plate section.) Photomicrographs of the six polymorphs of ROY **5-XII**, showing the different colour (R = Red, Y = yellow, O = orange) and morphology (P = plates, N = needles). From upper left, clockwise, OP, ON, YN, Y, R, ORP. (From Yu *et al.* 2000, with permission.)

any solid–solid transformation. A DSC trace of a mixture of the four (at a slower heating rate, Fig. 5.10) indicates subsequent recrystallization from the supercooled melt. An additional DSC experiment on YN (at $20\,^\circ$C min^{-1} (Fig. 5.11) shows the exothermic conversion to Y and (trace amounts of) R prior to melting. In other reported experiments, YN also converted to ON and ORP converted to ON or Y.

The data obtained from the DSC measurements could be used to determine free energy and entropy differences from expressions developed earlier by Yu (1995):

$$(G_j - G_i)_{Tmi} = \Delta H_{mj}(T_{mi} - T_{mj})/T_{mj} + \Delta C_{pmj}[T_{mj} - T_{mi}\ln(T_{mj}/T_{mi})], \tag{5.4}$$

$$(S_j - S_i)_{Tmi} = \Delta H_{mi}/T_{mi} - \Delta H_{mj}/T_{mj} + \Delta C_{pmj}\ln(T_{mj}/T_{mi}), \tag{5.5}$$

where T_{mi} and T_{mj} are the melting points of i and j, ΔH_{mi} and ΔH_{mj} are their enthalpies of melting, and ΔC_{pmj} is the heat capacity change upon melting j. The subscript indicates that $(G_j - G_i)$ and $(S_j - S_i)$ are evaluated at T_{mi}.

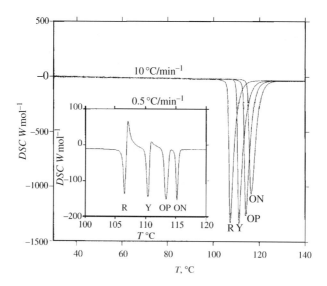

Fig. 5.10 DSC measurements on individual samples of polymorphs R, Y, OP and ON of **5-XII** (see caption for Fig. 5.9 for definition of terms), recorded at 10 °C min⁻¹, each exhibiting homogeneous melting. Inset: DSC trace of a mixture of the four modifications, recorded at 0.5 °C min⁻¹, showing better separated melting endotherms and exotherms resulting from crystallization from supercooled melts. (from Yu *et al.* 2000, with permission.)

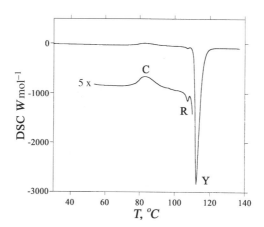

Fig. 5.11 DSC trace (at 20 °C min⁻¹) of polymorph YN of **5-XII** (see caption of Fig. 5.9 for definition of terms) showing an exothermic conversion (noted as C on the 5x expanded trace) and subsequent melting (noted as R, trace amount) and Y. The area under C gives an estimate of the enthalpy difference between YN and Y. (From Yu *et al.* 2000, with permission.) The authors attribute the differences between these traces and that in Fig. 5.10 to the difference in heating rate.

In many cases, data from DSC measurements are sufficient to determine the phase relationships among polymorphic modifications. However, due to the complexity of this system, the authors employed an older, but unfortunately rarely used technique of melting and eutectic melting data (McCrone 1957) to determine the stability relationship between the Y and ON forms. The procedure involves the preparation of mixtures of known composition of each polymorphic form under investigation with a series of common reference materials (McCrone 1957) to form eutectics that reduce the melting point by different degrees. For **5-XII** such mixtures were prepared with acetanilide, benzil, azobenzene, and thymol, and the resulting melting data are shown in Fig. 5.12. The data from such traces may be used to calculate $(G_j - G_i)$ at the eutectic melting temperatures by the relationship (Yu *et al.* 2000):

$$x_{ej}(G_j - G_i)_{Tei}$$
$$= \Delta H_{mej}(T_{ei} - T_{ej})/T_{ej} + \Delta C_{pej}[T_{ei} - T_{ej} - T_{ei}\ln(T_{ei}/T_{ej})]$$
$$+ RT_{ei}\{x_{ej}\ln(x_{ej}/x_{ei}) + (1 - x_{ej})\ln[(1 - x_{ej})/(1 - x_{ei})]\} \qquad (5.6)$$

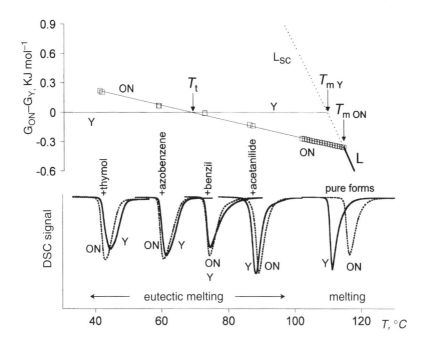

Fig. 5.12 Melting and eutectic data for determining the stability relationship between the Y and ON modifications of **5-XII**. Lower: melting endotherms of the two modifications as pure crystals, and in the eutectics with four different reference compounds. Upper: ΔG vs T relationship for the two forms based on data derived from eqns 5.4–5.6. (T_{mY} and T_{mON} indicate melting points of pure forms, T_t the transition temperature, L indicates the liquid phase and L_{SC} indicates the supercooled liquid phase.) (From Yu *et al.* 2000, with permission.)

where T_{ei} and T_{ej} are the melting points of i and j with a common reference compound, x_{ei} and x_{ej} are the eutectic compositions, ΔH_{mei} and ΔH_{mej} the enthalpies of eutectic melting, ΔC_{pej} the heat capacity change upon melting of the jth reference compound eutectic and R is the ideal gas constant.

The eutectic melting data for Y and ON with the reference compounds are given in Fig. 5.12, together with the thermodynamic data derived from eqns (5.4–5.6). From the relationship between the melting points of the eutectics in the lower part of the diagram, it is seen that at higher temperatures ON is more stable than Y, with a stability reversal at $\sim 70\,^{\circ}$C. The upper part of the diagram summarizes the free energy differences between ON and Y. The lines for Y and ON terminate at the line for the liquid.

Similar data obtained for other pairs of polymorphic modifications are presented in Fig. 5.13. The thermodynamic data and enantiotropic or monotropic relationship between modifications are given in Table 5.2, summarizing the relative stability of the phases from room temperature to the melting points (save the difficult to obtain ORP form).

The authors have made some interesting comments regarding these data in particular and the relative importance of energy and entropy contributions to the relative

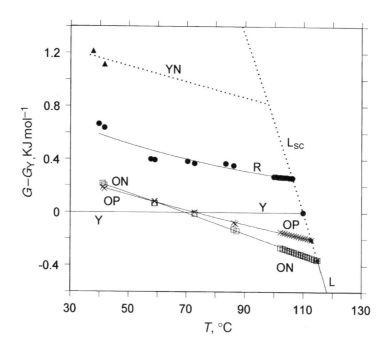

Fig. 5.13 Stability relationships between polymorphs of **5-XII**, constructed from melting and eutectic data as in Fig. 5.12. The symbols have the same meaning as in Fig. 5.12. (From Yu *et al.* 2000, with permission.)

Table 5.2 Summary of thermodynamic relationships[a] among modifications of **5-XII** (From Yu *et al*. 2000, with permission.)

Forms		ΔH, kJ mol^{-1}	ΔS, J K mol^{-1}	T_t, °C
Y = LT	ON = HT	2.6	7.7	70
Y = LT	OP = HT	1.9	5.3	72
Y = S	R = MS	1.4	3.0	c
Y = S	YN = MS	3.0[b]		c

[a] Enantiotropic systems: LT = low-temperature form, HT = high-temperature form. Monotropic systems: S = stable form, MS = metastable form.
[b] From YN–Y enthalpy of conversion.
[c] $T_t > T_m$ (virtual transition temperature of monotropic polymorphs).

stability of polymorphs in general. As noted in Section 2.2.2 the relative stability of polymorphs depends on the free energy ($\Delta G = \Delta H - T \Delta S$) between them. The relative importance of the two terms on the right can be measured by the ratio between them (say $T \Delta S / \Delta H$). As seen in Fig. 2.5 at absolute zero $T = 0$, $\Delta G = \Delta H$ and $T \Delta S / \Delta H = 0$. At a transition temperature between two polymorphic phases, $\Delta G = 0$ so the ratio $T \Delta S / \Delta H = 1$. Above a transition temperature this ratio will be > 1. Applied to some of the polymorphs of **5-XII**, for example, for the pair Y–R at the melting point of R the ratio is 0.85, which means that while Y is the more stable form at temperature, the entropy is an important contributor to the free energy. Other similar comparisons based on the data in Table 5.2 strengthen the notion of the importance of entropy in the consideration of thermodynamic relationships among polymorphs.

Having established the thermodynamic relationships among (most of) the polymorphs, the authors then carried out the X-ray crystal structure determination of all six of them. The most significant conformational variations are due to rotations about the N–C (thiophene) bond, ranging from 21.7° in the R polymorph to 104.7° in the Y polymorph. The nitro group is essentially planar with the phenyl ring in Y ($-1.8°$) and shows a maximum out-of-plane rotation of 18.7° in the OP modification, while the rotation of the phenyl group out of the C-N-C plane varies from $-150.0°$ to $-175.2°$. Hence there are experimentally significant conformational differences among all of the molecules in the six polymorphs.

The authors postulated that these differences in conformation should be manifested in the frequency of the CN stretch, which would also provide a measure of the molecular conformation in solution. The appropriate spectra, given in Fig. 5.14, clearly reflect the differences in molecular conformation, and indicate that both the melt and solution contain a mixture of the conformations found in the six polymorphic structures.

Essentially following the procedural stages outlined in Section 5.8, the authors then computationally investigated the energetics of the conformations observed in the six polymorphic forms. *Ab initio* calculations at the RHF/6-31G* level were used to explore the conformational energy profile about the N–C (thiophene) bond. The

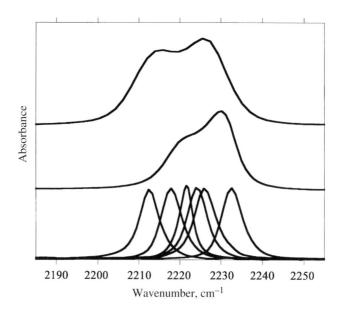

Fig. 5.14   Infrared absorption spectra in the region of the CN stretch for **5-XII**. Bottom, solid state spectra of forms (left to right) R, ORP, YN, ON, OP and Y. Middle, solution spectra in CCl$_4$ (0.90 mM). Top, supercooled melt at 22 °C. (From Yu *et al*. 2000, with permission.)

authors also considered the role of the (computed) molecular dipole moment (which changes with molecular conformation) in the stabilization of crystal energies (albeit while noting the evidence against the role of dipole moments in determining stability and space-group preference (Whitesell *et al*. 1991)). The combined results, presented in Fig. 5.15, indicate that a number of the structures are not in the conformational energy minimum (at ±90°), and that those away from the mimimum tend to have higher dipole moments. Both the *syn* planar and *trans* planar conformations are maxima, the former having a higher dipole moment.

Using the YN form as a reference, the authors then made a comparison of the computed conformational energies and the experimentally determined crystal energies, including noting the type of hydrogen bonding present in each of the structures (Fig. 5.16). The lowest energy modification (Y) is also unique in exhibiting an intermolecular hydrogen bond. The authors attribute the ordering for the other four polymorphs (again not including ORP) as due to two factors: favourable packing geometry of more planar conformers, and greater electrostatic interactions between larger dipoles. While these explanations are reasonable and consistent with other data, some additional evidence, provided say by some lattice energy calculations might strengthen them even more.

The study of the conformational polymorphism of **5-XII** by Yu *et al*. is one of most comprehensive reported to date. Similarly wide ranging studies, albeit with different emphases on the particular phenomena studied and the techniques employed have

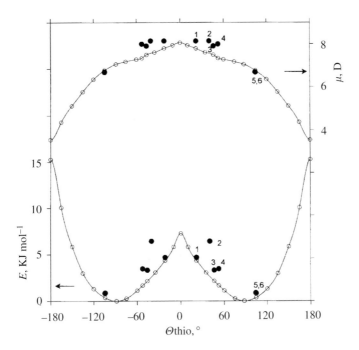

Fig. 5.15 Conformational energy (bottom, left scale) and dipole moment (top, right scale) calculated at the RHF/6-31G* level for **5-XII**. Solid circles correspond to conformers observed in the six crystal modifications: 1-R, 2-ORP, 3-OP, 4-ON, 5-Y, 6-YN. (From Yu *et al.* 2000, with permission.)

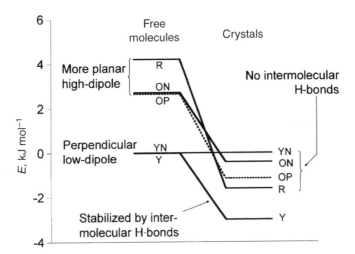

Fig. 5.16 Comparison of the computed conformational energies and experimentally determined crystal energies for **5-XII** (see text for discussion of the generation of these data points). Modification YN is the reference point (E = 0) for both comparisons. (From Yu *et al.* 2000, by permission.)

been appearing with increasing frequency and breadth, indicating concerted attempts to characterize and understand polymorphic systems, and these studies bode well for the increasing amount of information and understanding to be gained from such investigations. For instance, Stockton *et al.*'s (1998) study on pendimethalin **5-XIII** (with an orange and a yellow form) exhibits many of the features of the previous example, but in this case, the emphasis is on spectroscopic methods, with less emphasis on the thermodynamic properties. On the other hand, two different computational approaches were applied to lattice energy calculations (including optimization of the molecular geometry determined in the reported crystal structure determinations), both of which indicate that the orange form is stabilized by $\sim$ 4–5 kcal mol^{-1} relative to the yellow form.

5-XIII   5-XIV

Similarly, Näther *et al.* (1996), studied the dimorphic **5-XIV**, an analogue of the earlier rather extensively studied dichloro derivative (Byrn *et al.* 1972; Richardson *et al.* 1990). This is another excellent example of the irrelevance of the space group in determining the similarity or difference of polymorphs (see Section 2.4.2.1 and Bock *et al.* 1995). Form II, stable at room temperature, crystallizes in space group $P2_1/c$, and transforms to Form I at 391 K, in space group $P2_1/n$. The two space groups are simply cell transformations of the identical space group which is the most common one for organic and organometallic crystals (Baur and Kassner 1992; Padmaja *et al.* 1990; Brock and Dunitz 1994) listed as No. 14 in the International Tables for Crystallography (1983). The cell constants are totally different, as are the crystal structures, compared in Fig. 5.17.

The two forms of **5-XIV** often crystallize concomitantly, but may be separated manually by virtue of distinctions between habit and colour. A solvate containing two equivalents of ethanol is also obtained. The two unsolvated forms exhibit different molecular conformations and hydrogen bonding patterns. *Ab initio* calculations with a MIDI split-valence double-zeta basis set at the HF–SCF level indicated that the conformations observed in the unsolvated forms both correspond to local minima with the conformation observed in the room temperature Form II being energetically more favourable by 15 kJ mol^{-1}. (In Form I there are actually two crystallographically independent half molecules in the asymmetric unit, each lying on a crystallographic inversion centre, and they differ in computed conformational energy by 0.5 kJ mol^{-1}.) The (unminimized) lattice energies were computed using a 6-exp potential function (Gavezzotti and Filippini 1994), yielding $\sim$ −61 and $\sim$ −54 kJ mol^{-1} for Form II and Form I respectively, within the range expected for differences between polymorphs. The values of the densities (1.915 (II) and 1.899 (I) g cm^{-3}) and the packing coefficients (0.725 (II) and 0.717 (I)) (Kitaigorodskii 1961; Gavezzotti 1983, 1985) are

(a)

(b)

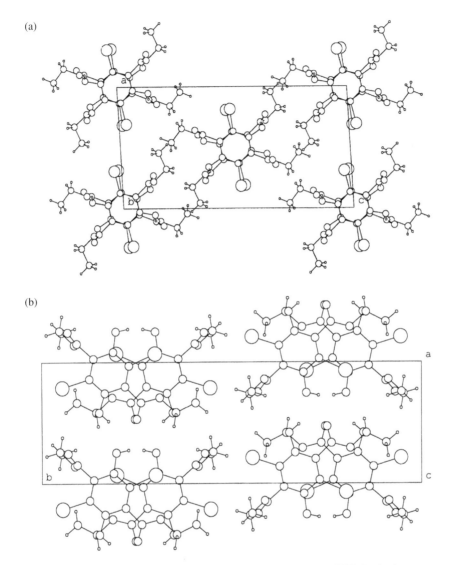

Fig. 5.17 Packing diagrams of the two polymorphic structures of **5-XIV**. Both views are on the shortest axis, which is approximately on the plane of the phenyl ring. Top, Form I, view along axis (7.85 Å). Bottom, view along the *c* axis (7.53 Å). Largest circles represent bromine substituents (Based on Näther *et al.* (1996*b*), with permission.)

similarly consistent with the ordering of lattice energies. The most stable computed molecular conformation was not found in any of the three crystal structures, leading the authors to propose the possible existence of an additional polymorphic modification below room temperature.

The computational aspects of both molecular conformation and lattice energetics were compared and evaluated in two studies by Buttar *et al.* (1998) and Starbuck *et al.* (1999). Once again, the molecules chosen for study **5-XV–5-XVIII** had a small number of conformational degrees of freedom. In the case of the dimorphic *o*-acetamidobenzamide **5-XV** (once again crystallizing in $P2_1/n$ (α form) and $P2_1/c$ (β form)), the former has an intramolecular hydrogen bond which is not observed in the latter, leading to significant differences in the two exocyclic torsion angles of over 100 degrees. The conformational energetics were investigated employing a variety of computational techniques, from molecular mechanics through semi-empirical and *ab initio* molecular orbital methods. The conformations observed in the crystal structures are near, but not at, the computed (gas phase) energy minima. In addition to those minima near the observed conformations, there are lower energy minima for (as yet not observed) conformations. While the computational methods employed at various levels of sophistication give some extreme values, all indicate that the conformation observed in the α form is more stable than that in the β form by an average 8.5 kcal mol^{-1}, a rather large value for conformational differences. In this system no overriding advantage was found for the use of high-level quantum mechanical computational methods over semi-empirical ones. The reason behind this may be that the molecule is one that contains well-parameterized atoms for semi-empirical methods. In the presence of less common valencies or third row atoms this result might not be obtained.

5-XV   5-XVI   5-XVII   5-XVIII

It has generally been assumed (e.g. see Bernstein and Hagler 1978) that lattice energies would be sufficient to stabilize otherwise unstable molecular conformations. In the case of **5-XV**, the lattice energies of the β polymorph are calculated to be more stable by about 2.6 kcal mol^{-1}, which means that the difference in the total of the inter- and intramolecular energy ($\sim$6 kcal mol^{-1}), is larger than the generally accepted values between polymorphs, and the energy deficit of the less stable conformation is not compensated for by the lattice energy. As an additional caveat, the authors reiterated the importance of optimizing the positions of hydrogen atoms as part of the computational protocol. While the hydrogen atom positions are not precisely determined by X-ray methods (and are often included in structure determinations in positions assumed from chemical and geometric principles), their location on the molecular periphery means that they make significant contributions both to non-bonded intramolecular interactions and to intermolecular interactions. It is therefore crucial to properly locate and optimize the hydrogen atom positions at the start of any computational procedure.

Another thorough study on the three diarylamines **5-XVI–5-XVIII** followed a similar strategy. **5-XVI** and **5-XVIII** are both dimorphic, while **5-XVII** exhibits three modifications with neutral molecules, and one modification with a zwitterion, a situation observed also in anthranilic acid (Ojala and Etter 1992, and references therein). The two major conformational parameters are the rotations about the N–C(phenyl) bonds, with the rotation of the carboxyl group comprising an additional torsional degree of freedom. Over the seven molecular structures studied, the former are in the range $\sim$ 0–180° for the rotation of the 'lower' ring, and in the range $\sim$ 0–37° for the 'upper' ring. Conformational preferences were studied by a variety of methods, again with the caveat that single point calculations without proper optimization of hydrogen positions do not give a good estimate of the relative energies of the structures observed in different polymorphs. In this case, the best computational approach for the torsional energetics was found to be 6-31G** basis set in *ab initio* optimizations, with other methods less successful in varying degrees, and the torsional potential energy surfaces were surveyed in considerable detail. In most cases, the conformations observed in the polymorphs appear at or near minima on the computed conformational potential energy surfaces. However, one of the observed conformations of **5-XVII** occurs at a maximum, while another appears between a minimum and a maximum.

The lattice energy calculations also were carried out with two programs, HABIT95 (Clydesdale *et al.* 1996) and (Cerius2 MSI) employing the SYBYL and Dreiding (Mayo *et al.* 1990) force fields. The calculations gave mixed results, in some cases conforming to experimental results and in other cases at variance with them. The authors also carried out a number of 'computational substitution' experiments (Hagler and Bernstein 1978) which consist of replacing molecules of with known structures by very similar molecules which do not exhibit the same crystal structures. Some of the resulting computed trial structures were sufficiently close in energy to observed structures that they might be expected to exist, although obviously, no means of preparing such possible polymorphic modifications can be extracted from such calculations.

The authors' conclusions from the two studies described briefly here indicate much regarding the state of the art of lattice energy calculations on polymorphic systems and their *predictive* capabilities. Not all the conformations observed in crystal structures appear at minima on computed gas phase potential energy surfaces; they also appear at maxima and non-stationary points. This raises questions about the widespread use of computed gas phase models as input for crystal structure solution or polymorph prediction for conformationally flexible molecules. In the calculation of possible crystal structures, not all the lowest energy minima are occupied by known crystal structures. In fact, some of the known ones appear above the generally accepted range of 1–2 kcal mol^{-1} for the distribution of polymorphic modifications. Hence, some known crystal structures would not have been predicted using these methods. The conformational energy surfaces of many of these molecules allow for considerable variability within a small energy range, further complicating the choice of starting model for structure solution or polymorph prediction. Clearly, systems of conformational polymorphs provide excellent test cases and benchmarks for the development and evaluation of these techniques, but there is much to be learned in this active and challenging aspect of polymorphism research (Reed *et al.* 2000). Various aspects of that activity are summarized in the next section.

5.10   The computational prediction of polymorphs

As described throughout this chapter, once the general features of molecular crystals became apparent through X-ray crystallographic methods (e.g. Pauling 1960, first edition 1939) there were attempts to model the structures, based on the notion of relative constancy of intermolecular contacts for specific atoms, and the tendency to minimize the amount of voided space in a crystal. This prompted the pioneering work by Kitaigorodskii in the late 1940s, summarized in his 1955 book (in Russian) (English translation: Kitaigorodskii 1961), which included the design and construction of a mechanical 'structure seeker' for physically modelling possible structures using hard sphere molecular models. Many of Kitaigorodskii's basic ideas have been incorporated into today's considerably more sophisticated computational approaches, but the original book still contains many useful insights into the nature of packing regularities in molecular crystals.

The ultimate goal is to be able to solve (see Timofeeva (1980) for a review of early approaches to this problem), or predict a crystal structure on the basis of the molecular structure alone. Clearly, for polymorphic systems, or potentially polymorphic systems, that goal includes some additional challenges: how many polymorphs can be expected to exist, what is the structure of each, and how might one proceed experimentally to obtain each of the polymorphs? As in most areas of chemistry, the modelling approaches range from empirical to *ab initio*, each with its advantages, limitations, successes, and failures. In this section, we will attempt to review those approaches as they relate to polymorphic structures in this dynamic and rapidly developing field (Gdanitz 1997, 1998; Lommerse *et al.* 2000; Mooij 2000; Verwer and Leusen 1998; Pillardy *et al.* 2000). While the emphasis here is on molecular crystals, similar efforts are being made on crystals of polymers (Ferro *et al.* 1992; Boyd 1994; Aerts 1996).

All the programs that attempt to predict the number and structures of polymorphs of course begin with a molecular structure. For some, a very rough estimate of the molecular shape and size is sufficient, while for others even precise details may not suffice, as described in the previous section. Moreover, in this still relatively early stage of development of all of these methods, it is usually not obvious *a priori* how much detail is required in the molecular model. At any rate, if the molecule does have conformational flexibility then choices must be made for the conformations of the starting model. Most current programs do not allow for the simultaneous variation of molecular conformation and crystal structure—although this is clearly a direction for urgent development. A promising development in the effort to combine intramolecular and intermolecular searches is the successful inclusion of the rotation of hydroxyl groups together with the search for possible crystal structures (van Eijck and Kroon 1999).

At the empirical end of the spectrum of methods employed for the theoretical/computational generation of possible polymorphic crystal structures is the approach taken by Gavezzotti (1991, 1994a, 1996, 1997) and co-workers (1995, 1997) and Filippini et al. (1999) as well as Holden et al. (1993). Trial structures with fixed molecular geometry are generated (Gavezzotti 1994a) using (singly, and in combination) the most commonly appearing symmetry elements in molecular crystals: inversion centres, two-fold screw axes, glide planes, and translation (sometimes appearing as centring) (Kitaigorodskii 1961). Of the 230 mathematically possible space groups, statistics indicate (Donohue 1985; Bauer and Kassner 1992) that organic crystals tend to crystallize in one of the following with a high probability: $P\bar{1}$, $P2_1$, $P2_1/c$, $C2/c$, $P2_12_12_1$, $Pbca$. The generated trial structures with the highest packing coefficients and substantially cohesive packing energy are accepted for further study, which involves optimization of the lattice energy by a suitable program (e.g. Williams 1983). The resulting structures must then be screened by the user, either visually, or with the aid of some quantitative criteria to choose reasonable structures. Of course, one quantitative measure is the lattice energy which should be at (or near) a minimum for the expected polymorphic structures. In many cases, however, the number of low-energy structures within the (rather narrow) energy range of $\sim 10\,\mathrm{kJ\,mol^{-1}}$ is much too large to permit a rational choice of a reasonable number of expected polymorphic structures even for relatively simple molecules (van Eijck et al. 1995; Mooij et al. 1998).

Another approach (common to many strategies of structure generation and 'polymorph prediction') is to attempt to match the powder diffraction pattern calculated from the generated trial structure with an experimental one, although this can no longer be justified as a truly predictive procedure.

In a strategy with a similar philosophy, Perlstein (1994a,b, 1996) has outlined an 'aufbau process' for building up trial structures, following the approach to understanding crystal packing originally advocated by Kitaigorodskii (1961). The process proceeds from one- and two-dimensional motifs using some common patterns revealed in the CSD (Perlstein 1992), which are then built up into a number of possible crystal structures, all of which are potential polymorphic structures (Perlstein 1999).

The CSD also provides information on the coordination spheres of molecules and the symmetries associated with them (e.g. Braga and Grepioni 1992; Allen and Kennard 1993, Dunitz 1996; Motherwell 1997). This information has been used by Holden et al. with a program called MOLPAK (1993) to generate trial structures on the basis of the densest packing of such coordination geometries. Using these model structures, other workers then minimized the resulting lattice energies. For instance, the group of Price has applied the quite sophisticated multipole electrostatic model for electrostatic contributions together with the Buckingham empirical atom···atom potentials in a program called DMAREL (Willcock et al. 1995) to successfully generate a number of crystal structures, including polymorphic ones containing rather complex hydrogen bonding patterns and to address a number of questions about competing forces (Coombes *et al.* 1997; Price and Wiley 1997; Aakeröy *et al.* 1998; Potter *et al.* 1999).

A more purist approach, which in principle ignores the accumulated structural data and associated structural trends, involves placing each independent molecule with a defined conformation at a certain location (X, Y, Z) with a certain orientation (ϕ, θ, φ) in a unit cell defined by its six parameters (a, b, c, α, β, γ). Each additional independent molecule in the asymmetric unit is defined with six additional location and orientation parameters. Each variation of conformation for one or all of the individual molecules constitutes a new starting model.

The starting models are then optimized by a variety of techniques in the search for the global energy minimum (for the computationally most stable structure) and energetically neighbouring structures (for possible less stable polymorphs). One of the fundamental problems in this process is the definition of the space group. In the procedure described in the previous, paragraph, the space group would be $P1$, rarely encountered for molecular materials. One solution is to add additional molecules, each with six parameters, which quickly taxes the power of any computer. A computationally more efficient approach is to check for the origin of symmetry as a result of the packing optimization process, a procedure incorporated into a number of standard programs for analysing the results of crystal structure determinations (e.g. PLATON program, Spek 1990). Of course, another solution is to adopt part of the earlier related strategy and incorporate accumulated information on space group frequencies, at the risk of not generating a true structure in a less commonly encountered space group. For instance, one of the two polymorphs of acetamide,

5-XIX

5-XX   5-XXI

5-XIX crystallizes in space group $R3c$ (Deene and Small 1971), one of the polymorphs of diethyl barbituric acid **5-XX** crystallizes in $R\bar{3}$ (Craven *et al.* 1969), and one of the modifications of aspartame **5-XXI** crystallizes in $P\bar{4}$. Additional considerations which must be included in developing a general strategy for generating trial structures include the possible presence of more than one molecule in the asymmetric unit (van Eijck and Kroon 1999), the location of a molecule on a crystallographic symmetry element, and the presence of solvent molecules. Currently available programs also have no facility for treating structures comprised of more than one molecular entity, such as organic salts host–guest complexes or molecular complexes (e.g. charge-transfer complexes). Furthermore, Dunitz *et al.* (2000) have recently shown that entropy differences among the computed polymorphs can be of the same order as differences in packing energy at 300 K (usually the temperature of greatest practical interest, if not of computational rigour). This suggests that a means for including lattice-vibrational entropy should also be included in these computational algorithims. The importance of the inclusion of dynamic effects into atom–atom computational procedures was demonstrated by Rovira and Novoa (2001).

In spite of these limitations—indeed, challenges for further development—there have been considerable activity and progress in generating possible crystal structures, and polymorphs, through the energy minimization of structures with one molecule in the asymmetric unit. A number of groups have developed algorithms and programs for the systematic search of packing arrangements. It is not clear yet which of these approaches will be the most fruitful and eventually the most useful. However the activity in this area is increasing rapidly, especially with access to increasing computer power, and developments should be closely monitored. Many choices have to be made as well in both the generation of the original model (e.g. geometry and calculation of partial atomic charges for use in the electrostatic terms), and in the details of the algorithms for methods employed in the actual calculations (e.g. rigid-body minimization or all atom minimization, Ewald summation for electrostatic terms). Subsequent lattice energy minimizations also involve a number of strategic options. Systematic searching of packing arrangements has been developed by the groups in Utrecht (van Eijck et al. 1995, 1998, 1999; Mooij et al. 1998), in Louisville (Williams 1996; Gao and Williams 1999) in Cleveland (Chaka et al. 1996) and in Ludwigshafen (Erk 1999). Random searching of possible structures is also an option in the former two approaches and has been adopted as well by Schmidt (1995; Schmidt and Englert 1996), by Dzyabchenko et al. (1984, 1985, 1987, 1996) by Gibson and Scheraga (1995), and by Pillardy et al. (2000), who have also studied the role of the dipole moment in considering the computed electrostatic energy of the crystal.

Another approach to generating possible crystal structures uses Monte Carlo methods, incorporating a minimum of chemical information (Gdanitz 1992; Karfunkel and Gdanitz 1992; Karfunkel and Leusen 1992; Karfunkel et al. 1994; Leusen 1996; Gdanitz 1998; Payne et al. 1998a,b, 1999) or incorporating the tendencies of certain groups to form patterns in crystals (Chin et al. 1999a; Chin 1999), such as hydrogen-bonded tapes (Chin et al. 1999b). The strategy of Karfunkel et al. for avoiding local minima is to carry out the initial calculations at high temperatures, and then 'cool' the system to lower temperatures, a computational process known as simulated annealing. They have noted the sensitivity of the method to the choice of partial atomic charges and also the difficulty in distinguishing among essentially equivalent structures. The resulting set of low-energy structures are then often compared with experimental X-ray powder diffraction patterns to help select probable solutions.

In the approach of Chin et al., the simulated annealing method was employed on molecules that tend to form hydrogen-bonded tapes (Schwiebert et al. 1995), thus providing a well-defined (but necessarily limited) starting model for the subsequent trial structures.

The problem of ranking or scoring the multiple solutions resulting from these approaches has been approached in a different manner by Hofmann and Lengauer (1997, 1999). Rather than base the scoring on energy, the trial structures, which are generated using a method similar to that of Gavezzotti (1994a), are ranked according to a function that is based on statistical trends of interatomic distances derived from the CSD. These statistical trends are applied with an inverse Boltzmann relationship to determine the connection between them and the energies. Motherwell has applied a similar approach, without invoking the energetic relationship derived from the reverse Boltzmann statistics, but with making a judicious choice of the statistical sample based on chemical and structural similarity between the structure under study and those in the CSD sample. This is a procedure that is considerably aided by the capability of newly developed software (Lommerse and Motherwell 1999) and intermolecular libraries (Bruno et al. 1997) derived from the CSD.

How well do these methods work individually and collectively? In 1999 the CCDC conducted a blind test of structure (including polymorph) prediction, in which most of the groups mentioned in this section participated (Lommerse et al. 2000).[5] Such tests should serve as a challenge, a model and a landmark for the development of these computational techniques.

| 5-XXII | 5-XXIII | 5-XXIV | 5-XXV |

Four compounds **5-XXII–5-XXV** with determined, but unpublished structures were chosen for testing with the following restrictions: $Z' = 1$; the space groups were not specified, but were limited to one of the more common ones, noted earlier. The molecules are essentially small and rigid, although two contain less common elements and **5-XXIV** has some torsional freedom.

Each of the participants was permitted to propose at most three structures for each compound. Four programs (MPA, Williams 1996; MSI-PP, Verwer and Leusen 1998; UPACK, van Eijck and Kroon 1999; Zip-Promet, Gavezzotti 1994b) correctly predicted the structure of **5-XXII**. One was successful for each of the other three [MSI-PP, UPACK, and UPACK/ ab initio (Mooij et al. 1998), respectively]. Unknown to the participants, one compound **5-XXII** had crystallized in two polymorphs. The structure that the forum successfully predicted was one that crystallized initially in space group *Pbca*. Following the experimental determination of its crystal structure at low temperature, it melted and could not be recovered; subsequent crystallizations led to a $P2_1/c$ structure that was not predicted by any of the participants, even though the experimental facts suggest that the $P2_1/c$ structure is a more stable form compared to the apparent disappearing *Pbca* polymorph (see Section 2.3).

---

[5] A second blind test was conducted in 2001 with many of the same participants. It is intended that the results will be published in a similar manner.

Of the seven correct predictions, five were the first choice (out of the three permitted) of the participants. As noted in the report, this suggests that 'the prediction problem, although beset with difficulties, is not hopeless'. On the other hand, the authors note that if the disappearing *Pbca* form of **5-XXII** had not been found at all, then there would have been only three successes overall, and none for that compound. The first sentence of the paper echoes the opening quotation of this chapter and perhaps best summarizes the current state of the art of the computational methods presented here: '... at the present stage of development, perhaps the best that can be expected from crystal structure prediction programs is to provide a list of possible candidates for experimentally observable polymorphs'. Clearly, additional similarly designed and executed blind tests will serve as milestones in the development of these computational techniques for the calculation of lattice energies of molecular crystals in general and those of known or expected polymorphic forms in particular.

6

Polymorphism and structure–property relations

Polymorphic modifications of a given compound show significant differences in chemical behavior. (Cohen and Schmidt 1964)

There are two strategies for studying structure/property relationships: maximize the number of observations or minimize the number of variables. (Willis 1987)

6.1 Introduction

The design and preparation of materials with desired properties are among the principal goals of chemists, physicists, materials scientists, and structural biologists. Achieving that goal depends critically on understanding the relationship between the structure of a material and the properties in question. To be most effective, studies of structure–property relationships generally require systematically eliminating as many as possible of the structural and composition variables that play the most important role in determining the particular property under investigation. For molecular materials, a typical strategy might involve, for instance, a methodical variation in the mode or type of substitution on one part of the molecule in order to test a particular hypothesis about the causal relationship between structures and properties. Variations in substituents, while they do often result in changes in structure, and the corresponding changes in properties, also lead to perturbations in the geometric and electronic structure of the molecules in question. In such cases, changes in properties cannot always be correlated directly with changes in structure. The existence of polymorphic forms provides a unique opportunity for the investigation of structure–property relationships, since by definition the only variable among polymorphs is that of structure, and one of the most effective strategies for studying structure–property relations has been to follow the behaviour of a physical property through a polymorphic phase change. The study of the thermodynamics, kinetics, and mechanism of phase transitions by most or all of these analytical or physical techniques is a discipline in itself (e.g. Verma and Krishna 1966; Rao 1984, 1987; Bayard et al. 1990; Rao and Gopalakrishnan 1997), and is not addressed specifically here; rather, the polymorphic systems are considered as points in the multidimensional phase space representing structures with different packing and possibly different molecular conformations for the utilization of the study of structure–property relations. For a polymorphic system, differences in properties among polymorphs must be due only to differences in structure. As a corollary to this principle, a constancy in properties for a polymorphic system indicates a lack of structural dependence on that property, at least within the limitation of the structural variation through that series of polymorphic structures.

For molecular materials, studies of structure–property relations fall into two broad categories. In the first category, the properties under investigation are due to strong interactions between neighbouring molecules, and one wishes to study the changes in bulk properties resulting from differences in the spatial relationships among molecules in the crystal. In the second category, we seek information on variations in properties related to differences in molecular structure, generally, for reasons noted in the previous chapter, in molecular conformation. Cases of conformational polymorphism, discussed in detail in the previous chapter, also provide opportunities for the study of the influence of crystal forces on molecular conformation. This chapter is devoted to describing representative examples of structure–property studies from both of these categories. It is not possible to be inclusive in this regard, but it is hoped that the description of a number of case studies will be sufficient to demonstrate the utility and advantages of choosing polymorphic systems, when possible, for the investigation of structure–property relationships. In addition to providing information on the structure–property relationships, many of these studies also reveal the experimental strategies and procedures required to obtain the polymorphic modification(s) with the specifically desired properties (Bernstein 1993).

6.2   Bulk properties

6.2.1   *Electrical conductivity*

Molecular materials are traditionally considered to be electrical insulators, but the discovery in the early 1970s of metallic conductivity in crystals of the π molecular complex of tetrathiafulvalene **6-I** and tetracyanoquinodimethane **6-II** (Ferraris *et al.* 1973, Coleman *et al.* 1973) led to a revolution in thinking about these materials in particular, and the potential for organic materials in general, as the basis for the next generation of electronic components (Wudl 1984; Williams *et al.* 1985), as well as the Nobel Prize in Chemistry for the year 2000. In contrast to the vast majority of previously known π molecular complexes that crystallize with plane-to-plane stacks of alternating donors and acceptors (Herbstein 1971) (Fig. 6.1), the complex of **6-I** and **6-II** crystallized with segregated stacks of molecules along the same crystallographic axis (but not mutually parallel molecular planes), each stack containing only one type of molecule.

6-I   6-II

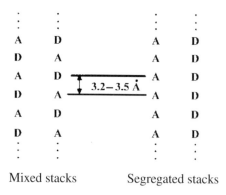

Fig. 6.1 Schematic diagram of the mixed-stack and segregated stack motifs for packing of π molecular charge-transfer complexes. (From Bernstein 1991b, with permission.)

This structural feature has been shown to be a necessary condition for electrical conductivity in these materials, although the mixed mode of stacking is generally considered to be the thermodynamically preferred one (Shaik 1982). Proof of the relative stability of the mixed and segregated stack motifs, and a recipe for obtaining crystals of the latter, came with the discovery of a pair of polymorphic 1 : 1 complexes of **6-II** with **6-III** (Bechgaard *et al.* 1977; Kistenmacher *et al.* 1982). The red, transparent, mixed-stack form of the complex is a semiconductor, while the black, opaque structure with segregated stacks is a conductor (Fig. 6.2).

6-III

This finding demonstrated that the presence of segregated stacks is a necessary condition for electrical conductivity. Reflecting the relative stabilities for the two stacking modes noted above, crystals of the red semiconductor form are obtained from a thermodynamic or equilibrium crystallization: equimolar solutions of the donor and acceptor in accetonitrile are mixed and allowed to evaporate slowly. On the other hand, crystals of the black form are obtained from a kinetic or non-equilibrium crystallization: hot equimolar solutions of the donor and acceptor in (the same) acetonitrile solvent are mixed and cooled rapidly. Some microcrystals of the resulting black powder are then used as seeds to obtain larger crystals of the segregated stack black form.

Non-equilibrium crystallization methods, in particular electrochemical techniques, have become standard procedure for obtaining crystals of organic conductors, in part because of the ability to control and reproduce the crystallization conditions

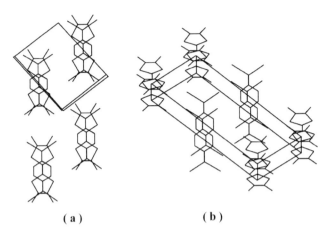

(a)   (b)

Fig. 6.2   Views of the two polymorphic structures of **6-II : 6-III**. In both cases the view is on the plane of the tetracyanoquinodimethane molecule **6-II**. (a) The red, transparent, mixed-stack complex, a semiconductor; (b) the black opaque, segregated stack complex, a conductor. (From Bernstein 1991*b*, with permission.)

(Williams *et al.* 1991). However, there are examples of cases of polymorphic charge-transfer complexes in which a mixed-stack structure and a segregated stack structure were obtained concomitantly under nearly equilibrium conditions of slow diffusion, although, not surprisingly, under these conditions, the segregated stack form was obtained 'with much difficulty' (Nakasuji *et al.* 1987). The segregated stack modification exhibits metallic conduction, while the mixed-stack form is a semiconductor (Imaeda *et al.* 1989). Similar principles may be applied and similar observations have been made when the structural units contain organometallic complexes (Legros and Valade 1988; Cornelissen *et al.* 1992, 1993; Almeida and Henriques 1997).

The utilization of polymorphism in understanding the connection between structure and electrical conducting properties is perhaps most poignantly represented by the so-called 'ET' salts of **6-IV** (Bechgaard *et al.* 1980). Generically designated (ET)$_2$X, they have generated intense activity since some exhibit the highest known T_c values for organic superconductors. Because the characterization of the solid state properties is the essence of the investigation of these materials, they have been intensively studied, and many have been found to be polymorphic (Kikuchi *et al.* 1988; Williams *et al.* 1991; Schlueter *et al.* 1994; Saito 1997). For X = I$_3^-$, there are at least fourteen known phases, and learning to define and control the growth conditions leading to a specific desired polymorph or the avoidance of an unwanted polymorph is one of the challenges facing workers in this field. The usual method of crystal growth for these compounds is electrocrystallization. Experiments can extend to periods up to months, with a number of polymorphs appearing on the same electrode, in many cases with indistinguishable colours or crystal habits. Only the physical characteristics (conductivity, spectral response, X-ray diffraction pattern, etc.) can be used to distinguish them, and often the identification of the various phases requires the examination and characterization of each individual crystal in a batch (Wang *et al.* 1989; Kato *et al.* 1987). For example, Kobayashi *et al.* (1987) identified the presence of the α, β, γ, δ, τ and κ polymorphs of $(ET)_2^+ I_3^-$ in the same experiment. Yoshimoto *et al.* (1999) have shown by solution mediated transformations that the β form is more stable than the α form in the temperature range 0–50 °C, leading to the selective and controlled growth of large crystals of each form.

6-IV

As an example of the type of variation observed, the β and κ phases can be compared (Fig. 6.3) along with the schematic packing motifs of other polymorphic forms. The former, apparently favoured by thermodynamic crystallization conditions (e.g. low current density) is a centrosymmetric triclinic structure with one formula unit of the salt in the unit cell. The symmetry arguments require that the anion lie on a

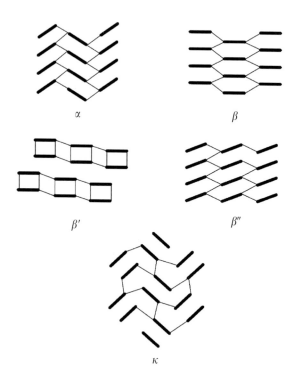

α

β

β'

β''

κ

Fig. 6.3 Schematic representation of the network of ET **6-IV** molecules in five phases of the salt $(ET)_2^+ I_3^-$. Thick lines indicate ET molecules, and thin lines indicate short intermolecular S$\cdots$S contacts. (A. Kini, personal communication.)

crystallographic inversion centre, and that the donor ET molecules all be parallel, as shown in Fig. 6.3. The structure is thus characterized by stacks along the diagonal of the unit cell, with strong intermolecular $S \cdots S$ interactions between stacks. T_c for this phase is 1.4 K. The κ phase may be obtained, concomitantly with the α and θ phases, in a tetrahydrofuran solution under N_2 with a mixed supporting electrolyte of $(n\text{-}C_4H_9)_4NI_3$ and $(n\text{-}C_4H_9)_4NAuI_2$ at 20 °C and constant current. It is characterized by the formation of dimers, crystallizing in a centrosymmetric crystal structure with one layer per unit cell in the $P2_1/c$ space group, and exhibiting a T_c at 3.6 K.

Understanding the structure–property relationship in these materials is crucial to the rational development of organic conductors and superconductors with increasingly higher T_c values. The plethora of polymorphic structures can easily lead the unwary investigator astray, but it provides an opportunity not available in many other systems for isolating the structural characteristics required for a very specific physical property. Williams *et al.* (1991) have also pointed out that the *isostructural* series of salts are important for the information they can yield. In this case, the *structural* parameter is kept nearly fixed (or only slightly perturbed) and the effect of *chemical* perturbations can then be evaluated. For instance, the ET salts tend to crystallize with the same structure, so the effect on the properties of varying the anion can be studied in a series of isostructural salts.

While polymorphic structures can reveal subtleties about the structure–property relationship, they do not necessarily provide the keys to this understanding, in particular when the structural changes are more subtle than their manifestations in particular physical properties. A recent study of the system of 1,6-diaminopyrene **6-V** and chloranil **6-VI** is edifying in this regard since it indicates both the utility and some of the limitations of studies of polymorphic systems. **6-V** is a relatively strong donor that was the subject of an early study as a component of potential electrically conducting complexes. (Kronick *et al.* 1961, 1964). Its complex with **6-VI** is polymorphic, with a dark brown α form and green β form. They may be obtained concomitantly (see Section 3.5) from benzene solution by both slow cooling and 'prolonged' slow evaporation (Inabe *et al.* 1996). Structurally, both forms are mixed-stack complexes, and on the basis of the discussion above might be expected to be electrical insulators. They do differ in crystal structure, however, as can be seen in Fig. 6.4. In the α form the stacks along the c axis are vertical, leading to a significant donor–acceptor overlap reminiscent of the earlier mentioned mixed-stack structure. In the β form the

6-V   6-VI

(a)

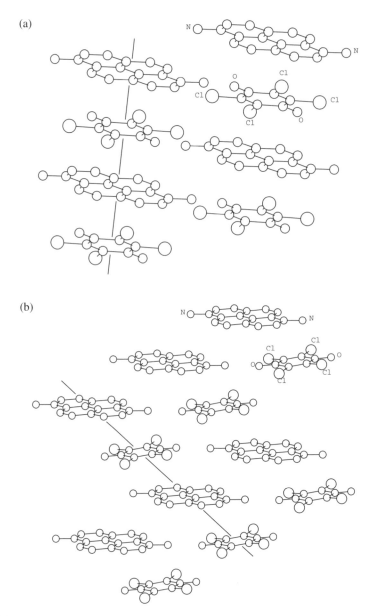

(b)

Fig. 6.4 Packing diagrams of the two forms of the complex formed by **6-V** and **6-VI**. (a) α form. The vertical mixed stacks are along the vertical c crystallographic axis. (b) β form. The mixed donor/acceptor stacks are along the diagonal indicated. The structure also exhibits donor/donor and acceptor/acceptor interactions in a direction approximately perpendicular to that of the mixed stacks. Nitrogen, oxygen and chlorine atoms are identified in representative molecules. (After Bernstein 1999, with permission.)

c-axis stacks are offset, leading to a reduction in the donor–acceptor overlap and an increase in both donor–donor and acceptor–acceptor overlap. Reiterating the meaning of the concomitant crystallization in practical terms, according to the authors the two polymorphs are nearly isoenergetic at room temperature.

Upon increasing the temperature of the α phase, the electrical conductivity rises by approximately seven orders of magnitude, although the onset of that increase varies from crystal to crystal. The same phenomenon is observed upon formation of a pellet prepared under pressure for conductivity measurements. In spite of the change in conductivity, the IR spectra are identical for the solids before and after the change. The full X-ray crystal structure determination of the highly conducting material indicates that no statistically significant structural change has taken place. Additional physical measurements (ESR, solid state CP/MAS NMR) on both the 'pristine' α form and the 'low resistance' α' form do not provide any clue as to the source of the vastly increased electrical conductivity (Goto *et al.* 1996). This system provides a poignant reminder that some physical phenomena are based on subtle or small perturbations which are not directly observable in structure determinations or are beyond the detection limits of other physical techniques. The closing statement of the authors that 'no ordinary conduction mechanism can rationalize the low resistivity value [of the low resistance α' form]' is a challenge to our ingenuity in measuring and interpreting physical phenomena.

6.2.1.1 *The neutral/ionic transition in charge-transfer complexes*

One particular phenomenon associated with mixed-stack charge-transfer complexes that has received considerable attention is the neutral-to-ionic (N–I) transition. As noted above, the vast majority of π charge-transfer complexes crystallize in the mixed-stack motif (Herbstein 1971). Except for cases in which the ionization potential of the donor is very low and the electron affinity of the acceptor is very high, the ground state of these complexes may be formally described as stacks of alternating neutral donors and acceptors. At the other extreme, in the formal definition of the ionic phase, each donor will have lost an electron and each acceptor will have gained one, and the structure will now consist of alternating anions and cations. In the range between these two extremes, there is a group of compounds forming mixed-stack complexes which can exist in both the ionic (I) and neutral (N) states, with transitions between them (Torrance *et al.* 1981; Jacobsen and Torrance 1983). In principle, this may also be considered a case of polymorphism, albeit with the definitional complication of whether the components are indeed the same for both phases. However, both dissolve to give the same solution, although the equilibrium distribution between ionic and neutral species may depend on the polarity of the solvent.

$$\cdots \mathbf{D^\circ A^\circ D^\circ A^\circ D^\circ A^\circ D^\circ A^\circ D^\circ A^\circ D^\circ A^\circ} \cdots \qquad \text{Neutral phase (N)}$$

$$\cdots \mathbf{(D^+A^-)(D^+A^-)(D^+A^-)(D^+A^-)(D^+A^-)} \cdots \qquad \text{Ionic phase (I)}$$

The prototypical N–I system is formed by tetrathafulvalene (TTF) **6-I** and chloranil (CA) **6-VI**, although others have also been reported (Katan and Koenig 1999). The

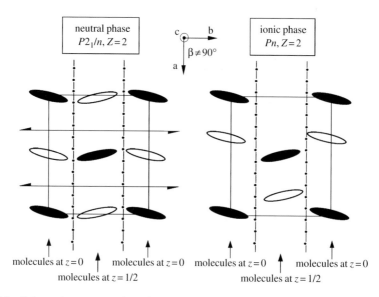

Fig. 6.5   Schematic representation of the structural features of the neutral and ionic phases of the complex of tetrathiafulvalene (TTF) **6-I** and chloranil (CA) **6-VI**. In both cases the view is on the *ab* plane. In the neutral phase the molecules lie on crystallographic inversion centres at (0,0,0) and (0, 1/2, 0), while in the ionic phase those points are no longer crystallographic inversion centres. (Based on Cailleau *et al.* 1997, with permission.)

nature of the structural change is represented in Fig. 6.5. At room temperature and atmospheric pressure, the neutral (monoclinic, $P2_1/n$) phase is stable. The asymmetric unit is composed of one half of each of the centrosymmetric donor and acceptor molecules. The crystal structure generated by the space group symmetry (with each molecule lying on a crystallographic inversion centre) leads to the mixed-stack structure with parallel stacks of regularly spaced alternating donor and acceptor molecules along the *a* crystallographic axis. Although formally represented as the neutral complex, the degree of charge transfer has been determined to be approximately 0.3 (Girlando *et al.* 1983; Jacobsen and Torrance 1983; Tokura *et al.* 1985). This means that over the space-averaged structure approximately 30 per cent of the TTF donors have lost electrons to the CA acceptors.

At 81 K, the first-order transition to the ionic state takes place leading to a degree of charge transfer of approximately 0.7, indicating that 30 per cent of the donors and acceptors are still neutral. The structural change involves the retention of the glide plane and virtual loss of the screw axis (with concomitant loss of the inversion centre), the space group being thus reduced to Pn (Le Cointe *et al.* 1995*a*), and the asymmetric unit being one full molecule each of TTF and CA. The lowered space group symmetry lifts the restrictions that led to the regular spacing and the structure is now characterized by pairs of molecules, rather than regular stacks, as indicated above by the parentheses in the schematic notation for the ionic structure. The change in the

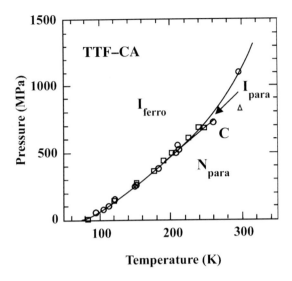

Fig. 6.6 (P,T) phase diagram of the TTF–CA system. Experimental points were obtained by a variety of methods, including neutron scattering $\bigcirc$, NQR $\square$, and conductivity $\triangle$. C indicates the estimated critical point. N_{para}, I_{para} and I_{ferro} indicate the neutral paraelectric, ionic paraelectric and ionic ferroelectric phases respectively. (From Cailleau *et al.* 1997, with permission.)

degree of charge transfer is also reflected in the molecular geometry of the molecular moieties (Le Cointe *et al.* 1995*a*) as well as in the ^{35}Cl Nuclear Quadrupole Resonance (NQR) response (Gallier *et al.* 1993; Le Cointe *et al.* 1995*b*).

The (P,T) phase diagram for the system has been prepared (Fig. 6.6) using a combination of data from neutron scattering, NQR, vibrational spectroscopy and conductivity measurements (Cailleau *et al.* 1997). The neutral phase is paraelectric, while the ionic phase has paraelectric and ferroelectric regions. The nature and mechanism of the phase transition has been reviewed and treated, as well as a number of other physical properties (Le Cointe *et al.* 1996; Cailleau *et al.* 1997).

6.2.2 Organic magnetic materials

Preparing magnetic materials from organic molecules is no less a challenge than preparing electrically conducting organic materials, although activity in this field started somewhat later and has not been quite as widespread. The subject has been reviewed recently (Kinoshita 1994; Miller and Epstein 1994, 1996; Miller 1998; Kahn *et al.* 1999), and activity has increased significantly in recent years. The study of magnetism has traditionally been closely but indirectly related to that of polymorphism, in particular with the changes in magnetic behaviour resulting from phase changes. The area has been dominated very much by physics and physical measurements and the emphasis has been phenomenological, focusing more on the temperature of phase transitions and the macroscopic nature of the changes in magnetic behaviour associated with those phase transitions, rather than on the structural basis for those changes or differences in behaviour. For examples of some studies on the magnetic properties of polymorphic transition metal complexes see, for instance, Boyd *et al.* (1981), Scheidt *et al.* (1983) and Decurtins *et al.* (1983). The development of molecular magnetic materials by chemists has expanded the field considerably to include an interest in the fundamental individual molecular basis for potential magnetic behaviour, as well as the geometric and spatial requirements of the crystalline solids comprised of those molecular bases for different magnetic behaviour (Kahn *et al.* 1999).

A design requirement of magnetism in molecular materials is the presence of unpaired electrons on the molecular species. The magnetic behaviour is then determined by the environment of the electrons and the nature of the interactions among them. Thus, polymorphic structures can provide particularly useful information about the changes in magnetic behaviour resulting from changes in molecular environment, and a number of case studies of magnetic behaviour of polymorphic materials demonstrate the utility as well as some of the problems inherent in such studies.

One of the few purely organic materials to have been shown to undergo a well characterized ferromagnetic transition is the azaadamantane derivative **6-VII** (Dromzee *et al.* 1996). The material is dimorphic, and both forms appear concomitantly by evaporation from a diethyl ether solution at room temperature, suggesting that they are thermodynamically nearly equivalent. In fact, the authors note that they have not succeeded in determining the conditions for selectively crystallizing either form. The monoclinic α form undergoes a ferromagnetic transition at 1.48 K (Chiarelli and Rassat 1991). Views of the α and β forms are shown in Fig. 6.7. It can be seen that there is a subtle difference in the environment of the oxygen atoms on which (at least formally) the unpaired electrons are located (as indicated in **6-VII**), and the crystal structure analysis suggests that the oxygen is disordered in the β phase. As yet, no detailed analysis of the structural basis for the magnetic behaviour has been given, although Miller (1998) has noted that in the monoclinic phase the intermolecular NO$^\bullet$ distances are shorter than in the orthorhombic phase. This is a system that can potentially provide additional useful information.

6-VII   6-VIII

(a)   (b)

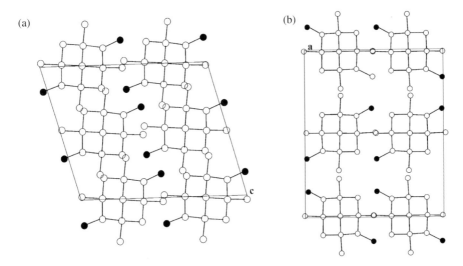

Fig. 6.7 Representation of the packing of the two forms of **6-VII**. Oxygen atoms are denoted by solid circles. (a) The monoclinic α form; (b) the orthorhombic β form. (From Dromzee 1996, with permission.)

The sensitivity of magnetic properties to changes in polymorphic structure is demonstrated in the trimorphic system of decamethylferrocene **6-VIII** with TCNQ **6-II**. Two forms were originally prepared by Miller *et al.* (1987). The 'thermodynamically favoured' form is prepared from an acetonitrile solution refrigerated at $-35\,^{\circ}$C for more than two weeks, or electrochemically generated in 24 h. It appears as purple plates which behave as a paramagnetic material. The crystal structure contains 'dimeric' units (Fig. 6.8) which are stoichiometrically represented as $[Fc^*]_2^{\bullet+}[TCNQ]_2^-$, $(Fc^* = Fe(C_5Me_5)_2)$ and form the building blocks of a classic herring-bone structure of a molecular crystal (Desiraju 1989). The 'kinetically favoured' phase is prepared from warm solutions of donor and acceptor in dichloroethane or acetonitrile, from which the salt rapidly precipitates, yielding air-sensitive chunky green crystals. In this case, the basic building block is a one-dimensional mixed-stack chain (hence referred to as the '1-D' phase by the authors) Fig. 6.8, and the crystals are metamagnetic. Form III was obtained by recrystallization of the complex from acetonitrile at $-20\,^{\circ}$C. It is also purple in colour but 'upon close inspection' (Broderick *et al.* 1995) is distinctly different from Form I, appearing as parallelepipeds rather than as plates. It is an air-sensitive ferromagnetic material, which crystallizes in one-dimensional chains as in Form II but exhibits a slightly different colour. There are a number of structural differences, though. The two Cp* groups within a donor are staggered in Form II, and eclipsed in Form III. Within a stack in Form II, all the acceptors are oriented in the same direction; in Form III they form a zigzag arrangement along the b axis. The interstack arrangements are also different.

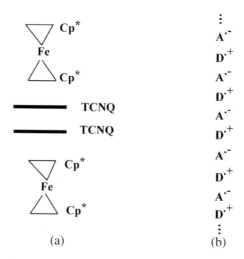

Fig. 6.8 (a) Schematic representation of the 'dimeric' units which are the structural build-ing blocks for the herringbone motif in the crystal structure of Form I of the complex $[Fc^*]_2^{\bullet+}[TCNQ]_2^{\bullet-}$. $C_p^* = C_5Me_5$ (b) schematic representation of the mixed stacks of anion and cation radicals in Forms II and III of the complex. (From Bernstein 1999, with permission.)

6-IX   6-X

A number of additional polymorphic molecular materials exhibiting variations in magnetic behaviour warrant mention here. The stable free radicals **6-IX** and **6-X** are known to be at least dimorphic (Banister *et al.* 1995, 1996; Palacio *et al.* 1997*a,b*) and tetramorphic respectively (Turek *et al.* 1991; Nakazawa *et al.* 1992). Both forms of **6-IX** may be obtained by sublimation, the thermodynamically stable β form being produced slowly, while the triclinic α form is generated via a more rapid sublimation. They are conformational polymorphs (Section 5.4) differing in the dihedral angle between the rings ($\alpha = 32.2°$; $\beta = 57.8°$) (Banister *et al.* 1995, 1996). Both struc-tures are characterized by the formation of head-to-tail chains, all parallel in the β structure, but antiparallel in the α structure. The former undergoes weak ferromag-netic ordering below 35.5 K, whereas the latter orders antiferromagnetically below 8 K, and attempts have been made to attribute these differences in behaviour to the structural features distinguishing the two modifications.

The four forms of **6-X** have very different crystal structures (Miller 1998), and correspondingly different magnetic properties. At room temperature, the thermodynamically stable orthorhombic β form is obtained by slow evaporation from benzene at 20 °C. Other solvents and crystallization temperatures are used to generate the three additional forms (α: 1,1,2-trichloroethane at ~ 30 °C; γ: acetonitrile or chloroform at 65 ± 5 °C; δ: chlorobenzene at 132 °C). The α- and δ-phases are paramagnetic (Allemand *et al.* 1991; Tamura *et al.* 1993), while the γ-phase orders antiferromagnetically at $T_N = 0.65$ K (Kinoshita 1994) and the β-phase undergoes a ferromagnetic transition $T_c = 0.6$ K (Tamura *et al.* 1991). Three polymorphs (α, β, and δ) of the compound have been prepared as a thin film (Fraxedas *et al.* 1999a) and the conditions and mechanism of the transition of the α to the β form have recently been studied in some detail (Fraxedas *et al.* 1999b).

The closely related molecule **6-XI** has been shown to be dimorphic (Sugawara *et al.* 1994). Both forms can be crystallized from diethyl ether, but at different temperatures: the bluish-purple blocks of the α phase above 4 °C, and the blue needles of the β phase below 0 °C. Selective crystallization can be accomplished by seeding. The α modification, structurally characterized by both intramolecular and intermolecular hydrogen bonds, is a three-dimensional ferromagnet. On the other hand, the β form exhibits intermolecular hydrogen bonds and π–π stacking, and its magnetic behaviour was interpreted by the ferromagnetic ST model, which was accompanied at lower temperatures by a weak antiferromagnetic interaction (Matsushita *et al.* 1997). The authors concluded that the hydrogen bond is not only a structural feature, but also plays a role in the transmission of spin polarization between molecules. The structures and properties of different polymorphic forms, as well as the role and potential use of hydrogen bonds in generating structures of these (and other) materials as organic magnets was reviewed by Veciana *et al.* (1996).

6-XI

6-XII

6-XIII

Two additional dimorphic molecular systems that have been studied recently for their magnetic behaviour include the copper complex [Cu(cyclam)(TCNQ)$_2$]TCNQ **6-XII** (Ballester *et al.* 1997) and the gold complex **6-XIII** (Leznoff *et al.* 1999). The common structural feature in the former is the presence of parallel chains of copper macrocyclic units linked by dimeric (TCNQ)$_2^{2-}$ units which are coordinated to the metal. The chains are connected by neutral TCNQ units. In one polymorph the nature of the connection leads to a two-dimensional network while in the second modification a three-dimensional network is generated. The magnetic behaviour of the two forms is similar, but not identical. In both there is strong antiferromagnetic coupling between the organic radicals, and the dimeric (TCNQ)$_2^{2-}$ anion behaves diamagnetically below room temperature. On the basis of analysis of the distance between neutral and anionic TCNQ species (~ 0.2 Å shorter in Form 2 than in Form 1) the authors suggest that the formally neutral TCNQ is less so (i.e. more changed) in the former than in the latter, indicating more delocalization and hence accounting for a lower antiferromagnetic coupling between radicals in accord with the magnetic susceptibility measurements.

In the gold complex **6-XIII**, there are differences in the packing modes which are particularly manifested in the intermolecular contacts involving the radical-bearing aminoxyl N-oxide fragments. This leads to differences in the (N–O)$\cdots$(N–O) interactions between the two structures (α and β) which results in differences in the magnetic behaviour. Both exhibit intermolecular antiferromagnetic interactions. The β modification shows no maximum in that behaviour, the magnetic susceptibility rising with decreasing temperature. On the other hand, the α form goes through a maximum in the magnetic susceptibility at 3.5 K, a behavioural effect which could be successfully modelled based on the observed crystal structure.

A number of other systems involving magnetic properties of organic polymorphs are worthy of brief note here, to demonstrate the variety of issues that have been raised, if not always resolved, and the kinds of opportunities that polymorphic systems provide for investigating these systems.

Compound **6-XIV** (tanane) has a high temperature tetragonal phase that transforms at around 14 °C to an orthorhombic phase, which is both ferroelastic and ferroelectric, and a monoclinic phase has also been reported and prepared (Capiomont *et al.* 1972; Jang *et al.* 1980). Structural data (Capiomont *et al.* 1981) were used to model the transition (Legrand *et al.* 1982).

6-XIV   6-XV

Banister *et al.* (1996) prepared an organic free radical **6-XV** not based on NO. It is concomitantly dimorphic by sublimation, and varying the conditions can lead to either the centrosymmetric $P\bar{1}$ α form (Banister *et al.* 1995) or the non-centrosymmetric

Fdd2 β form (Banister *et al.* 1996). The result is that in the α structure neighbouring chains are antiparallel, while in β they are all parallel. Due to these symmetry considerations the α modification cannot exhibit a so-called 'spin-canting' mechanism (Carlin 1989) which can account for the transition from a low-dimensional antiferromagnet to a weakly ferromagnetic state. However, these restrictions are absent in the β modification, and this mechanism, based to a large extent on the structural distinction between the two polymorphs, has been used to account for the fact that this β form is the first example of an open-shell molecule to exhibit spontaneous magnetization above liquid helium temperature (36 K).

Finally, the stable free radical, diphenylpicrylhydrazyl **6-XVI**, is the classic reference standard for ESR measurements, with solid samples showing stability over as long as thirty years (Yordanov and Christova 1997). Despite the common use of this material and a number of crystallographic studies, the polymorphic and solvate behaviour apparently still have not been fully characterized. A variety of crystal modifications, both solvates and solvent free, have been reported (Weil and Anderson 1965). Among the solvent-free modifications were included an amorphous form with a melting point (*sic*) of 137 °C, an orthorhombic form with a melting point of 106 °C (Williams 1965) and a triclinic form with melting point 128–129 °C (Williams 1965). Williams (1967) reported the structure determination of a benzene solvate, which was repeated by Boucherle *et al.* in 1987. The structure reported by Kiers *et al.* (1976) was originally thought to be the triclinic form of Williams (1965) but turned out to be monoclinic (*Pc*) and to be an acetone solvate. A monoclinic form has been reported by Ellison and Holmber (1960) and another modification, with similar but apparently not identical cell constants and unidentified space group has been studied by Prokop'eva and Davidov (1975). The solid state magnetic properties, presumably of the orthorhombic form, were studied by two groups (Burg *et al.* 1982; Boon and Vangerven 1992) but the apparent structural variety and the variation of magnetic properties with crystal structure still have not been fully elucidated.

6-XVI

Also worthy of note here is the fact that both a yellow and an orange modification of a 2,2-di(*p*-nitrophenyl) hydrazine derivative of **6-XVI** crystallize concomitantly from dichloromethane, with very similar molecular conformations (Hong *et al.* 1991). The colour difference (*vide infra*) has been attributed to a difference in molecular packing.

Which of these structural factors, if any, determine the differences in magnetic properties? What kinds of experiments can be designed to make these distinctions? Are there additional polymorphs that reveal the subtleties of structural differences and physical properties? There are many questions, challenges and opportunities for discovery and understanding in systems of this type that exhibit distinguishable structures and properties, and it is clear that the growing awareness of the valuable information gained by studying polymorphic materials will lead to increasing study and understanding of these systems.

6.2.3   *Photovoltaicity and photoconductivity*

The oil crisis of the middle 1970s generated a great deal of activity in the search for alternate energy sources, including organic photovoltaic and photoconducting materials. That activity slackened off during the 1980s, but the development of new technologies to generate and structurally characterize materials, surfaces and structures of bulk and monomolecular thick films, often at the atomic level, has led to a resurgence of activity in this field (Law 1993), especially in the evolution of advanced photocopier and laser printer technology. While much direct information is becoming available, as with any correlation of structure and properties, the information provided by polymorphic systems can still prove very useful. Many of the materials studied for photovoltaic or photoconductive properties are also used as pigments and dyes, and the more general discussion of the polymorphic behaviour of that class of compounds is found in Chapter 8. We will concentrate here on those two related photophysical phenomena.

Much of the early and continuing activity on photovoltaic materials was centered about squarylium **6-XVII** and cyanine dyes **6-XVIII** (Merritt 1978; Morel *et al.* 1984; Law 1993), but there has been continuing and intensive work on other promising classes of candidate materials such as azo pigments and perylene pigments (Borsenberger and Weiss 1993) as well as a variety of other more specific molecular systems (Youming *et al.* 1986; Law *et al.* 1994). In spite of the increasing interest in these two photoeffects and the effort that has been made in developing and studying these materials, there have been few direct studies of photovoltaic or photoconducting behaviour as a function of polymorphism. This is in part due to the lack of proclivity of many these materials to crystallize at all, to say nothing of crystallizing as polymorphs, so that relatively few crystal structures have been determined. Also, since the potential applications would involve nanodimensional devices, much of the effort in the study of the materials has been concentrated on films, composites, etc.,

6-XVII   6-XVIII

although there is also evidence for multiplicity of structural motifs in these media as well (Ashwell *et al.* 1997; Wojtyk *et al.* 1999; Stawasz *et al.* 2000).

Nevertheless among the well-studied photoconductive molecular materials, polymorphism is most prevalent in the phthalocyanines, perhaps reflecting McCrone's (1965) adage about the relationship between the time and energy spent on the study of a compound and the number of polymorphs known. There are also some examples from the squaraines, apparently a few instances among the perylenes, but as yet apparently no reported examples among the photoconducting azo dyes. This situation is bound to change as intensive work continues in this field.

In Section 3.5.3 we discussed the structural aspects of the dimorphic green and violet forms of the concomitant crystallizing squaryllium dye **6-XVII (R = C$_2$H$_5$, R' = OH)** (Bernstein and Goldstein 1988). For another derivative **6-XVII (R = C$_4$H$_9$, R' = H)**, Ashwell *et al.* (1996) have reported the crystal structures of two polymorphs, which coincidentally are also green and violet. There are no studies of the photoconductivity of either of these structurally characterized polymorphic systems, although Tristani-Kendra and Eckhardt (1984) did carry out a thorough spectroscopic investigation of the nature and manifestation of the intermolecular coupling on the solid state optical spectra of the two forms of the former. These investigations also provide information on the important question of the direct structural relationship between dye aggregates in solution and macroscopic crystals, a question that has been addressed by Marchetti in a number of spectroscopic studies (Marchetti *et al.* 1976; Young *et al.* 1989).

Law (1993) has reviewed the widespread polymorphism in phthalocyanines and its relationship to their photoconductivity and use in xerographic applications. These are best demonstrated by the prototypical copper phthalocyanine **CuPc (6-XIX, M = Cu)**, whose structural chemistry is discussed in detail in Section 8.3.3.1. The polarized absorption spectra of five forms are given in Fig. 6.9. One outstanding feature of these spectra is the rather intense red shifted band at ~770 nm in the δ and ε modifications compared to the solution absorption at ~678 nm (Law 1993). This

6-XIX

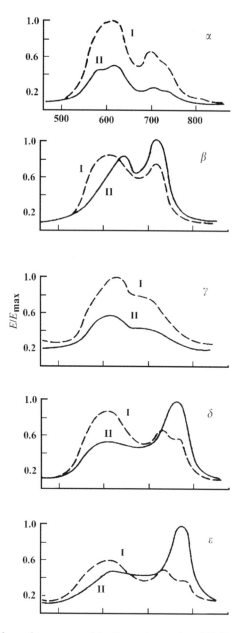

Fig. 6.9 Polarized absorption spectra of the five polymorphs of **CuPc**. The directions of the polarized light are perpendicular (I, - - -) and parallel (II, —) to the long axis of the pigment aggregates. (Adapted from Sappock 1978, with permission.)

band has been interpreted as arising from an intermolecular charge-transfer interaction, the latter being required for high photoconductivity. In fact, it has been reported that the xerographic performance of the ε form is superior to the other crystalline modifications (Yagishita *et al.* 1984; Enokida and Hirohashi 1992).

Law (1993) has pointed out the difficulties in comparing the photoconductivity data of different compounds and even of different polymorphs. Many of these difficulties arise from the variations in processing procedures and conditions in various laboratories, and the possibility of polymorphic changes during processing. Moreover, the direct relationship between the crystal structure and the photoconductivity has not been firmly established, because of difficulties caused by impurities, crystal size, etc. Nevertheless, the importance of that relationship is clearly recognized, and the results on photoconductive polymorphic phthalocyanines that have been intensively studied for their xerographic response are **6-XIX (M = 2H)**, and **6-XIX (M = TiO)**, and **6-XIX (M = VO)** (and derivatives thereof) led Law (1993) to state that 'the precise stacking arrangement is important. There are plenty of examples to support this view indirectly', among them the data cited above for **CuPc**—namely, the common feature of all photosensitive phthalocyanines is the presence of the red shifted absorption, found in the aggregate and absent in solution. The structural conclusions are that the photoconductivity is indeed a solid state phenomenon requiring strong intermolecular interactions, short (~ 3.5 Å) interplanar distances and large charge mobility upon excitation (i.e. the presence of charge-transfer states). There is clearly much still be learned here, which can be revealed by combined structural and spectroscopic studies on the various polymorphs.

6.2.4 *Nonlinear optical activity and second harmonic generation*

The transformation of data transmission systems from analogue to digital means, more specifically to optically based digital systems, has led to intense interest in the development of optical switching devices. Molecular materials comprise a promising group of potential candidate systems for those switching devices (Kolinsky 1992) and much work has been done over the past two decades in preparing and characterizing potentially useful compounds for these applications (Munn and Ironside 1993; Zyss 1994).

The principal structural requirement for second order nonlinear effects in assemblies of molecules is the lack of a centre of symmetry, and considerable efforts have been expended in trying to induce potentially useful molecular entities to crystallize in non-centrosymmetric or polar crystals (Curtin and Paul 1981; Etter *et al.* 1991). As demonstrated below, this is a necessary, but not sufficient condition for obtaining nonlinear effects. True to form, the variety of crystallization experiments has led to a number of polymorphic structures, and to information about the relationship between the properties of these materials and their structures, as well as useful guidelines for attempting to obtain the desired non-centrosymmetric crystal structures.

Perhaps the seminal work in this regard is that by Hall *et al.* (1986). This study summarizes work on three widely studied molecules whose crystals have been shown to have nonlinear optically active molecules **6-XX–6-XXII**, acronymically designated as PAN, FNBH and MNP, respectively, and demonstrates the importance of space group symmetry as well as the details of packing in determining the nonlinear optical activity of molecular crystals. The first two are known to be dimorphic while four polymorphs have been reported for the third.[1] The two known forms of PAN crystallize in the non-centrosymmetric space group $P2_1$. Both are grown from solutions of water and absolute alcohol. The one which does not exhibit activity may be obtained with platelike morphology by 'equilibrium' methods such as programmed decrease of solution temperature or evaporation. The form exhibiting activity may be obtained as dendritic needle-like crystals by 'non-equilibrium' methods, for instance rapid crystallization from a highly supersaturated solution. In general, dendritic growth often results from kinetic crystallization, since many nuclei are generated, and the high degree of supersaturation means that growth continues from them essentially unabated. Under less kinetic conditions, the larger crystals grow at the expense of the smaller ones. In fact, it was noted in this system that upon standing, solutions containing the nonlinear optically active dendritic crystals tended to convert to the inactive plate-like crystals, consistent with the relative stability implied by the growth conditions. Such a solvent-assisted conversion process also indicates that the inactive form is more stable. The crystal structures of the two forms are compared in Fig. 6.10.

PAN
6-XX

FNBH
6-XXI

MNP
6-XXII

It can be seen that in both forms the molecular dipole moment has a similar projection on the polar b axis. In the active form, the molecular dipole vectors of the two molecules in the unit cell are oriented so that they add to give a net component along the b axis, while the pseudocentrosymmetric arrangement of the molecular dipoles in the inactive form (albeit in the non-centrosymmetric space group) leads to an effective cancellation of the components along that axis. Note that the inactive form contains two pseudocentrosymmetrically related molecules in the asymmetric unit, leading to an overall Z-value of 4. This is the crystallographic situation that permits the cancellation of the dipoles. With one molecule per asymmetric unit in the active form, the screw axis must generate a net dipole moment. Using the structural

[1] The crystallographic data for these polymorphic structures do not yet appear in the CSD, except for the inactive form of FNBH (CSD Refcode SEFBEK) (Aldoshin *et al.* 1988). We are grateful to Prof. M. Hursthouse (Southampton) for providing those data.

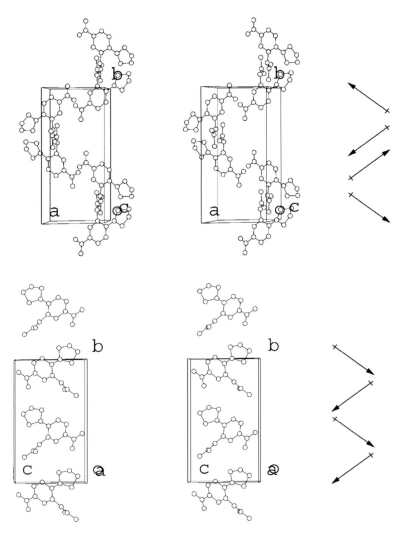

Fig. 6.10 Stereo views of the two forms of **6-XX** (PAN). Both views are oriented so that the *b* crystallographic axis is vertical. On the right approximate orientations for the molecular dipole moments are also shown (see text). Lower, crystal form exhibiting nonlinear optical activity; upper, inactive form. For clarity, some molecules in the unit cell, related by symmetry to those shown, have not been included.

information and knowledge of the crystal faces that dominate the non-polar modi-fication, Popovitz-Biro *et al.* (1991) designed a polymeric crystallization inhibitor for that form, resulting in the preferential crystallization of the polar polymorph (see Section 3.7).

In the case of **6-XXI** (FNBH) different crystallization conditions from absolute alcohol led to the two forms, equilibrium conditions leading to crystals of high optical quality of the nonlinear optically inactive form, while a kinetic crystallization from highly supersaturated solution led to poor quality crystals of the active form. Here, the space group symmetries clearly distinguish between the two forms. The inactive form crystallizes in the centrosymmetric space group $P2_1/c$, with eight molecules in the unit cell, the asymmetric unit being composed of an approximately centrosymmetrically related pair of crystallographically independent molecules, as shown in Fig. 6.11. On the other hand the active form is in (the rarely encountered, at least for molecular crystals) space group $I4_1cd$, in which the c axis is the polar direction (Fig. 6.11).

In **6-XXII** (MNP) the four different forms may be obtained from different solvents under different rates of cooling and evaporation. The space groups for the two forms for which crystal structures have been reported (Forms 2 and 3) are again both noncentrosymmetric monoclinic $P2_1$. The needles of Form 2 obtained by rapid cooling

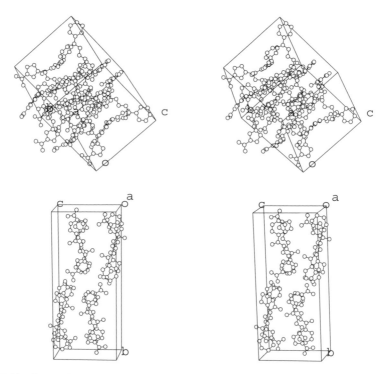

Fig. 6.11   Stereo views of the two forms of **6-XXI** (FNBH), upper, tetragonal form exhibiting nonlinear optical activity; lower, inactive monoclinic form, showing the presence of centrosymmetrically related dimers of crystallographically independent molecules. The tetragonal structure appears particularly complex because of the presence of 16 molecules in the unit cell, all of which are shown. When viewed in stereo individual molecules, and the relationships between them, may be discerned.

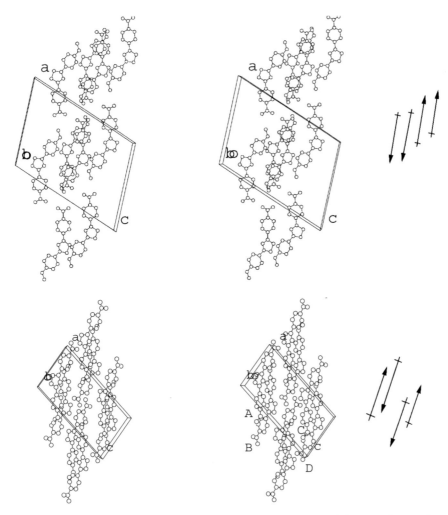

Fig. 6.12   Stereo views of the two forms of **6-XXII** (MNP). On the right approximate orientations for the molecular dipole moments are also shown (see text). Note that in both cases the molecular dipole moments are nearly perpendicular to the polar *b* axis, which is perpendicular to the plane of the paper. Top, Form 2, with two molecules in the asymmetric unit; bottom, Form 3, with four molecules in the asymmetric unit, designated by capital letters A, B, C, D. A and B are pseudocentrosymmetrically related, as are C and D.

are unstable in the presence of Form 3. The structures of both forms are shown in Fig. 6.12. In Form 2, the asymmetric unit is again a pair of crystallographically inequivalent but pseudocentrosymmetrically related molecules. In Form 3, there are two such pairs. In both cases all the molecular dipoles are approximately perpendicular to the polar *b* axis, so that they essentially cancel upon the action of the twofold screw parallel to this axis. The necessary condition of a non-centrosymmetric space group is met, but the sufficiency condition of favourable orientation of molecular dipoles is not satisfied and both cases lead to a near cancellation of optical coefficients and a very small nonlinear effect.

These three examples clearly demonstrate the physical principle of the necessity for a polar structure to obtain nonlinear optical effects, the difficulty in obtaining that polar structure, and possible ways to overcome those difficulties and the utility of studying polymorphic structures to establish these principles. The crystallographically centrosymmetric arrangement is clearly preferred under equilibrium conditions, but even under kinetic conditions molecules apparently tend to arrange themselves with spatially opposing dipoles. The not-necessarily foolproof recipe for obtaining nonlinear optical activity would therefore strongly suggest the use of kinetic conditions of crystallization. There have also been recent attempts to associate particular molecular conformations with the tendency to pack in non-centrosymmetric structures (Huang *et al.* 1996).

The recognition of these structural features associated with nonlinear optical activity in fact led to the discovery of a new polymorph of *m*-nitrophenol **6-XXIII**. The description of the material appears in Groth's compendium (1919), indicating two melting points, but the interfacial angles determined by Barker (1908) and Steinmetz (1915) were consistent, not hinting at polymorphism. However, in a 1934 determination of the index of refraction, Davies and Hartshorne noted 'occasional individuals (of crystals) with more or less rhombic outlines' amid monoclinic prismatic crystals. In 1972, Shigorin and Shipulo reported that the compound exhibited a strong second harmonic generation, but Groth's morphological description indicated a centrosymmetric space group. In an attempt to resolve this discrepancy, Pandarese *et al.* (1975) carried out the crystal structure determination of the material described by Groth and others from the melt and from benzene. The crystals were indeed centrosymmetric in space group $P2_1/n$ leaving the nonlinear optical activity unaccounted for.

6-XXIII

With the dilemma not resolved, Pandarese *et al.* carefully reexamined many individual crystals from the benzene-grown batch, and discovered that approximately 20 per cent of them belonged to the non-centrosymmetric orthorhombic space group $P2_12_12_1$, thus explaining the source of the nonlinear effects. A method for purifying *m*-nitrophenol and growing crystals was developed by Wojcik and Marqueton in 1989 and the structure of the non-centrosymmetric orthorhombic form was published by Hamzaoui *et al.* in 1996.

Subsequent to the report by Hall *et al.* (1986), a number of other groups have utilized the information on the kinetic vs thermodynamic crystallizations in polymorphic potential nonlinear optical materials to generate active materials. Black *et al.* (1993)

utilized selective anisotropic solvent adsorption on specific crystal faces to favour the growth of morphologically polar crystals. Some additional reports of the study of crystal modification and nonlinear optical activity include those on anhydrous and hydrated sodium *p*-nitrophenolate (Brahadeeswaran *et al.* 1999), derivatives of 2-benzylideneindan-1,3-dione (Matsushima *et al.* 1992), straight-chain carbamyl compounds (Francis and Tiers 1992), benzophenone derivatives (Terao *et al.* 1990), a 1,3-dithiole derivative (Nakatsu *et al.* 1990), α-[(4′-methoxyphenyl)methylene]-4-nitro-benzene-acetonitrile (Oliver *et al.* 1990) and so-called 'lambda shaped molecules' (Yamamoto *et al.* 1992). Hall *et al.* (1988) followed the thermal conversion of the centrosymmetric ($P2_1/c$) form of 2,3-dichloroquinazirin to the non-centrosymmetric Pc form by monitoring the development of an SHG signal. Consistent with the earlier observation, the centrosymmetric form was obtained under equilibrium conditions, while the non-centrosymmetric one could be obtained under more kinetic conditions.

A change of solvent of crystallization may also generate polymorphs with different SHG responses, as observed in 2-adamantylamino-5-nitropyrimine (Antipin *et al.* 2001) and in the trimorphic 4-methoxy-4′-nitrotolane system (Tabei *et al.* 1987). In one case at least, it has been demonstrated that the noncentrosymmetric polymorph is preferred in crystallizations from non-polar solvents, while the centrosymmetric polymorph results from crystallizations from polar solvents (Sharma and Radhakrishnan 2000). In another case, a centrosymmetric and a non-centrosymmetric form were found to crystallize concomitantly (Timofeeva *et al.* 2000). The two forms have almost identical layer structures differing in the superposition of successive layers; hence they may be considered molecular polytypes (Verma and Krishna 1966). Serbutoviez *et al.* (1994) also demonstrated for two molecules that the addition of one equivalent of pyrrolidine to 2 : 1 ethanol/water solvent mixtures could also be used to generate SHG active crystal modifications in zwitterionic substances. In the case of **6-XXIV** (R = SCH₃) two polymorphic monohydrates were obtained with and without the addition of pyrrolidine; upon heating, both converted to the same inactive anhydrate. For **6-XXIV** (R = H), the monohydrate obtained in the absence of pyrrolidine was inactive, the dihydrate obtained from the addition of pyrrolidine was active, and again, both converted to an inactive anhydrate upon heating.

6-XXIV

6.2.5 *Chromisomerism, photochromism, thermochromism, mechanochromism, etc.*

Changes in the colour of a substance resulting from perturbations which result in changes in structure or environment—so-called 'chromogenic effects' (Nassau

1983)—have fascinated chemists for centuries (Kahr and McBride 1992). The observation that solutions of a pure substance could lead to concomitantly crystallizing crystals of different colour led chemists in the first two decades of the twentieth century to coin the term *chromoisomerism* (Hantszch 1907*a*,*b*, 1908; Toma *et al.* 1994). The crystals may undergo changes in colour due to exposure to light (photochromism), to changes in temperature (thermochromism) or to mechanical shock (mechanochromism). In some cases, the structural changes associated with the colour change are at the molecular level, while in other cases they may be due to subtle or significant changes in the crystal structure; a combination of molecular and bulk structural changes is also possible (e.g. Reetz *et al.* 1994; Reed *et al.* 2000). If the molecular integrity is maintained, then such changes involve polymorphic systems. When the chemical nature of the molecular entity changes, one is again faced with the definitional difficulties of polymorphism (see Section 1.2). With those caveats in mind we describe a number of systems that demonstrate some of the principles and challenges in these systems involving changes in colour. Many other systems described in this book exhibit colour differences among polymorphs—in fact, that is often the reason for the recognition of the polymorphism—and those colour differences are often noted, although the reasons for the differences in colour are not further investigated or specifically described. In view of the fact that for some of the chromogenic phenomena thousands of chemical systems have been recognized and studied, the coverage here can serve only as a brief introduction to the important connection between these effects and polymorphism. We will concentrate here on systems chosen to study the specific physical phenomenon and, where possible, the attempts to relate the consequences of the physical perturbation to changes in structure and colour. All of these involve monomolecular species or salts but they serve equally well as prototypes for polymorphic polymeric systems (e.g. Chumwachirasiri *et al.* 2000).

6.2.5.1   *Chromoisomerism*

The rather thorough study by Toma *et al.* (1994) of the red and green polymorphs of 9-phenylacridinium hydrogen sulfate **6-XXV** demonstrates many aspects of the current state of the art in studying these closely related phenomena (along with an historical summary of the debate surrounding the origin of chromoisomerism). The two forms of **6-XXV** crystallize concomitantly, although selective crystallization may be accomplished by appropriate seeding; they do not undergo thermal interconversion. The molecular structures found in the two modifications are essentially identical, with the torsion angle of the exocyclic phenyl ring differing by one degree, also in agreement with the geometry expected from computed molecular energetics. The hydrogen-bonding patterns are also identical ($\mathbf{D}_3^3(9)$ $\mathbf{R}_2^2(8)$, see Section 2.4.2.2.1) (involving centrosymmetric dimeric hydrogen sulphate units flanked by (also centrosymmetrically related) phenylacridinium units, although the packing of these units is quite different (Fig. 6.13). In spite of the fact that this was a rather thorough study, it was not possible to determine the physical basis for the difference in colour between the two polymorphs.

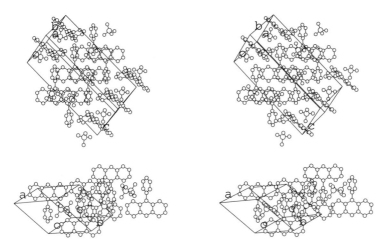

Fig. 6.13 Stereoviews of the red (upper) and green (lower) polymorphs of **6-XXII**. In order to facilitate comparison, in both cases the view is on the best plane of a centrally located anthracene moiety. Hydrogens have been deleted for clarity.

6-XXV

6-XXVI

Toma *et al.* refer also to the studies of **6-XXVI** (Byrn *et al.* 1972; Yang *et al.* 1989; Richardson *et al.* 1990) as an example of a case in which the source of differences in colour were explained. While these systems were both very thoroughly studied and structural changes were *correlated* with differences in colour (for another excellent example, see, for instance, Desiraju *et al.* 1977), the *origin* of the differences in colour, which is based on the details of the electronic structure, has for all intents and purposes remained elusive in all of the systems studied. One notable exception is indeed the case of **6-XXVI** in which the colour change could be associated with a proton transfer (Richardson *et al.* 1990).

There may be a number of reasons for the difficulties in determining the origin of colour differences. First, the colour of a crystal actually observed by a human being results from a variety of optical phenomena (scattering, absorption, reflection, etc.) (Nassau 1983), which are not generally simultaneously measured by any particular analytical optical or spectroscopic technique. Also, as noted for magnetic phenomena in Section 6.2.2, the origins of the changes in colour may be beyond the detection limits of the techniques currently in use. The human eye is a rather sensitive detector of colour and variations in colour. While we can quite confidently determine differences in structure to less than one part per 100 (generally better than 0.01 Å for a nominal bond length of 1.5 Å) by a variety of analytical methods, the small differences in extinction coefficient or subtle changes in the overlap of absorption bands can lead to optically recognizable changes in shade or colour which are often difficult to quantify experimentally.

6.2.5.2   Photochromism

The photochromic process involves the formation, destruction or change of colour due to light. The fact that such phenomena can be useful for optical devices information storage has made photochromism a subject of considerable interest (Bouas Laurent and Durr 2001; Maeda 1991; Weber *et al.* 1998).

Compound **6-XXVII** can be obtained from xylene, toluene or benzene as 'superb, long, very dark blue-black needles with metallic lustre'. A 'brilliant scarlet form' can be obtained from polar solvents such as acetic acid, DMF or DMSO. The black form can be converted to the red form both photophysically and thermally (abruptly at 139–142°C) (Begley *et al.* 1981). The two structures are shown in Fig. 6.14. In the black form, the molecules form infinite stacks ('in the manner of roof tiles') in the *ac* plane with intermolecular contacts of 3.31 Å which the authors attribute to charge-transfer interactions. The stacking is absent in the scarlet form, and the molecular conformation is considerably more non-planar, making this pair of structures also conformational polymorphs. No mechanism for the conversion or colour difference has been proposed.

6-XXVII

In another example, the photochromic transformation was carried out in solution and the starting material and product were investigated in the solid in order to establish the structural nature of the process (Burns *et al.* 1988). The compound **6-XXVIII** is a formazan derivative that can adopt a number of conformations. The orange, light-stable form crystallizes from 1 : 1 ethanol–water, while the red form is obtained by dissolving the orange form in the dark and evaporating the solvent under vacuum. The crystal structures, Fig. 6.15, clearly show that in the red form, the molecule adopts the *syn,* s-*trans* **6-XXVIIIa** conformation, while in the light stable orange form

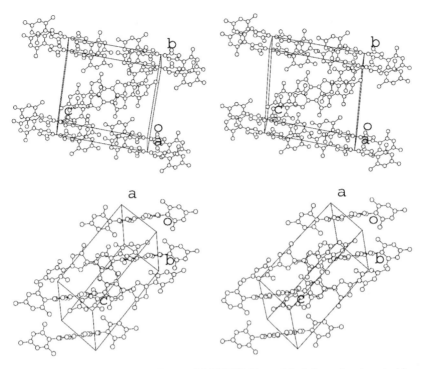

Fig. 6.14 Stereoviews of the two forms of **6-XXVII**. Upper, black form showing stacking of neighbouring molecules. Lower, scarlet form. In both structures the view is on the best plane of the essentially planar central portion of the reference molecule near the middle of the diagram, shown with its surroundings.

exhibits the *anti, s-trans* conformation **6-XXVIIIb** is adopted. Thus, this could also be considered an example of conformational polymorphism, although the solution composition of isomers apparently depends on the light level. The source of the difference in *solid state* colour has not been investigated.

6-XXVIIIa   6-XXVIIIb

6.2.5.3 *Thermochromism*

Thermochromism has been widely studied, and Nassau noted even in 1983 the existence of several thousand examples. Thermal transformations of molecular crystals have also been studied in some detail (Paul and Curtin 1973), and many of these are

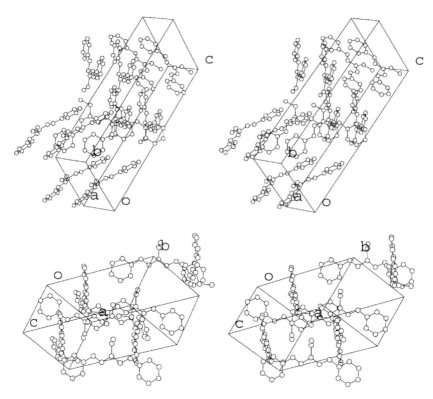

Fig. 6.15 Stereo packing diagrams of the red form (upper) and the light stable orange form (lower) of **6-XXVIII**. Both structures are plotted on the same reference plane which is the N–C(Et) = N atomic triplet in a molecule located near the centre of the cell.

accompanied by changes in colour. The salicylideneanilines (Cohen *et al.* 1964*a–c*) and the substituted salicylideneamines (Carles *et al.* 1987) have provided a source of a variety of compounds, exhibiting both photochromic and thermochromic crystal modifications.

Reetz *et al.* (1994) reported the preparation of concomitant polymorphs of the 4,4′-bypyridium salt of squaric acid **6-XXIX**. The major product is a monoclinic ochre modification, while 'a few individual crystals' of a slightly darker triclinic form were also isolated and characterized. Upon heating to 180 °C the monoclinic form undergoes a single-crystal to single-crystal thermochromic change to a red crystal. The change takes place via a reaction front that moves rapidly (i.e. a few seconds) along the needle axis, which corresponds to a short (~3.8 Å) crystallographic axis in the starting material. The transformation is reversible, but is accompanied by hysteresis, occurring in the opposite direction only at 150 °C. Despite the fact that the process appears reversible there are some subtle changes that occur upon cycling. Upon returning to the low-temperature ochre modification the relative intensities of the UV maximum (at 240 nm) and the charge-transfer band (at 390 nm) are different from the starting material. The locations of the lines in the X-ray powder diffraction pattern (reflecting the unit cell dimensions) are not perturbed, but the intensities (reflecting the unit cell contents and orientation) are altered. The same effects were observed when the transformation was carried out under hydrostatic pressure. The colour change has been attributed to a proton transfer mechanism along the chains of the interleaved bipyridinium and squaric acid moieties.

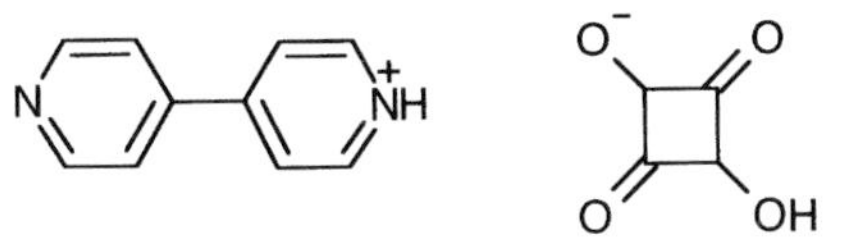

6-XXIX

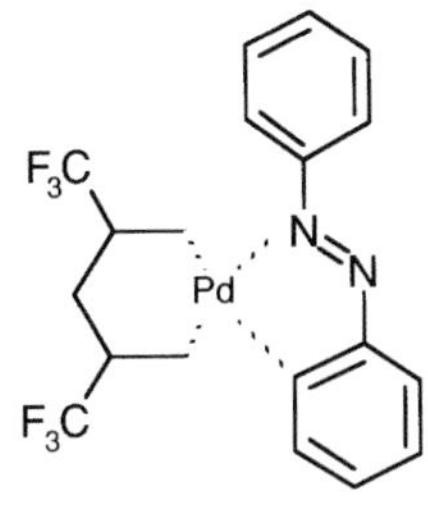

6-XXX

The thermochromic transition of an organometallic compound **6-XXX**, reported by Etter and Siedle (1983), exhibits a number of additional noteworthy features and riddles. Yellow triclinic needles obtained from hexane expand suddenly along the needle axis when heated to $90 \pm 10\,^\circ$C. If they are heated only on one face, they exhibit the thermosalient (i.e. 'hopping crystal') effect (Steiner *et al.* 1993; Zamir *et al.* 1994; Lieberman *et al.* 2000) (Section 6.2.6). X-ray photographs indicate that the original lattice is maintained at this point, but disorder is increased. Following the expansion, a red phase develops at one end of the needle and a front progresses along the needle as the temperature is raised. The transition is not reversible and at $157\,^\circ$C the red crystals melt. The red material, identical by X-ray diffraction with the transformed phase, may be obtained independently from hot xylene. The crystal structures of the two forms indicate that they are conformational polymorphs, with very different packing (Fig. 6.16). Etter and Siedle attempted to characterize the yellow to red transition structurally and thermodynamically, but noted then that significant questions remain unanswered, in particular why the colour changes along a front *after* the thermal expansion.

Other systems raise similar questions. Crystals of the overcrowded molecule **6-XXXI** undergo a reversible orange to red conversion repeatedly, apparently without

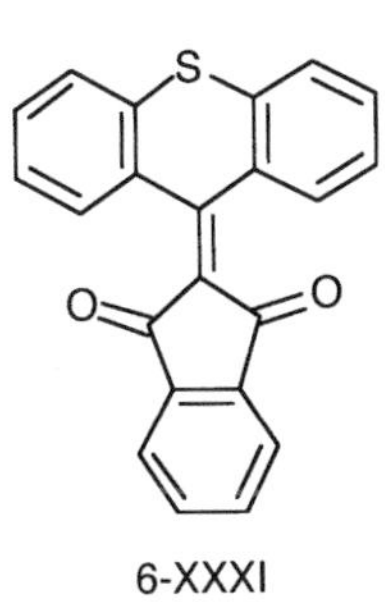

6-XXXI

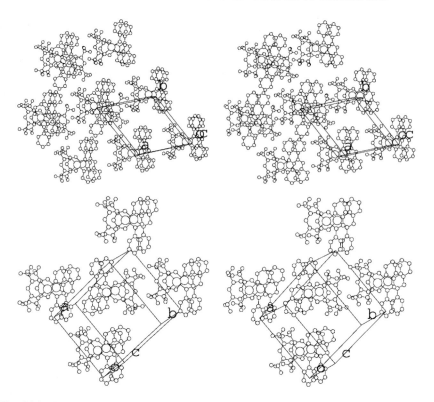

Fig. 6.16 Stereo packing diagrams of the low temperature yellow (upper) and the high temperature red (lower) forms of **6-XXX**. In both cases the view is on the best molecular plane, which is approximately along the *c* axis, measuring 8.44 Å in the yellow form and 6.99 Å in the red form. Large circles depict the Pd atoms.

any sign of visible degradation. Yet a phase change is not detectable by DSC measurements (Stezowski *et al.* 1993). Finally, in a remarkable system studied by Katrusiak and Szafranski (1996), guanidinium nitrate undergoes a first-order phase transition at 296 K, followed by another continuous phase transition at 384 K. The lower temperature transition is accompanied by large crystal strain and a physical lengthening of the crystal by over 44 per cent, which is not reported specifically to lead to physical movement of the crystal, but can hardly be ignored. The mechanism, based on structural and calorimetric data and lattice energy calculations, has been interpreted in terms of the rearrangement of the hydrogen-bonding patterns in the course of the transition.

6.2.5.4 *Mechanochromism and triboluminescence*

These are two closely related phenomena that may not warrant individual definitions. Mechanochromism describes a change in colour resulting from a physical perturbation involving pressure such as crushing or grinding, while triboluminescence is defined as the generation of light due to friction. In a demonstration of the mechanochromic effect for polymorphic structures, **6-XXXII** was found to crystallize in four modifications, in some cases concomitantly, depending on the solvent and conditions; three structures have been reported (Mataka *et al.* 1996). Colours vary from colourless (the form whose structure has not been reported) through pale yellow to dark orange. They differ slightly but experimentally significantly (by a few degrees) in the rotations of the phenyl groups and the carbonyls and so may be considered conformational polymorphs, but the authors attribute differences in colour more to intermolecular interactions than to conformational variations. Stereo views of a pair of molecules in the three structures are shown in Fig. 6.17.

6-XXXII

The mechanochromism resulting from grinding or crushing was monitored using visible fluorescence resulting from 365 nm incident radiation. The blue fluorescence is exhibited by the pale yellow form, which upon crushing becomes green with green fluorescence. The dark orange form which fluoresces yellow-green is also transformed to a green solid with green fluorescence. The chromatic changes are less dramatic for the yellow form, which shows a change in the shade of yellow and fluoresces yellow-green. Upon grinding, the colourless form with blue fluorescence, turns yellow green with yellow-green fluorescence. The X-ray powder patterns of the yellow and dark orange forms change as a result of the grinding or crushing; the pale yellow and colourless ones do not change. The original colours can be restored by washing the products or exposing them to solvent vapour. Except for this phenomenological description, no structural or physical basis for the mechanochromic effects has been proposed.

During the 1970s, there was rather intense activity surrounding the triboluminescent effect, which can readily be demonstrated by chewing Wintergreen Life Savers® in front of a mirror in a darkened room. Much of the work was summarized in a review by Zink (1978). The prevailing model at the time was that the triboluminescent effect required a non-centrosymmetric crystal (such as the sugars in the Life Savers®) which developed fluorescent charged states as a result of the friction generated by chewing. The study of a polymorphic system with a non-triboluminescent centrosymmetric form and a triboluminescent non-centrosymmetric form (Hardy *et al.* 1978) tended to confirm such a model. In subsequent work there were a number of reports

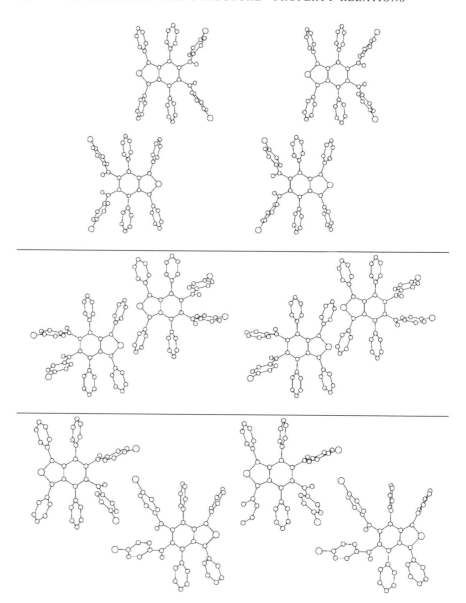

Fig. 6.17 Stereoviews of the molecular conformation and pairwise relationship of three of the four modifications of **6-XXXII**. All three structures are viewed on the same reference plane, which is the benzothiophene ring. Upper, pale yellow form; middle, light yellow form; lower, dark orange form.

of centrosymmetric crystals that nonetheless exhibited triboluminescence, which in some cases was attributed to disorder (e.g. Chen *et al.* 1998), meaning that local regions of the crystal may not have been strictly centrosymmetric. Sweeting *et al.* (1997) carried out a study of twelve 9-anthracene-carboxylic acids and their esters to correlate their triboluminescent activity with crystal structure, purity and photo-luminescence. Their conclusion was that a non-centrosymmetric crystal structure is a necessary, but not sufficient condition for triboluminescence, with charge separation and recombination actually being responsible for the triboluminescent effect. Clearly, the identification and study of some non-disordered polymorphic structures could provide additional edification in this matter.

6.2.6 Thermal phase changes and the thermosalient effect— 'hopping' or 'jumping' crystals

The field of thermally induced phase transitions is rather vast, and while involving transformations among polymorphic forms, is beyond the scope of this work (see, e.g. Rao and Rao 1978; Bruce and Cowley 1981; Salje 1990). However, as polymorphs are simply different phases, there has been increasing interest in the study of the structural mechanism (in chemists' terms the 'reaction coordinate') of these transitions in molecular crystals. As Dunitz (1991) has pointed out, there is no clear boundary between solid state transitions and chemical reactions. Only a few of these investigations, all emphasizing different aspects of such studies, are noted here. They all have one feature in common; that is, the structures of the starting and product phases are known, and attempts are made to deduce the nature of the intra- and intermolecular motions during the course of the thermal transformation. These motions may lead to thermally induced reactions (Paul and Curtin 1973) for which the kinetics can be quite complex (Shalaev and Zografi 1996), and the current status of this aspect of solid state reactions and the benefits that can be derived by a multidisciplinary investigative approach have been given in a recent review (Even and Bertault 2000). The development of time-resolved crystallography, especially with synchrotron radiation, promises to shed much light on the details of these reactions.

There have been occasional reports of crystals that jump in the course of these thermal phase transitions—the thermosalient effect—in organic crystals (Gigg et al. 1987; Ding et al. 1991; Steiner et al. 1993; Zamir et al. 1994; Lieberman et al. 2000), and inorganic crystals (Crottaz et al. 1997). The hopping or jumping of the crystal, visible on a macro scale, is usually associated with a thermally induced polymorphic phase change. In virtually all of these reports, the starting and final structures are known, some of them are reversible (e.g. Zamir et al. 1994), but determining the precise relationship between the microstructural changes (detectable in the structure determinations) and the macrostructural manifestations of those changes remains a considerable challenge. There are undoubtedly many more systems that exhibit thermosalient phase changes which can be readily observed on the hot stage microscope (see Section 4.2).

6.3 Molecular properties

The manifestation of changes in molecular structure among polymorphic modifications as detected by a number of physical and analytical techniques was discussed in

Chapter 4. The primary aim there was to describe how those techniques can be used to detect and characterize polymorphic forms. In this subsection, we wish to demonstrate how those techniques can be used to investigate structure–property relations on a molecular level utilizing the polymorphic nature of the compound. The distinction between these two approaches is not always sharp, as suggested also by the discussion of bulk properties. Again, the coverage here is not meant to be inclusive, but rather to demonstrate how the utilization of polymorphic structures can often provide the key to unraveling structure–property relations, regardless of the analytical technique employed.

6.3.1 *Infrared and Raman spectroscopy*

Since IR spectroscopy is a standard, and perhaps currently the most widely used tool in the search for and characterization of polymorphs, there are likely to be thousands of references to the use of the technique in connection with polymorphs. The vast majority of these deal with the determination of the IR fingerprint of a polymorphic modification. In this section, we wish to note a few cases in which the IR and Raman techniques were employed to obtain chemical information somewhat beyond the mere identification of a particular crystal modification. For instance, Mathieu (1973) showed for a number of chiral compounds that it is possible to distinguish between a *dl* racemate and a conglomerate of *d* and *l* crystals by use of IR and/or Raman spectroscopy, even when it may not be possible to make such a distinction by physical or visual means.

As in many studies of polymorphism the combination of techniques is particularly effective for obtaining structure–property relationships (Yu *et al.* 1998). Combining IR with SSNMR and X-ray crystallography to study the two polymorphs of **6-XXXIII**, Fletton *et al.* (1986) were able to correlate the IR frequencies with the number of molecules in the unit cell as well as the intramolecular and intermolecular hydrogen bonding.

6-XXXIII   6-XXXIV

Perhaps a classic case of the application of Raman spectroscopy on polymorphs involved the resolution of a controversy surrounding the correlation of geometry and stretching frequency near $500 \, cm^{-1}$ of the biologically important disulphide bridge, $-S-S-$ (Lord and Yu 1970*a*,*b*). Dithioglycolic acid, **6-XXXIV**, may be considered a model compound for this important chemical linkage. In a series of papers, Sugeta and coworkers had suggested that bands showing large shifts to higher frequencies indicate one (for $\sim 525\,\mathrm{cm}^{-1}$) or two (for $\sim 540\,\mathrm{cm}^{-1}$) C–C–S–S– torsion angles approximating the *trans* ($\sim 180°$) conformation (Sugeta *et al.* 1972, 1973; Sugeta 1975). Van Wart and Scheraga had argued contrarily that the same frequency shifts were characteristic of considerably smaller torsion angles, in the range 20–50° (van Wart and Scheraga 1976a,b, 1977).

The resolution of the controversy was provided by Nash *et al.* (1985) who prepared two polymorphs of **6-XXXIV**. Form I was obtained only when an aqueous solution evaporated slowly (through an orifice) over a period of weeks. On the other hand, Form II was obtained by a more rapid evaporation of an aqueous solution or from a variety of organic solvents. The latter thus appears to be the kinetic form; the former the thermodynamic form. The two forms crystallize concomitantly when aqueous solutions are allowed to evaporate more rapidly. In Form I, the molecule lies on a crystallographic twofold axis, requiring equality of the two –CCSS– angles ($-167.4°$) and a value of $-86.3°$ for the –CSSC– torsion angle. Molecules in Form II lie on crystallographic general positions, the two –CCSS– angles being $-76.2°$ and $-64.6°$, with a –CSSC– angle of $-92.3°$. There are some other geometric parameters that differ by experimentally significant amounts. Thus these are also conformational polymorphs, with variations in torsion angles, suitable for addressing the spectroscopic points of the controversy.

As expected, the Raman spectra of the two polymorphs also differ. In the region under dispute, Form I has a very strong line at $536\,\mathrm{cm}^{-1}$, while the corresponding line for Form II appears at $510\,\mathrm{cm}^{-1}$. Van Wart and Scheraga had reported a very strong line at $508\,\mathrm{cm}^{-1}$, which means that they must have obtained Form II, not Form I as they contended. On the other hand, the $536\,\mathrm{cm}^{-1}$ value for Form I does correspond to that of Sugeta *et al.*, confirming their conclusions. Nash *et al.* also performed normal coordinate calculations that were consistent with these findings, which were subsequently confirmed by Van Wart and Scheraga (1986).

6.3.2 *UV/Vis absorption spectroscopy*

Most interactions of electromagnetic radiation with matter contain a geometric, as well as an energetic component. For visible and ultraviolet absorption this is because the fundamental relationship governing the absorption of light is the transition moment integral, (6.1), in which **r** is the transition moment *vector* defining

$$f = \int \Psi \mathbf{e} \cdot \mathbf{r} \Psi^* \mathbf{d}\tau \qquad (6.1)$$

the direction along which the electric vector **e** of the light must operate for the transition to occur from the ground state, defined by the wave function Ψ, to the excited state, Ψ^*. While many, indeed most, attempts at reconciling absorption spectral observations with theory concentrate on the energetics, the directional properties of the spectral features are certainly no less important, and in some instances may even be more critical than the energies in determining the extent to which experimental results correspond to theoretical models. In solution the directional properties are randomized, but crystals provide an essentially fixed matrix for orienting the molecules in a way potentially known through the crystal structure determination. It is not sufficient, however, to fix the molecules in a crystal structure. The direction of the electric vector of the incident light must be fixed by using polarized light. Then, by carrying out spectroscopic studies using polarized light on single crystals for which the crystal structure has been determined, and with a known orientation of the crystal with respect to the polarization of the electric vector of the incident light, the directional properties of the observed spectral transitions can be determined. These can then be compared directly with the symmetry properties of the wave functions Ψ, Ψ^* which form the basis of the electronic model or theory.

A complicating factor arises when the oscillator strength of the transition is large (e.g. has a large extinction coefficient) as it is in many materials of spectroscopic interest. In such cases it may not be possible to prepare single crystals or films of known orientation sufficiently thin to allow the passage of light for an absorption measurement. However, because of a complementary relationship between the absorption and reflection of light, it is possible to measure the polarized reflection spectra of single crystals, and these spectra may be converted to the equivalent absorption spectra through a Kramers–Kronig transformation (Anex and Simpson 1960; Anex 1966; Anex and Fratini 1964). The application of such techniques to single crystals of polymorphic materials can give direct and rich information on the origin of optical effects resulting from molecular as well as bulk processes.

The utility of such an approach is demonstrated on the molecular level with the case of benzylideneaniline **6-XXXV** (R–H), which occupied spectroscopists for nearly three decades (Haselbach and Heilbronner 1968, and references therein). This material is isoelectronic with azobenzene **6-XXXVI** and stilbene **6-XXXVII**, but its solution absorption spectrum differs significantly from them (Fig. 6.18).

6-XXXV   6-XXXVI   6-XXXVII

The difference in the absorption spectra between **6-XXXVI** and **6-XXXVII** on the one hand and **6-XXXV** on the other hand has been attributed to a difference in molecular conformation in solution: the former two were believed to be essentially planar on the average, while the latter is not. The non-planarity of **6-XXXV** is due to the repulsion between the hydrogen on the bridge and one of the *ortho* hydrogens on the aniline ring. The bridge hydrogen is absent in **6-XXXVI** and the increased length of the –C=C– bond in **6-XXXVII** alleviates that steric effect. For **6-XXXV** the tendency towards non-planarity due to the hydrogen repulsion is balanced in part by

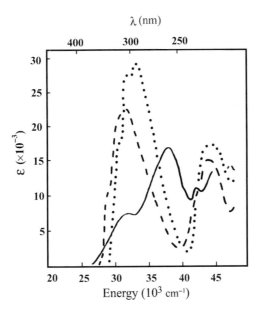

Fig. 6.18   Solution absorption spectra of benzylideneaniline **6-XXXV** (R–H) (—), azoben-zene **6-XXXVI** (---) and stilbene **6-XXXVII** (· · ·) in ethanolic solutions. (From Bernstein *et al.* 1979, with permission.)

the π-electron conjugation, leading to the minimum energy conformation of $\alpha \approx 50°$ (Bernstein *et al.* 1981).

The existence of two polymorphic structures of the dichloro derivative of **6-XXXV** (R = Cl) (Bernstein and Izak 1976) provided an opportunity for the direct examination of the relationship between the molecular structure and the electronic spectrum. The two structures are conformational polymorphs, with the metastable very pale yellow triclinic needle form exhibiting a planar molecular conformation ($\alpha = \beta = 0°$) (Bernstein and Schmidt 1972) and the stable yellow orthorhombic form (with chunky rhombic crystals) exhibiting a non-planar conformation ($\alpha = 25°$; $\beta = -25°$) (Bernstein and Izak 1976). Assuming that the two crystal structures merely serve to hold the molecule in the two different conformations (i.e. the 'oriented gas' model), the absorption spectra should reflect the difference in conformation: that measured on the triclinic structure, with a planar conformation, should closely resemble the spectra of **6-XXXVI** and **6-XXXVII**, while that for the orthorhombic structure, with the nonplanar molecular conformation, should retain the characteristics of **6-XXXV** (R = H) in solution.

Polarized normal incidence specular reflection spectroscopy on single crystals was used to investigate this system (Bernstein *et al.* 1979). Faces of the two polymorphs were chosen for study on the basis of a favourable projection of the long axis of the molecule (Fig. 6.19), which was believed to be the polarization direction of the lowest

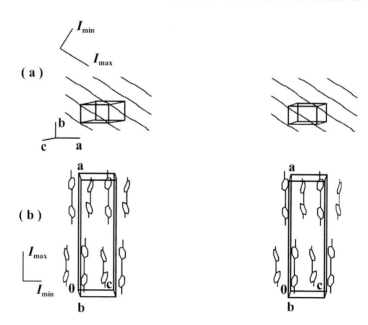

Fig. 6.19 Projections of the crystal structures of **6-XXXV** (R = Cl) onto the faces studied spectroscopically: (a) the (001) face of the triclinic form; (b) the (110) face of the orthorhombic form. I_{min} and I_{max} are the extinction directions that were oriented parallel to the electric vector of the incident light for the measurement of the reflection spectra. (From Bernstein *et al.* 1979, with permission.)

energy transition (Bally *et al.* 1976). The reflection spectra with the light polarized nearly along the long axis of the molecule (Fig. 6.20) clearly indicate a significant difference, with that of the triclinic form even exhibiting vibronic structure reminiscent of the stilbene and azobenzene solution spectra. The corresponding absorption spectra obtained through the Kramers–Kronig transform (Anex and Fratini 1964) are given in Fig. 6.21. These can be considered equivalent to the absorption spectra that would be obtained if the measurement could be made through that particular crystal face.

The absorption spectra differ in the same manner as do the reflection spectra. The similarity to the expected solution spectra is indicative of weak interactions between the molecules, hence confirming the assumption that these crystals may simply be considered to be oriented gases. The spectrum of the planar molecule exhibits a single low-energy band, which as noted above, is punctuated by reproducible vibronic structure due to the superposition of the $-C{=}N-$ stretching mode ($\sim$1350 cm^{-1}) on the long-axis transition. In the non-planar conformation found in the orthorhombic structure the conjugation of the π system is broken, so the long-axis transition associated with that system no longer exists and consequently no vibronic structure is expected or observed. These observations provide direct confirmation of the interpretations of spectroscopists on the origins of the spectral differences among benzylideneanilines and stilbenes and azobenzenes. This system is a good example of how polymorphic

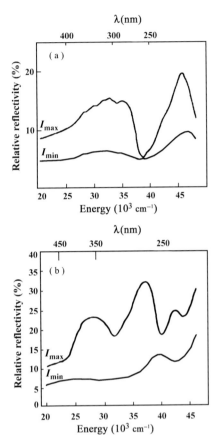

Fig. 6.20 Normal incidence polarized reflection spectra for the two forms of **6-XXXV** (R = Cl) I_{max} and I_{min} correspond to the directions indicated on Fig. 6.19: (a) the (001) face of the triclinic structure; (b) the (110) face of the orthorhombic form. (From Bernstein *et al.* 1979, with permission.)

structures, in particular those which are conformational polymorphs, can provide excellent systems for the study of the relationship between molecular structure and spectral properties. The assumption that these crystals behave as oriented gases is borne out by the similarity between the Kramers–Kronig derived absorption spectra and the solution spectra.

At the opposite extreme from the oriented gas model for molecular crystals, the neighbouring molecules do interact with each other resulting in spectral properties of the bulk that differ considerably from those of the individual molecule. Interacting molecules of this type often tend to form aggregates even in solution, a phenomenon that has been exploited by the photographic industry for the tuning of the spectral response of silver halide emulsions (Herz 1974; Smith 1974; Nassau 1983). Aggregate formation can lead to the development of new, and often quite intense absorption bands

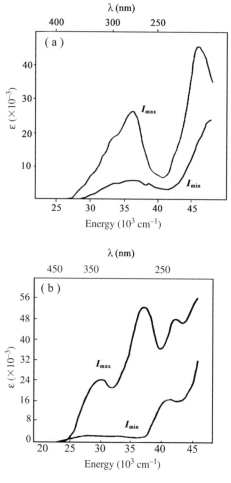

Fig. 6.21   Kramers–Kronig transformed spectra of **6-XXXV** (R = Cl) obtained from the reflection spectra of Fig. 6.20. I_{max} and I_{min} correspond to the directions indicated on Fig. 6.19. (From Bernstein *et al.* 1979.)

(e.g. Jelley 1936), and the nature of both the aggregates and their spectral response has been a matter of long-standing interest and study.

Polymorphism may be utilized to study this phenomenon of dye aggregation, employing similar experimental spectroscopic techniques as above. As noted and discussed earlier (Sections 3.5.3 and 6.2.3) the dye molecule **6-XVII** (R = Et, R′ = OH) with the solution spectrum shown in Fig. 6.22, has been shown to concomitantly crystallize as violet and green crystals (Tristani-Kendra *et al.* 1983). Of interest here are the spectral manifestations of those structural differences.

The molecule is essentially flat in both structures, so differences in the spectroscopic properties must be an expression of the intermolecular relationships in the two

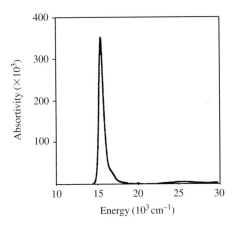

Fig. 6.22   Solution absorption spectrum of molecule **6-XVII** (R = Et, R′ = OH) in methylene chloride. (From Tristani-Kendra and Eckhardt 1984, with permission.)

structures. The polarized normal incidence reflection spectra of the two crystals are given in Fig. 6.23, and it can be readily seen that the spectra, representing extremes for quasimetallically reflecting crystals (Fanconi *et al.* 1969; Anex and Simpson 1960) are significantly different between the two forms for light polarized essentially along the long axis of the molecules. This must be a consequence of the difference in the interaction of a single molecule with its surroundings. The reflection spectrum of the band in the monoclinic form has been interpreted as being composed of two transition dipole oscillators, while that of the triclinic form contains three, or possibly four, oscillators. The interpretation of these spectra was one of the first attempts to investigate the nature of dye aggregation by studying polymorphic systems (Tristani-Kendra and Eckhardt 1984).

Similar studies employing polarized specular reflection spectroscopy on single crystals have been carried out on a polymorphic cyanine (Sano and Tanaka 1986) dye and a polymorphic acridine (Mizuguchi *et al.* 1994) dye. As many as seven polymorphic oriented layers of a polymethine dye have been prepared and characterized, the different polymorphs serving as different models for states of aggregation (Dähne and Biller 1998a,b).

6.3.3   *Excimer emission*

In 1954 Förster and Kasper reported that in some cases the emission of light following excitation of a molecule can be attributed to fluorescence from a dimer rather than from the individual molecule that initially absorbed the radiation; the emitting moiety was termed an 'excimer', and the determination of the relationship between its structure and emitting properties was a matter of considerable interest (Chandra *et al.* 1958; Murrell and Tanaka 1964; Smith *et al.* 1966).

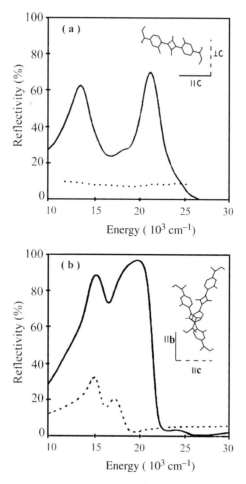

Fig. 6.23 Normal incidence polarized reflection spectra of the two forms molecule **6-XVII** (R = Et, R′ = OH). For each crystal modification there are two spectra measured with the light polarized along each of the two directions (the so-called principal directions), as indicated in the upper right hand corner, which also shows the projection of the molecule(s) onto the crystal face studied. (a) Triclinic polymorph, (100) face; (b) monoclinic polymorph, (100) face. (From Tristani-Kendra and Eckhardt 1984, with permission.)

The discovery of two excimer emitting polymorphs of the stilbene derivative **6-XXXVIII** (Cohen *et al.* 1975) provided an opportunity to determine directly the relationship between excimer structure and its emission properties. The structure of Form A was determined crystallographically and indicated a stacking arrangement, while the structure of Form B, inferred from the cell constants and the photochemistry consistent with the topochemical principles (Section 6.4), was a pairwise arrangement

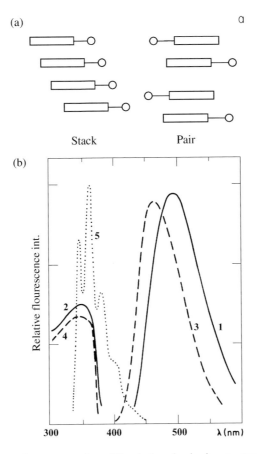

Fig. 6.24   (a) Schematic representation of the stack and pairwise structures for the two polymorphs of **6-XXXVIII**. (b) Emission and excitation spectra of **6-XVIII** 77 K. 1,2 : emission
and excitation spectra of modification A; 3,4 : emission and excitation spectra of modification
B; 5 : emission spectra of glassy ethanol solution. (From Cohen *et al.* 1975, with permission.)

(Fig. 6.24). The emission and excitation spectra for the two polymorphs are compared to that for a glassy ethanol solution in Fig. 6.24. There are clear differences that
the authors have semi-quantitatively attributed to differences in the potential energy
curves of the ground state for the stack and pair structures ($c.1000\,\mathrm{cm}^{-1}$).

6-XXXVIII   6-XXXIX

In a later study Theocharis *et al.* (1984), were able to make a direct determination of the relationship between intermolecular registry and excimer behaviour through the study of two conformational polymorphs of bis(*p*-methoxy)-*trans*-stilbene **6-XXXIX**. Irradiation of the orthorhombic form with a planar conformation is likely to form excimers which requires the approach of two non-parallel neighbouring molecules. Apparently, relaxation to the original molecular positions following excimer emission may not be possible, leading to a topotactic transformation to a monoclinic form, in which the molecules are not planar, but are parallel, since they are related by translation.

6.3.4 *Excited state diffraction studies*

As with analytical methods employed to characterize polymorphs, the combination of techniques in studying structure–property relations is also seeing increased use. With the increasing power and speed of flash photo and synchrotron sources it is now possible to study the structural changes that take place in transient states, in particular in excited-state species (Coppens *et al.* 1998; Ozawa *et al.* 1998). Using the combination of flash laser techniques and synchroton radiation, Zhang *et al.* (1999) studied the excited state of 4,4′-dihydroxybenzophenone in two polymorphs of its complex with 4,13-diaza-18-crown-6. In its own crystal, the benzophenone molecules are in close proximity, leading to triplet–triplet annihilation. The formation of a complex leads to greater separation between individual benzophenone molecules, and the existence of two polymorphs means that the triplet lifetimes can be studied as a function of separation of carbonyl groups. The carbonyls are more widely separated in the triclinic modification than in the monoclinic one, leading to corresponding triplet lifetimes of 49.2 ± 0.5 and 44.2 ± 1.2 µs, respectively. The difference makes these promising candidates for diffraction studies of excited-state species.

6.4 Photochemical reactions

In 1964, Cohen and Schmidt formulated the topochemical principles of reactions in the solid state, in particular photochemical reactions, very much on the basis of the different photochemical behaviour of polymorphic substances (Hertel 1931). Their investigations arose from the rather long-standing controversy over the nature of the solid state dimerization of *trans* cinnamic acid **6-XL** (Störmer and Förster 1919; Störmer and Laage 1921; Stobbe and Steinberger 1922; Stobbe and Lehfeldt 1925; de Jong 1922*a,b*). The material was long known to be polymorphic (Lehmann 1885; Erlenmeyer *et al.* 1907), and considerable light was shed on the problem by Bernstein and Quimby (1943), but it remained for the combined application of organic photochemistry and X-ray crystallography to these polymorphic systems to define the topochemical principles (Cohen *et al.* 1964*d*; Schmidt 1964).

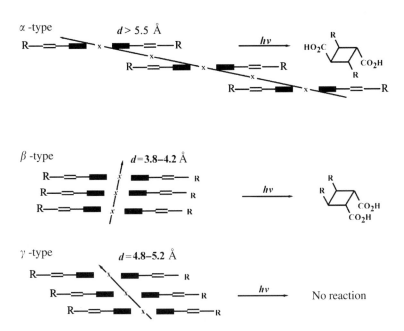

6-XL

The principles derived from these experiments are summarized in Fig. 6.25. In the β-type crystal, the molecules are arranged in a translationally equivalent manner along a short ($< \sim 4.2$ Å) crystallographic axis, so that the necessarily parallel reactant double bonds are within an appropriate distance and in the proper orientation to undergo a [2 + 2] addition; the resulting photochemical product, with molecular mirror symmetry, reflects the preregistry of the reactant molecules in the β structure. In the prototypical α modification, neighbouring molecules are related by an inversion centre, which again results in a parallel registration of the reactive double bonds, but the crystallographic axis is approximately doubled to 7.5–8.0 Å. The photochemical [2 + 2] product also has inversion symmetry. In the absence of a short contact and/or suitable orientation of the neighbouring reactive centers, the reaction does not take place; this is the case for γ-type crystals.

A great deal of solid state chemistry has been carried out and understood using these topochemical principles (Schmidt 1971; Thomas *et al.* 1977; Lahav *et al.* 1979;

α -type   *d* > 5.5 Å

β -type   *d* = 3.8–4.2 Å

γ -type   *d* = 4.8–5.2 Å

No reaction

Fig. 6.25 The topochemical principles, as demonstrated by the photochemical behaviour of polymorphic cinnamic acids. The filled rectangles in the reactant molecules represent carboxyl groups.

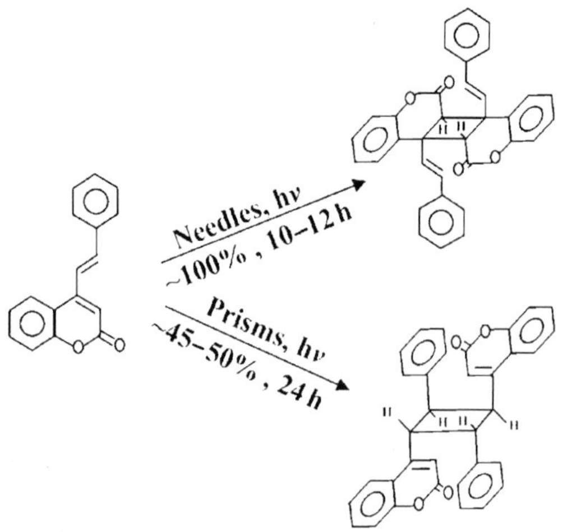

Fig. 6.26   Schematic diagram of the different photoreaction pathways for the triclinic needles and monoclinic prisms of 4-styrylcoumarin **6-XLI**. (From Moorthy and Venkatesan 1994, with permission.)

Thomas 1979; Hasegawa 1986; Ramamurthy and Venkatesan 1987; Venkatesan and Ramamurthy 1991; Singh *et al.* 1994). Some polymorphism-based variations on this basic theme are worthy of note. Moorthy and Venkatesan (1994) studied the concomitantly crystallizing dimorphs of 4-styrylcoumarin **6-XLI**, which has two reactive double bonds. The triclinic needles and the monoclinic prisms, which are easily distinguished and separated, yield two different photoproducts at different rates and with different yields (Fig. 6.26) depending on which of the two double bonds participate in the [2 + 2] photoreaction. The potential limited variety of photoproducts from these systems depends on the polymorphic form, and the desire to control the nature of the product has led to attempts to generate polymorphs of styrylcoumarins with intermolecular relations that would lead to the preferred products (Vishnimurthy *et al.* 1996; Row 1999).

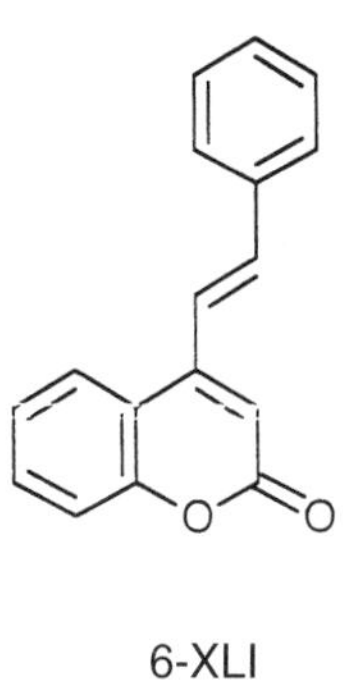

6-XLI 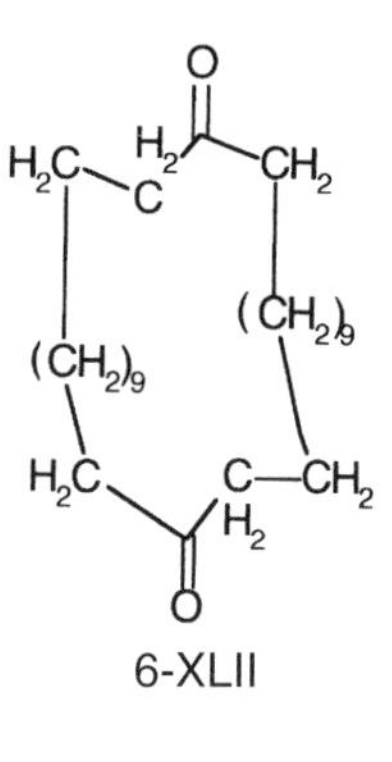   6-XLII

The topochemical principles have also been applied to the interpretation of *intra*molecular photochemical reactions of polymorphic materials depending on polymorphic form. 1,14-cyclohexacosanedione **6-XLII** can be prepared as conformational dimorphs that undergo Norrish type II photochemistry which can be correlated with the molecular conformations in the two modifications, one leading to a *cis*-butanol, while the other leads to a *trans* product (Gudmunsdottir *et al.* 1996). Similarly, these principles can be invoked to understand the relative photochemical stability of two polymorphs of the active ingredient in a compound developed for the treatment of psoriasis (Lewis 2000).

In the study of the four conformational polymorphs of a cobaloxime complex, Sawada *et al.* (1996) were able to establish a quantitative relationship between the size of the reaction cavity for an intramolecular photoisomerization reaction and the rate constant for the process in three of the modifications. Such studies provide a great deal of direct detailed information about the relationship between the environmental influences on the mechanism of a reaction at essentially the atomic level, information which is much more difficult to obtain from studies on a single crystalline form.

6.5   Thermal reactions and gas–solid reactions

Thermally induced phase transformations and reactions in the solid state also provide a potentially rich source of information on structure–property relations and often have practical economic implications (e.g. Loisel *et al.* 1998). As noted earlier, in the past the investigation of such phenomena relied to a large extent on the determination of the structures (and properties) of the starting materials and products (e.g. Jameson *et al.* 1984; Theocaris and Jones 1984). Two reviews based on that principle surveyed the literature through the 1970s (Paul and Curtin 1973; Curtin *et al.* 1979), with a more recent one by Even and Bertault (2000). Hager *et al.* (1998) have carried out a survey of the CSD (October 1996 release) to determine the number of temperature-dependent phase transitions for which structures of all resulting forms had been reported. Sixty-five of the 187 compounds for which the presence of a phase transition was recorded in the CSD have crystal structure determinations for more than one phase. The current ability to control and program the crystal temperature, combined with increasingly rapid means for determining crystal structures, and the development of techniques for the study of surfaces that also reveal details of internal structure all serve to provide information that can be used to study stages along the reaction coordinate, rather than just at its extremities (Boese *et al.* 1987; Sanchis *et al.* 1997). Individual crystal structures determined as the reaction proceeds provide snapshots of the changes in molecular structure, and in combination produce a representation of the dynamics of the transition or reaction (e.g. Katrusiak 2000). As noted elsewhere in this book, such studies often indicate the lack of a clear distinction between polymorphic transitions and thermal reactions in the solid state (Dunitz 1991).

On the other hand, reactions between solids and gases involve two chemical species and such a distinction is much clearer. The 1975 and 1987 reviews by Paul and Curtin remain two of the key sources of information on the subject, and both contain a number of examples in which the study of polymorphic systems shed considerable light on the mechanisms by which such reactions do or do not proceed (Perrin and Lamartine 1990). One of the important features of gas–solid reactions is the anisotropy of the reaction, which depends on the fact that on the various faces of a crystal different parts of a molecule may be exposed to the reagent gas, and the reaction will proceed at various rates (or not at all) depending on the functionality of the exposed part of the molecule. Similarly, different structures lead to different crystal habits, but also to different intermolecular relationships that may allow or prevent the initial surface reaction to proceed through the bulk of the crystal.

Two classic studies on polymorphic systems are worthy of note here. One is the difference in the reactivity of the β- and γ-polymorphic forms of *meta* substituted cinnamic acids **6-XL**, in which the instantaneous rates and product distributions varied with time (Hadjoudis *et al.* 1972). A second example involves the compound **6-XLIII** (Voght *et al.* 1963) that has been shown to exist in a number of different crystalline modifications. A chloroform solvate (Schaeffer and Marsh 1969) is reactive towards oxygen, while Brückner *et al.* (1971) have reported the structure of an inactive form. The ability to probe the details of the reactions at the gas–solid interface with atomic force microscopy (Kaupp *et al.* 1996) and in the bulk by ^{1}H NMR imaging (Butler *et al.* 1992) has led to increased understanding and renewed interest in these systems that have considerable industrial promise especially due to the avoidance of solvents that otherwise would require disposal or recycling (Dittmer 1997).

6-XLIII

6.6   Pressure studies

Although pressure is one of the fundamental physical constants, the experimental challenges of pressure-dependent investigations compared with, say, temperature-dependent ones, is still rather daunting. This is in spite of the fact that some quite simple devices were developed for high-pressure crystallographic studies on molecular crystals over a quarter of a century ago (Piermarini *et al.* 1969; Block *et al.* 1970; Bassett and Takahashi 1974). On the other hand the combination of pressure, temperature and crystallographic studies can lead to detailed understanding of the structural basis of the phase diagram and the mechanisms of the phase transitions (Szafranski and Katrusiak 2000). As with the 'hyphenated' analytical techniques described in Chapter 4, the advantages to be gained by carrying out those measurements at a variety of pressures carries great potential for the study of polymorphic systems and their utilization for structure–property relationships (Szafranski *et al.* 1992; Busing *et al.* 1995; Boldyreva 1999).

It is natural to expect pressure to induce polymorphic phase changes, and considerable progress has been made recently in developing the techniques to make pressure-dependent crystallographic measurements more readily accessible (Katrusiak 1991). This has led to structural studies of pressure-induced phase transitions (Katrusiak 1990, 1995). A number of recent structural investigations as a function of pressure have also been carried out by Boldyreva *et al.* 2000 (and references therein). One of these was aimed at comparing the anisotropy of the pressure-induced lattice compression of two polymorphs of the complex salt $[Co(NH_3)_5NO_2]I_2$. The two forms (I and II) are concomitant (Ephraim 1923), but show no evidence of a solid to solid phase transformation. In a rather detailed study over a number of pressures, Boldyreva *et al.* found a non-reversible transition from I to a new third phase (III), which appears when the hydrostatic pressure is applied in the presence of a methanol–ethanol–water mixture. Phase I is more compressible than II or III with increasing pressure. The structural distortion of the unit cells for all three polymorphs is anisotropic, with the major distortion occurring along the shortest axis, and on the molecular level these distortions could be related to hydrogen-bonding networks and iodine–iodine interactions.

6.7 Concluding remarks

Polymorphic systems present unique opportunities for the study of structure–property relations. Just as virtually any property of a material may vary from polymorph to polymorph, so any structure–property relationship may be studied by utilizing polymorphic systems. In this chapter we have attempted to present some representative examples, with perhaps some prejudice resulting from our own experience and interests. But for the limitation on space many more examples could have been cited, and it is hoped that the selection made demonstrates the advantages to be gained by investigating such systems and provides the impetus to give polymorphic systems a high priority in choosing materials for structure–property studies.

7

Polymorphism of pharmaceuticals

After discovery of the first cases of polymorphism with dramatic differences in biological activity between two forms of the same drug . . . no pharmaceutical manufacturer could neglect the problem. (Borka 1991)

There are many mysteries of nature that we have not yet solved. Hurricanes, for example continue to occur and often cause massive devastation. Meteorologists cannot predict months in advance when and with what velocity a hurricane will strike a specific community. Polymorphism is a parallel phenomenon. We know that it will probably happen. But not why or when. Unfortunately, there is nothing we can do today to prevent a hurricane from striking any community or polymorphism from striking any drug. (Sun 1998)

7.1 Introduction

The increasing awareness and importance of polymorphism in the past 30 years or so is perhaps nowhere more evident than in the field of pharmaceuticals (Bavin 1989). The landmark chapters of Buerger and Bloom (1937) and McCrone (1965) did not place any special emphasis on pharmaceutical materials. One outstanding exception was the book of Kofler and Kofler (1954), whose authors were members of an academic department of pharmacognosy (Webster: the branch of pharmacology that treats or considers the natural and chemical history of unprepared medicines). The seminal paper on the subject of polymorphism of pharmaceuticals was the review of Haleblian and McCrone (1969), which set the scope and the standards for many subsequent works. The literature on the polymorphism of pharmaceuticals is now best described as vast. A multiauthored monograph has appeared covering various aspects of the subject (Brittain 1999*b*), along with major sections of other books (Byrn 1982; Byrn *et al.* 1999), and an ever increasing number of reviews (Haleblian 1975; Kuhnert-Brandstätter 1975; Bouche and Draguet-Brughmans 1977; Giron 1981; Burger 1983; Threlfall 1995; Streng 1997; Caira 1998; Yu *et al.* 1998; Winter 1999; Vippagunta *et al.* 2001) covering various aspects of polymorphism as related directly to problems in the pharmaceutical field and/or pharmaceutical compounds, including some of the economic and intellectual property implications (Henck *et al.* 1997). Therefore, it would be foolhardy to attempt to present a comprehensive review of the subject here. Rather, in keeping with the general philosophy of this book, the aim is to provide a general introduction to the subject, with sufficient examples to demonstrate the points raised, and commensurate relevant references for the reader to seek further details and information.

7.2   Occurrence of polymorphism in pharmaceuticals

7.2.1   *Drug substances*

The development of a new drug from a promising lead compound to a marketed product is a long and expensive process, with odds of success estimated at 1 in 10 000 (Yevich 1991). The strict quality control requirements and the intellectual property implications of the drug industry lead to thorough and intensive investigations of the formation and properties of solid substances intended for the use in pharmaceutical formulations, both active ingredients and excipients. These efforts, often extending over long periods of time and with many potential experimental and environmental variables, can create conditions that can lead to the appearance of polymorphic forms, intentionally or serendipitously. While it may not be surprising that many pharmaceutically important materials have been found to be polymorphic, or that any particular compound may turn out to be polymorphic, every compound is essentially a new situation, and the state of our knowledge and understanding of the phenomenon of polymorphism is still such that we cannot predict with any degree of confidence if a compound will be polymorphic, prescribe how to make possible (unknown) polymorphs, or predict what their properties might be (Beyer *et al.* 2001).

There have been a few attempts to compile instances of polymorphism in pharmaceutically important materials. Since even the definition of what comprises 'pharmaceutically important materials' is itself subject to debate it is difficult to judge how comprehensive such compilations might be. However, generally they serve as useful references and are given here. One of the first organized attempts at such a compilation for steroids, sulphonamides and barbiturates was by Kuhnert-Brandstätter (1965) (see also Kuhnert-Brandstätter and Martinek 1965). Much of those data may be found in her subsequent book (Kuhnert-Brandstätter 1971), which also contains a compilation of many of the thermal studies of pharmaceutical compounds that revealed polymorphic behaviour. Numerous additional reports of studies by the Innsbruck school of polymorphic pharmaceutical compounds (using thermomicroscopy and IR spectroscopy) have appeared in the literature since the middle 1960s; many of those have been listed by Byrn *et al.* (1999). Borka and Haleblian (1990) compiled a list of over 500 references to reports of polymorphism in over 470 pharmaceutically important compounds.[1] This was shortly followed (Borka 1991) by a review of polymorphic substances included in Fasciculae 1–12 of the European Pharmacopoeia (EP), including a comparison of melting points in the EP and the original literature. The latter review was subsequently updated in 1995 by including EP entries for Fasciculae 13–19 (Borka 1995).

Griesser and Burger (1999) compiled the information regarding 559 polymorphic forms, solvates (including hydrates) of drug solids at 25 °C in the 1997 edition of

---

[1] Dr Borka has communicated with this author that he and Dr Haleblian did not receive galley proofs of this paper, which unfortunately contains 'numerous printing errors'. The list of errata actually contains 48 of them. Even with that cautionary note, it is a useful compilation.

the EP. They also noted that of the 10 330 compounds in the 1997 edition Merck Index only 140 (1.4 per cent) are specifically noted as polymorphic, 540 (5 per cent) are noted as hydrates and 55 (0.5 per cent) have been specified as solvates. These numbers reflect a failure to report or to include these phenomena rather than representative statistics, and may suggest the current state of awareness of polymorphism on the part of compilers of such compendia and reference works.

A survey by this author of the 1 October 2000 release of the CSD (~225 000 entries) yielded 6353 hits for the qualifier 'form', 1045 hits for the qualifier 'phase' 528 hits for the qualifier 'polymorph', 201 hits for the qualifier 'modification', 28 342 hits for the qualifier 'solvate' and 21 132 hits for the qualifier 'hydrate'. Of course, the last two numbers give no indication of whether the materials are polymorphic and none of these statistics indicate the instances for which the structures of more than one form have been determined. The data for hydrates and solvates from the CSD may be considered quite reliable, since molecules of solvation are usually positively identified in the course of a crystal structure determination. The frequency of polymorphs, however, is likely to be underestimated, since many crystal structures of polymorphic systems, or what later are discovered to be polymorphic systems, are reported without making note of that fact. If the structure of only one member of a polymorphic family has been reported, then there may not be any reference to the polymorphism in the CSD. References to drugs discovered prior to 1971 that form solvates have been compiled, along with a separate summary of the thermomicroscopic behaviour of drug hydrates by Byrn *et al.* (1999).

Because of the rather select nature of the sample set and the distinct possibility that not all forms are always reported caution must be exercised in drawing conclusions from such statistical surveys. Nevertheless, according to Griesser and Burger (1999) there may be some apparent tendencies which should be monitored as data continue to accumulate: polymorphism seems to be more common for compounds with molecular weight below 350; polymorphism seems to be more common for compounds with low solubility in water; for organic salts, the formation of hydrates appears to be more common among larger molecules; organic solvates appear to be more common among neutral compounds with higher molecular weights.

7.2.2 *Excipients*

Pharmaceutical formulations contain the active drug ingredient(s) as well as excipients that serve a variety of purposes: fillers, stabilizers, coatings, drying agents, etc. As solid materials excipients exhibit varying degrees of crystallinity, from the highly crystalline calcium hydrogen phosphate to nearly amorphous derivatives of cellulose. These materials can also exhibit polymorphism which may influence their performance in the formulation. Giron (1995, 1997) has listed many of the excipients that are known to exhibit a number of forms (polymorphs, solvates and amorphous). They include many of those that are widely used: lactose, sorbitol, glucose, sucrose, magnesium stearate, various calcium phosphates and mannitol (Burger *et al.* 2000).

The solid nature of the excipient may influence the final physical form of the tablet (Byrn *et al.* 2001), such as a tendency to stick (Schmid *et al.* 2000), or may induce a polymorphic conversion of the active ingredient (Kitamura *et al.* 1994). Hence, there have been attempts to develop protocols for the selection of compatible active ingredient–excipient compositions (Serajuddin *et al.* 1999). For instance, nuclear magnetic resonance spectroscopy has been employed to study the structural changes in epichlorohydrin cross-linked high amylose starch excipient (Shiftan *et al.* 2000), and has also been used to discriminate between two polymorphs of prednisolone present in tablets with excipients, even at low concentrations (5 per cent w/w) of the active ingredient (Saindon *et al.* 1993). The characterization of excipients by thermal methods has also been reviewed by Giron (1997).

7.3   Importance of polymorphism in pharmaceuticals

Polymorphism can influence every aspect of the solid state properties of a drug. Many of the examples given in preceding chapters on the preparation of different crystal modifications, on analytical methods to determine the existence of polymorphs and to characterize them and to study structure/property relations, were taken from the pharmaceutical industry, in part because there is a vast and growing body of literature to provide examples. One of the important aspects of polymorphism in pharmaceuticals is the possibility of interconversion among polymorphic forms, whether by design or happenstance. This topic has also been recently reviewed (Byrn *et al.* 1999, especially Chapter 13) and will not be covered here. Rather, in this section, we will present some additional examples of the variation of properties relevant to the use, efficacy, stability, etc. of pharmaceutically important compounds that have been shown to vary among different crystal modifications.

7.3.1   *Dissolution rate and solubility*

The dissolution properties and solubility are often crucial factors in the choice of the crystalline form for formulation of a drug product (Carstensen 1977). In general, these two factors play a major, if not over-riding role in determining the bioavailability of the drug substance (see also Section 7.3.2). The physiological absorption of a solid dosage form usually involves the dissolution of the solid in the stomach and the rate and extent of that dissolution is often the rate-determining step in the overall absorption process. Since different crystalline forms can exhibit different dissolution kinetics and limits, these properties are routinely studied in great detail for any drug substance, whether polymorphic or not; clearly, characterization for polymorphic substances is even more critical. As a result, there is extensive literature covering such studies. An early review contained a compilation of references to many of the previously studied materials (Kuhnert-Brandstätter 1973).

The fundamentals of the measurements of solubility and dissolution properties of pharamceuticals are given elsewhere in considerable detail (Vachon and Grant 1987;

Byrn *et al.* 1999, esp. Chapter 6; Grant and Brittain 1995). Reviews of the physical principles of dissolution and solubility behaviour of organic materials (Grant and Higuchi 1990) and pharmaceuticals (Grant and Brittain 1995) have been given and, particularly relevant to this chapter, typical examples of the variation of these processes among crystalline forms of pharmaceuticals are comprehensively described by Brittain and Grant (1999).

7.3.2   *Bioavailability*

The rate and extent of the physiological absorption of an active drug substance are decisive factors in its overall efficacy (e.g. carbamazepine, Zannikos *et al.* 1991). These can vary among different crystal modifications, and they have become an important scientific and regulatory issue (Ahr *et al.* 2000).

While a number of studies of the connection between the crystal modification and bioavailability have been published, it is reasonable to assume that many more remain the intellectual property of pharmaceutical companies or in confidential documentation submitted to regulatory agencies. Also many *in vitro* studies, especially those of extent or rate of dissolution, are used to extrapolate to expected bioavailabilites (e.g. Chikaraishi *et al.* 1995; Shah *et al.* 1999), which is indeed proven in some cases such as the tetramorphic tolbutamide (Kimura *et al.* 1999). We cite here a few additional examples from the open literature, limiting ourselves (except for one example below) to cases in which the bioavailability does differ among crystal modifications.

The two polymorphs of the barbiturate pentobarbital exhibit significantly different rates of absorption and area under the curve for oral administration (Fig. 7.1) (Draguet-Brughmans *et al.* 1979). Comparison of rectal absorption of suppositories containing the two polymorphs of indomethacin showed higher levels for the α

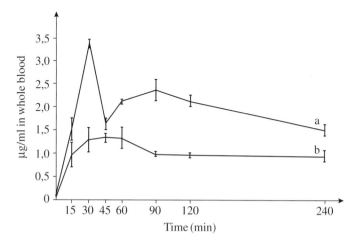

Fig. 7.1   Rate of absorption of the two polymorphs of orally administered pentobarbital. a, Form I; b, Form II. (From Draguet-Brughmans *et al.* 1979, with permission.)

form than for the γ form (Yokoyama *et al.* 1979). For the trimorphic antileukemic mercaptopurine Form III was found to be 6–7 times more soluble than Form I, but the bioavailability (in rabbits) was approximately 1.5 times greater for Form III than for Form I (Yokoyama *et al.* 1980). The antianxiety agent nabilone has at least four polymorphs, designated A, B, C and D, which appear to be equally hydrophobic and insoluble. Forms B and D are bioavailable in dogs, while A and C are not. On prolonged storage, heating or grinding, all convert to the nonbioavailable Form A (Thakar *et al.* 1977).

The bioavailablity can also vary between crystalline and amorphous modifications (Section 7.7), as well as among polymorphic forms. The amorphous form of the antibacterial azlocillin sodium has more activity than the crystalline form (Kalinkova and Stoeva 1996). Similarly, an antinematode drug, code named PF1022A exists in four modifications, designated α (amorphous) and crystalline I, II and III. α and III have higher solubility and are more effective than I and II against tissue-dwelling nematodes (Kachi *et al.* 1998). An amorphous form of another code-name drug (YM022), generated by spray drying, has enhanced bioavailability over two crystalline forms (Yanu *et al.* 1996).

The degree of hydration of different modifications also plays a role in the bioavailability. One of the most studied systems in this regard is the anhydrate/trihydrate of the antibiotic ampicillin, although the results have not always led to consistent conclusions. Early *in vitro* solubilities (Poole *et al.* 1968) and rates of dissolution (Poole and Bahal 1968) were shown to differ. *In vivo* the anhydrous form reaches a maximum concentration. *In vivo* bioavailability studies by Ali and Farouk (1972) (Fig. 7.2)

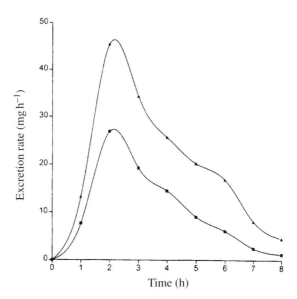

Fig. 7.2   Urinary excretion rates following administration of the anhydrate (▲) and trihydrate (■) forms of ampicillin. (From Brittain and Grant 1999, with permission.)

clearly indicated differences between the two. Other studies, however, indicated similar bioavailabilities (Cabana *et al.* 1969; Mayersohn and Endrenyi 1973; Hill *et al.* 1975). Brittain and Grant (1999) have suggested that the differences reported by various groups indicate that the bioavailability in the case of ampicillin is strongly influenced by the nature of the formulation of the dosage form.

There are also some examples of solvates (as opposed to hydrates) that have been reported to exhibit differing bioavailability. For instance, for implants of the methanol solvate of predisolone *tert*-butyl acetate the mean absorption rate was found to be 4.7 times that of the anhydrous form, which in turn was similar to the hemiacetone solvate (Ballard and Biles 1964). The same authors also reported that the mean absorption rate for all solvates of cortisol *tert*-butyl acetate were significantly different from that of the anhydrous form.

While the *rate* of absorption may differ, the *extent* of absorption, may still be equivalent. Such is the case for sulphameter (sulphamethoxydiazine). The different polymorphic forms have been shown to exhibit different equilibrium solubilities and dissolution rates (Moustafa *et al.* 1971). Form II is thermodynamically less stable and has an absorption rate about 1.4 times that of Form III, which is more stable in water (Khalil *et al.* 1972). This group determined that the two forms have different absorption, but using 72-hour excretion data showed that the extent of the absorption of the two forms was equivalent (Khalafallah *et al.* 1974).

Perhaps the classic example of the dependence of bioavailability on polymorphic form is chloramphenicol palmitate. Chloramphenicol is a broad spectrum antibiotic and antirickettsial which was developed in the 1960s and had a significant portion of the market until the appearance of side effects limited its use to topical application. The exceedingly bitter taste of the active chloramphenicol led to its formulation as an oral suspension of the tasteless 3-palmitate (CAPP). The early physical and physiological studies on the material were summarized by Aguiar *et al.* (1967) (see also Mitra *et al.* 1993). There are three polymorphic forms (A, B and C) in addition to an amorphous form. The characterization of the various forms by melting point and IR analyses proved problematic, even inconclusive, due to polymorphic transitions during grinding for sample preparation (Borka and Backe-Hansen 1968).

The A form is the most stable, but only the B and amorphous forms are biologically active. Aguiar *et al.* determined the physiological absorption rate as a function of the A and B polymorphs, as shown in Fig. 7.3. The suspension containing only the metastable B form gives higher blood levels following oral doses than those containing only Form A, by nearly an order of magnitude. Since particle size was shown to have little effect on blood levels, it was concluded that the structure of the solid plays an intimate role in determining the physiological absorption rate. As a result of this finding, the mechanism of this absorption and its connection with the polymorphism were investigated in considerable detail.

CAPP is nearly insoluble in water; hence, it must be hydrolysed by enzymes in the small intestine before absorption can take place. According to one possible proposed mechanism (Aguiar *et al.* 1967), the first and rate determining step in the total process

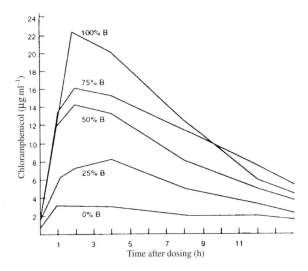

Fig. 7.3 Peak blood serum levels of chloramphenicol following dosing for pure polymorphs A and B and various mixtures for a single oral dose equivalent to 1.5 g of chloramphenicol palmitate. (From Haleblian and McCrone 1969 (after Aguiar *et al.* 1967) with permission.)

is a dissolution of the ester followed by enzymatic hydrolysis of CAPP. Such a mechanism is consistent with the generally accepted dissolution/absorption mechanism for most solid drugs. However, Andersgaard *et al.* (1974) proposed a second mechanism in which *solid* CAPP is enzymatically attacked in the small intestine. If dissolution is the first and rate determining step of the total process, then there should be a close relationship between the rates of dissolution and the rates of enzymatic hydrolysis of polymorphs A and B. On the other hand, no relationship of this sort is expected if CAPP is attacked in the undissolved state.

The rates of dissolution on the one hand and *in vitro* hydrolysis of the solid by the enzyme pancreatic lipase on the other hand are given in Fig. 7.4. If dissolution is the first step in the total hydrolysis process, the reaction scheme may be written as

$$\text{undissolved CAPP} \quad \Rightarrow \quad \text{dissolved CAPP} \quad \Rightarrow \quad \text{hydrolyzed CAPP}.$$

Since the rate of the second step of this process must be the same for Forms A and B of CAPP, this model for the absorption process leads to the conclusion that any differences in the rate of formation of hydrolysed CAPP must be due to a difference in the rate of dissolution of the two polymorphs. The data presented in Fig. 7.4 are not compatible with the assumption that dissolution is the first and rate determining step, since the slopes at time zero are significantly different from those expected for such a mechanism. Rather, Andersgaard *et al.* claim that it is more reasonable to assume that CAPP is attacked in the undissolved state, probably by pancreatic lipase, which is known to act on substances insoluble in water (Waki 1970). Further studies on CAPP and the analogous stearate (Cameroni *et al.* 1976) apparently corroborate

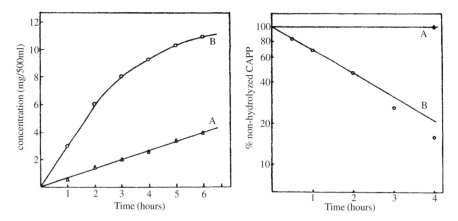

Fig. 7.4 Left: Rates of dissolution of polymorphs A and B of CAPP. Right: Rates of *in vitro* enzymatic hydrolysis by pancreatic lipase of polymorphs A and B of CAPP. (After Andersgaard *et al.* 1974, with permission.)

this assumption, indicating that there is a direct connection between the solid state structure and the enzymatic hydrolysis of chloramphenicol palmitate. It has been found that storage of the B form at the relatively high temperature of 75 °C leads to conversion to the inactive A form in a matter of hours, suggesting that extended storage at higher temperatures could lead to reduced efficacy of a formulated product (Devilliers *et al.* 1991).

The effect of the transformation of crystal form after ingestion has been of increasing interest in the past decade (see also Chapter 10). In a recent study of the two anhydrous Forms (I and III) and a dihydrate of carbamazepine (Kobayashi *et al.* 2000), it was shown that the initial *in vitro* dissolution rate was of the order III > I > dihydrate, with Form III being transformed more rapidly to dihydrate than Form I. The solubilities of the two anhydrous forms were 1.5–1.6 times that of the hydrated form. When dosed to dogs at 40 mg per body weight there were no significant differences between for the area under the curve. However, at doses of 200 mg per body weight significant differences in plasma concentration versus time were observed. It was suggested that the difference was due to rapid transformation from Form III to dihydrate in the gastrointestinal fluids.

Two recent examples involving commercially successful drugs indicate the diversity in bioavailability among different polymorphic systems. At one extreme, the two polymorphic forms of ranitidine hydrochloride (Glaxo SmithKline's H_2-antagonist Zantac®) have been shown to be bioequivalent, which is one of the reasons that ethical and generic companies were involved in litigations over the two forms (see Section 10.2). At the other end of the spectrum is Abbot's protease inhibitor Norvir®, generically ritonavir. After approximately two years on the market a new thermodynamically stable polymorph began to precipitate out of the semisolid formulated product. This proved to have lower solubility with greatly reduced bioavailability, resulting in removal of the drug from the market for almost a year, until a new gel capsule (i.e. 'solution') formulation could be developed (Chemburkar *et al.* 2000). The apparently greater stability of a newly prepared crystal form of glibenclamide has also been attributed to reduction in dissolution and bioavailability of its tablets (Panagopoulou-Kaplani and Malamataris 2000).

7.4   Microscopy and thermomicroscopy of pharmaceuticals

In Chapter 4, we noted the efficiency, simplicity, and utility of microscopy and thermomicroscopy in the characterization and study of polymorphs. Perhaps the most widespread and systematic application of these techniques has been in the field of pharmaceutical materials, as promoted particularly by the Institute of Pharmacognosy at the University of Innsbruck, and summarized in the (now out of print) book by Kuhnert-Brandstätter (1971). In addition to introductory chapters on techniques of hot stage microscopy, the book contains a compilation of the results of hot stage studies on approximately 1000 pharmaceutically important compounds. Many of these results were reported in more detail in an extensive series of papers beginning in the 1960s (Kuhnert-Brandstätter 1962) and continuing through the 1980s by Kuhnert-Brandstätter and her late successor, A. Burger. Later contributions often dealt with compounds actually listed in a pharmacopoeia (e.g. Burger *et al.* 1986).

Recent descriptions of microscopy and thermomicroscopy applied to pharmaceutically important compounds, along with quite comprehensive references to standard texts on the fundamentals and apparatus of the technique may be found in Byrn *et al.* (1999) and Brittain (1999*b*).

Prior to the development of routine and increasingly sophisticated analytical methods, hot stage microscopy competed as one of the principal tools for polymorph characterization and classification. As noted in Chapter 4, successful strategies for the investigation of polymorphs require the application of as many analytical techniques as possible, and hot stage microscopy should be considered as one of the first, if not *the* first to be employed in a comprehensive characterization of a compound (Morris *et al.* 1998). In spite of our current ability to generate a great deal of precise analytical data, there is often no substitute for physically observing solid materials and their behaviour as a function of temperature, preferably on the polarizing microscope. Evidence for the success of this approach has been provided in a number of recent studies.

In some cases, new phases that may not be detectable by other methods may be detected optically (Chang *et al.* 1995). Solid state conversions and their monotropic (Burger *et al.* 1997) or enantiotropic nature (Henck *et al.* 2000), or the products of desolvations may be easily recognized (Schinzer *et al.* 1997). Intimate processes of polymorphic behaviour, such as nucleation, crystal growth, habit transformation, sublimation and properties of the melt (e.g. degradation) may be readily observed and video recorded (de Wet *et al.* 1998).

As noted at the conclusion of Chapter 4, the amalgamation of a number of analytical techniques into a single instrument considerably expands the possibilities for detecting and characterizing polymorphs. This is a particularly powerful combination when optical/thermomicroscopy (preferably with video recording) is combined with other analytical methods. Thermomicroscopy combined with DTA led to the verification of a monotropic transformation between two forms of ibopamin (Laine *et al.* 1995). Combining hot stage microscopy with Raman spectroscopy led to the direct characterization and correlation of the thermal and spectroscopic information on three polymorphs of paracetamol, as well as the first report of lufenuron, a chitin synthesis inhibitor used in pest control and crop protection, and the identity of the polymorphic form in the marketed tablets (Szelagiewicz *et al.* 1999). FTIR was incorporated into a hot stage microscope to simultaneously obtain the visible and spectroscopic characterization of the three polymorphic forms of carbamazepine (Rustichelli *et al.* 2000). Finally, microspectroscopic FTIR and FT-Raman were combined with hot stage microscopy to study the polymorphism in (R, S)-proxyphyilline, including the production and characterization of a new, kinetically very unstable form that most likely could not be detected or analysed by any other technique (Griesser *et al.* 2000).

7.5   Thermal analysis of pharmaceuticals

The fundamentals of the application of thermal analysis in the study and characterization of polymorphs are given in Section 4.3, and many of the examples presented there are on pharmaceutically important compounds. The use of thermal analysis in the pharmaceutical area, including specific applications to polymorphic materials was reviewed in the early 1980s (Wollmann and Braun 1983; Giron-Forest 1984), the former reference containing a specific list (with citations) of pharmaceutically important compounds (both actives and excipients) that had been studied using thermal methods. The rapid technological developments in this area have led to increasingly wider use in the pharmaceutical industry in general (Giron and Goldbronn 1997; Clas *et al.* 1999; Thompson 2000), with a number of more recent reviews covering various aspects of polymorphism (Giron 1990, 1995, 1997, 1998; Kuhnert-Brandstätter 1996).

Many of the new developments in thermal methods as applied to pharmaceuticals in general and polymorphic systems in particular have been reviewed by Giron (1999). A study combining the use of thermal methods with a variety of additional techniques to characterize the polymorphic behaviour of a purine derivative (designated MKS 492 by Novartis) again demonstrates the complementarity of these methods and the advantages to be obtained by applying as many of these as possible (Giron *et al.* 1999). The study revealed an amorphous form in addition to six crystalline modifications of which four are pure crystalline (i.e. unsolvated) forms. Temperature resolved X-ray diffraction was used in conjunction with differential scanning calorimetry (DSC) to aid in interpretation of thermal events, leading to the identity of the most stable form at room temperature, which was chosen for further development. Using microcalorimetry, the authors developed a protocol for quantitative determination of the amorphous material as a contaminant in the desired polymorphic form. The method was particularly useful at amorphous content <10 per cent with a limit of detection of about 1 per cent.

The development of new thermal analytical techniques has naturally led to the reinvestigation of previously studied systems. The thermal behaviour of sulphapyridine was used to demonstrate many aspects of 'traditional' DSC described in Chapter 4, and a thorough study leading to the energy–temperature diagram was published by Burger *et al.* (1980). Nevertheless, there was still considerable confusion among various authors as to the naming and identity of the five polymorphs (Bar and Bernstein 1985). The compound was recently reinvestigated by Bottom (1999) combining the traditional DSC with the recently developed modulated temperature DSC (MTDSC) monitored by thermomicroscopy, resulting in considerably more understanding and insight into the nature of the solid–solid transitions, in particular the glass transition, all of which were more easily interpretable using the MTDSC technique.

Tetracaine hydrochloride was also revisited using thermal methods, revealing a complicated system containing six anhydrous crystalline forms, an amorphous form, a hemihydrate, a monohydrate and a tetrahydrate. As with the purine derivative noted above, the identification and classification of these modifications was considerably aided by the use of temperature resolved X-ray diffraction (Giron *et al.* 1997).

We mention briefly here some additional recent developments in the application of thermal methods to polymorphism in pharmaceuticals. Subambient DSC has been used to determine the melting behaviour of dosage forms which may vary depending on the polymorph that might crystallize out below room temperature (Schwarz and Pfeffer 1997). Pressure differential scanning calorimetry (PDSC) has been used in conjunction with variable temperature X-ray diffraction to quantify the relative amounts of the anhydrate and trihydrate of ampicillin in mixtures of the two. PDSC could also be used to detect changes in crystallinity due to milling on a quantitative basis with a high degree of precision (Han *et al.* 1998).

The so-called 'hyphenated techniques', incorporating thermal methods as one of the combined analytical techniques are sure to play an increasing role in the identification and characterization of crystalline forms of pharmaceutical substances. The combination of TGA with FTIR allows the simultaneous quantitative analysis of weight changes during thermal processes with the IR identification of the decomposition products (e.g. solvent) resulting from those processes (Materazzi 1997). For substances with low volatility, the FTIR analysis may be replaced with mass spectroscopy (Materazzi 1998).

7.6   The importance of metastable forms

As the energy–temperature phase diagram (Section 2.2.3) indicates, at any particular temperature only one polymorphic form is the thermodynamically stable one (except, of course at the temperature of a transition point). The stable form is also the least soluble form at a given temperature (Byrn *et al.* 1999). All other phases are higher in energy and metastable with respect to the most stable phase. A metastable phase may be fleeting in nature (e.g. benzamide, Bernstein and Davis 1999) or may coexist indefinitely with its more stable counterpart in the absence of any perturbations (e.g. ranitidine hydrochloride (Cholerton *et al.* 1984; see also Giron 1988; Giron *et al.* 1990)).

The lower solubility of stable forms may limit their pharmacological utility (e.g. ritonavir (Chemburkar *et al.* 2000; Bauer *et al.* 2001)), so that it may be advantageous to selectively obtain and maintain a metastable form in a formulation (e.g. Shah *et al.* 1999). In such cases, crystallization strategies may be designed on the basis of the principles derived from the energy–temperature or pressure–temperature diagrams (Toscani 1998), as described in Chapter 3. It will be recalled that even if qualitative in many aspects, such diagrams serve to summarize a great deal of information in a very compact manner. For instance, characterization of the two polymorphs of taltireline, a central nervous system activating agent, indicated that they were enantiotropic, but the α form, metastable at its crystallization temperature of 10 °C, was preferred for formulation. Critical evaluation of the crystallization parameters isolated the factors that led to conversion to the stable form, and these were controlled to prevent conversion (Maruyama *et al.* 1999). For a two-component system generation of the phase diagram can also prove very useful in developing strategies for obtaining a number of crystal modifications, including a metastable one (Henck *et al.* 2001).

Together with knowledge of the phase diagram an increasing variety of techniques have been designed and employed to generate metastable modifications. Seeding of course, is one of those strategies, and Beckmann *et al.* (1998) developed a seeding strategy for a batch cooling crystallization to obtain quantitatively and reproducibly a metastable form of abecarnil, regardless of the purity of the material. In another approach, after thorough characterization of three polymorphic modifications by a variety of analytical methods, a desired metastable form of (R,S)-proxyphylline was crystallized in gram quantities from the supercooled melt, and proved to have considerable kinetic stability under dry atmospheric conditions (Griesser *et al.* 2000). A variation on that same theme was the successful high-temperature crystallization from the amorphous material of the metastable α form of indomethacin, whereas the low temperature crystallization yielded the stable γ form (Andronis and Zografi 2000).

One traditional strategy for screening a compound for polymorphic behaviour involves the trial of a variety of solvents and solvent mixtures. Our understanding of the role and choice of solvent has improved considerably and this information, combined with a knowledge of zones of stability can aid in determining crystallization conditions for obtaining metastable forms (Threlfall 2000). In addition, there has also been considerable progress in understanding and utilizing the interactions of solvent with the growing crystal (Weissbuch *et al.* 1991; Lahav and Leiserowitz 1993). Combining the detailed structural information available from the single crystal structure determinations of polymorphs with crystal morphological data (i.e. crystal habit, and the orientation of molecules projecting from the particular faces exposed) and with known intermolecular interactions between solute molecules and solvent functional groups allows the rational choice of solvent to select a particular polymorphic form (Weissbuch *et al.* 1995). An analysis of this nature was carried out and experimentally confirmed by Blagden *et al.* (1998*a*,*b*) for polymorphic modifications of sulphathiazole.

Another manifestation of the strategy of using detailed structural information to steer a crystallization involves the design of tailor-made additives (Weissbuch *et al.* 1995) which can be used to inhibit the growth of the stable form in order to favour the growth of a metastable form. In the classic examples employing this strategy, the additives are chosen to favourably interact with the molecules or parts of molecules projecting from particular crystal faces. Upon interacting they create a crystal surface which is no longer suitable for successive growth, thus leading to the inhibition. In a variation on this theme, when the crystal forms involved are conformational polymorphs, the choice of additive can be aided by considering the conformational differences between the polymorphic forms, a strategy that was successfully followed to inhibit the growth of the stable (β) form of L-glutamic acid to obtain the desired metastable α form (Davey *et al.* 1997). Another group, faced with the same problem at the batch level, found that seeding with the α form (seeds that do not contain any other solid phase) and increasing the cooling rate could lead to the selective crystallization of the desired α phase (Yokota *et al.* 1999).

This is clearly an area where the combination of thermodynamic, kinetic and structural information potentially can lead to successful strategies for controlling the polymorphic form obtained, in specific instances a metastable form, and as the means for obtaining these data become more sophisticated the approaches described here are sure to be developed and expanded (see also Section 3.7).

7.7   The importance of amorphous forms

Although polymorphism generally refers to different *crystalline* forms, no real crystal is perfect. The lack of perfection is manifested in disorder, and in the extreme when the entire material lacks long range order (although it may maintain some short range order), the result is an amorphous material (Klug and Alexander 1974). Amorphous materials are generally more energetic than crystalline materials; hence, they tend to have higher solubilities and rates of dissolution, properties which may even make an amorphous form advantageous over a crystalline one in a pharmaceutical formulation (Hancock and Parks 2000; Yu 2001). By the same token, the presence of some amorphous material in a crystalline sample may profoundly influence the properties of the material.

Because of their metastable state, amorphous materials are often difficult to prepare and handle; moreover, there is always the possibility that a stable crystalline state will crystallize from the amorphous material, leading to changes in properties and possibly rendering the resulting form unsuitable for pharmaceutical use (Craig *et al.* 1999). Nevertheless, quite a few pharmaceutical products are marketed with amorphous material as the active ingredient (e.g. Byrn *et al.* 1995; Giron 1997) and increasing attention is being paid towards the preparation, detection, characterization, and stabilization of amorphous forms (Threlfall 1995; Hancock and Zografi 1997;

Byrn *et al.* 1999; Yu 2001). One increasingly used method to stabilize the amorphous form is to disperse it in a polymer matrix such as polyvinylpyrrolidone, although there is still the possibility that crystalline material will crystallize out (Kimura *et al.* 2000).

Amorphous pharmaceutical materials are typically obtained by employing crystallization procedures far from equilibrium such as (rapid) solidification from the melt, lyophillization (freeze drying) or spray drying, removal of solvent from a solvate, precipitation by changing pH, or by mechanical processing such as granulation, grinding or milling (Guillory 1999). Recently developed supercritical fluid technology is certain to play a role in preparation of amorphous and metastable forms (Shekunov and York 2000). The interest in understanding the processes involved in preparing amorphous materials or those that often result from the inherent tendency of amorphous materials to undergo changes to energetically lower states has led to some comprehensive studies. Many of these involve the same analytical techniques used to characterize crystal modifications. For example, the *formation* of amorphous anhydrous carbamazepine from the crystalline dihydrate was studied *in situ* by DSC, TGA and variable temperature X-ray diffraction (Li *et al.* 2000). As demonstrated by Ceolin *et al.* (1995) in another study, the amorphous state may also be obtained by quenching the melt. On the other hand, the *degradation* of amorphous quinapril hydrochloride, prepared by two methods, was followed by DSC, TGA, X-ray powder diffraction, polarized optical microscopy, scanning electron microscopy and IR spectroscopy (Guo *et al.* 2000).

Since amorphous material may act as a contaminant, it is often necessary to quantitatively determine its concentration in a crystalline sample. A variety of techniques of analysis for the quantitiative detection of an amorphous phase in a crystalline phase have been developed, and these are undergoing rapid change and improvement (Buckton and Darcy 1999; Stephenson *et al.* 2001). As always, the choice of method will depend on the material and the circumstances, but a few examples are noted here to demonstrate some recent developments. Taylor and Zografi (1998) employed FT-Raman spectroscopy to quantify the degree of crystallinity of indomethacin. They found a linear correlation over a 0–100 per cent range of crystallinity with a limit of detection of 1 per cent crystalline or amorphous content. Diffuse near-infrared reflectance spectroscopy (NIRS) was employed on the same compound together with mixtures of amorphous/crystalline sucrose (a potential excipient) (Seyer *et al.* 2000). The method was found to be slightly more precise than analyses carried out by the more traditional X-ray (Klug and Alexander 1974; Zevin and Kimmel 1995) and DSC techniques.

The evolution in calorimetry technology has also led to the development of protocols for quantitative analysis (Buckton and Darcy 1999). Fiebich and Mutz (1999) determined the amorphous content of desferal using both isothermal microcalorimetry and water vapour sorption gravimetry with a level of detection of less than 1 per cent amorphous material. The heat capacity jump associated with the glass transition of amorphous materials MTDSC was used to quantify the amorphous content of a micronised drug substance with a limit of detection of 3 per cent w/w of amorphous substance (Guinot and Leveiller 1999). Significant variations in the chemical stability profiles of an experimental drug could be attributed to differences in amorphous content of around 5 per cent, which was also quantified by MTDSC (Saklatvala *et al.* 1999).

7.8   Concluding remarks

The development of a pharmaceutical product can be a long, arduous and expensive process. For those that are marketed in solid dosage forms that process deals with many matters of solid state chemistry, among them the detection, characterization, and preparation of various modifications. We have attempted here to touch on many of those issues, and as noted at the beginning of the chapter, examples of these various aspects of polymorphism and their manifestations in the pharmaceutical industry may be found throughout this book.

For scientific, regulatory and intellectual property reasons recent years have witnessed increasing efforts to discover and characterize as many different crystal modifications as possible of substances associated with a pharmaceutical product. This effort has been aided and enhanced by the rapid developments in instrumentation and analytical techniques, so that often old compounds have been revisited for study to the same extent that new ones have been subject to investigation. In light of the new and often more sensitive techniques available, many questions and problems noted 20 or more years ago warrant reinvestigation. One of these involves the large numbers of crystal modifications reported for some substances, for instance, phenobarbitone (thirteen modifications) (Cleverly and Williams 1959; Mesley *et al.* 1968; Stanley-Wood and Riley 1972) or cimetidine (seven modifications) (Hegedus and Görög 1985; Sudo *et al.* 1991). In a recent reexamination of a number of the sulpha drugs Threlfall reported discovering a new polymorph of sulphathiazole, two new polymorphs of sulphapyridine, 120 solvates of sulphathizole and 30 solvates of sulphapyridine (Threlfall 1999, 2000). Complete structural studies of diverse crystal modifications of a single substance are beginning to appear, for instance the seven solvated forms of 3,5-dinitrosalicylic acid (Kumar *et al.* 1999). In some cases of historically well documented multifarious polymorphism, systematic attempts have recently been made to prepare the various forms for single crystal analysis. For instance, p'-methylchalcone has been shown to exist in 13 different polymorphic forms (Weygand and Baumgärtel 1929; Eistert *et al.* 1952). Five of the forms that are metastable at room temperature have been prepared; three crystal structures have been carried out, while the other two underwent transformations during the data collection (Bernstein and Henck 1998).

Earlier reports of complex energy–temperature relationships (e.g. dapsone and ethambutol chloride (Kuhnert-Brandstätter and Moser 1979)) or unusual crystal chemistry (e.g. oxyclozanide (Pearson and Varney 1973)) also warrant serious consideration for reinvestigation. The formulation implications of chemical reactivity in solid-state pharmaceuticals have recently been reviewed by Byrn *et al.* (2001).

One issue that we have not touched on is that of the difficulties that are often encountered in process of scaling up from laboratory quantities and procedures, through the pilot plant and into full production. Equipment changes, differences between the quality of laboratory grade and bulk chemicals, variations in heating/cooling rates, stirring procedures (Genck 2000), seeding (Brittain and Fiese 1999; Beckmann 2000) etc., can all influence the result of a crystallization procedure and the polymorph obtained (Morris *et al*. 2001). Little of this information is documented in the literature (although for exceptions see Wirth and Stephenson 1997), for it is often a matter of empirical testing and development maintained as trade secrets or incorporated into the collective memory of an industrial concern, although passing mention may appear in the literature (Yazawa and Momonaga 1994; Giron 1995; Giron *et al*. 1999; Rodriguez-Horneido and Murphy 1999). Scale up and subsequent formulation also involve more complex transfer and processing procedures, which are affected by the physical and mechanical properties of the solids, likewise a function of crystal modification (Sun and Grant 2001). Some of these have been described by Hulliger (1994), including tensile strength (Summers *et al*. 1977), compression (e.g. DiMartino *et al*. 1996, 2001; Suihko *et al*. 2000), flow properties (Beach *et al*. 1999), filtration and drying characteristics (Crookes 1985; Crosby *et al*. 1999).

We have mentioned in passing the regulatory aspects of polymorphism in the pharmaceutical industry. Increasing attention is being paid by regulatory agencies to the preparation, identity, characterization, purity, and properties of the crystal form used in pharmaceutical products (Byrn *et al*. 1999). Such information is now required by the U.S. Food and Drug Administration in a New Drug Application, and Byrn *et al*. (1995) have presented a set of decision trees to aid in presenting the data on different crystal forms (polymorphs, solvates, desolvated solvates, and amorphous forms) to regulatory agencies. The International Committee on Harmonization has set up guidelines (Q6A) for addressing the issue of polymorphism in pharmaceuticals (Federal Register 2000).

8

Polymorphism of dyes and pigments

The representatives of organic chemistry are coming more and more to the conclusion that the formulae of limiting states which we have been using so far fail to reflect the real conditions prevailing in nature, for there exist such subtle differences in the state of matter compared to which our methods of description that still are very simple sometimes appear wholly inefficient. (Ismailsky 1913)

8.1 Introduction

Man has always been fascinated and charmed with colour. The dye indigo has been in use for over four millennia, and frequently is the source of the colour of the ubiquitous blue jeans around the world. It is probably not an exaggeration to state that dyes and pigments formed the basis of modern industrial organic chemistry. The synthesis of 'aniline purple' (mauvein) by the 18-year-old Perkin in 1856 catalysed a revolution in the chemical industry, with widespread commercial and even cultural ramifications. The dye and pigment industry quickly became the cornerstone of the chemical industry in both England and Germany and generated many additional new technologies and new industries. As of 1995 the worldwide consumption of organic pigments was estimated at 210 000 tonnes, valued at approximately 4 billion dollars,[1] although reliable data are notoriously difficult to obtain (Herbst and Hunger 1997). According to Hao and Iqbal (1997) about half of this market is based on the azo chromophore, while another quarter is based on phthalocyanines (mainly copper). While considerably less than the pharmaceutical industry, this is still a significant commercial enterprise, and the understanding of, and control over the nature of the materials in question is no less important.

The basis of the theory of all aspects of the sources and perception of colour, including an excellent discussion on the physics and chemistry of dyes and pigments has been given by Nassau (1983) (see also Tilley 1999). Actually, the precise distinction between dyes and pigments is still a matter of discussion, even controversy, much the same as the definition of polymorphism (see Section 1.2.1), but as in the latter case the working definition is generally accepted by most workers in the field. Most dyes are generally soluble, while pigments are regarded as insoluble in the medium being considered. Hence, the physics of the colour discerned by the viewer is different for the two classes (Evans 1974). As Nassau (1983) has noted, one feature common to dyes

---

[1] The value of all coloured goods is estimated (2001) at $300 bn, with $120 bn derived from azo dyes, $80 bn from phthalocyanines, and $5 bn from tetraphenylporphyrins.

Table 8.1 Collection of some references to polymorphic behavior of pigments[a]

Pigment	Number of forms	Reference or page number[b]
Quinacridone pigments	At least 3	41
Acid blue 324	2	Sandefur and Thomas 1984
Acid Orange 156	2	Höhener and Smith 1987
Dioxazine violet	2	Hayashi and Sakaguchi 1981, 1982
Disperse Brown 1	α, β	Kruse and Sommer 1976
Disperse Orange 5	2	Ghinescu et al. 1984
Disperse Orange 29	3	Hähnle and Opitz 1976
Disperse Red 65	α, β	Sommer and Kruse 1979
Disperse Red 73	3	von Rambach et al. 1974; Wolf et al. 1986a
Disperse Yellow 23	2	Koch et al. 1987a,b
Disperse Yellow 42	2	Burkhard et al. 1968; Flores and Jones 1972; Koch et al. 1987a,b
Disperse Yellow 68	3	Sommer et al. 1974; Wolf et al. 1986b
Pigment Red 9	At least 2	285
Pigment Red 12	At least 2	285
Pigment Red 49	Several modifications	317
Benzimidazolone pigments	'polymorphism common'	351
'several azomethine pigments'	?	411
'Polyster red A' $C_{25}H_{23}N_5O_4S$	3	Kuhnert-Brandstätter and Reidmann 1989
'Polyester red B' $C_{27}H_{25}N_5O_5S$	3	Kuhnert-Brandstätter and Reidmann 1989
Linear trans acridones	Multiple (α, β, γ)	461
"	β, two more γ modifications	462
Pigment Black 31	I and II	Hädicke and Graser (1986a); Mizuguchi (1998b)
Pigment Violet 19	5	465–6
Pigment Violet 23	2	Sakaguchi and Hayashi 1981, 1982; Curry et al. 1982
Pigment Violet 122	4	465, 469
Pigment Orange 36	2	Dainippon 1982
Pigment Red 1	3	Grainger and McConnell 1969 Whitaker 1979, 1980a,b, 1981, 1982
Pigment Red 31	3	Griffiths and Monahan 1976
Pigment Red 53 : 1 (Ba^{++} salt)	α and β	Schui et al. 1983; Hoechst 1982; Duebel et al. 1984
Pigment Red 53 : 2 (Ca^{++} salt)	15 polymorphic and pseudopolymorphic forms	Schmidt 1999a,b, 2000; Schmidt and Metz 1998a,b, 1999a,b; Farbwerke 1902

Table 8.1   (*Continued*)

Pigment	Number of forms	Reference or page number[b]
Pigment Red 53 : 3 (Sr^{++} salt)	4	Dainippon 1996a–d
Pigment Red 57 (Ba^{++} salt)	α and β	Dainippon 1988
Pigment Red 57 (Sr^{++} salt)	α and β	Danippon 1986
Pigment Red 57 : 1 (Ca^{++} salt)	α and β	Chen 1990
Pigment Red 57 : 2	4	Kobayashi and Ando 1988a,b,c
Pigment Red 122	4 (?)	Eshkova *et al.* 1976; Kelly and Giambalvo 1966
Pigment Red 149	α, β and γ	Bäbler 1983; Imahori and Hirako 1976; Spietschka and Tröster 1988a,b
Pigment Red 170	3	305 Ribka 1969, 1970
Pigment Red 177	3	Kosheleve *et al.* 1987
Pigment Red 187	At least 2	258, 308 Ribka 1961
Pigment Red 194	2	Shtanov *et al.* 1980, 1981; Pushkina and Shelyapin 1988; Shelyapin *et al.* 1987
Pigment Red 202	3	465, 470
Pigment Red 207	4 (?)	Wagener and Meisters 1970a,b
Pigment Red 209	3 or 4	Curry *et al.* 1982; Deuschel *et al.* 1964
Pigment Red 224	α and β	Ogawa *et al.* 1999
Pigment Red 247 (Ca^{++} salt)	α and β	341 Hoechst 1988; Froelich 1989
Pigment Red 254	α and β	Hao *et al.* 1997
Pigment Red 255	2	Ruch and Wallquist 1997
Pigment Yellow 5	α and β	Whitaker 1985a,b
Pigment Yellow 10	4 or 5	Momoi *et al.* 1976; Bäbler 1978; Fuji *et al.* 1980
Pigment Yellow 12	3	Kawamura 1987; Tuck *et al.* 1997; Curry *et al.* 1982
Pigment Yellow 16	5	Rieper and Baier 1992
Pigment Yellow 17	α and β	Shenmin *et al.* 1992
Zapon Fast Yellow C2G	3	Susich 1950
Solvent Orange 63	2	Shimura *et al.* 1988
Solvent Yellow 18	3	Whitaker 1990, 1992
Sudan Orange R	4	Susich 1950

Table 8.1   (*Continued*)

Pigment	Number of forms	Reference or page number[b]
Vat Green 1	2	Popov *et al.* 1981
Pigment Blue 60 (Indanthrone)	4	516; FIAT 1948; Susich 1950; Dainippon 1994
Dioxazine	More than one	533
Indigo	A and B	von Eller 1955; Süsse and Wolf 1980
Cu phthalocyanine	$\alpha, \beta, \gamma, \delta, \varepsilon$	41, 425, 434–443
	As many as 9	Erk 1998
Metal phthalocyanines	2–4	Law 1993
Metal-free phthalocyanine	5	439
	7	Whitaker 1995

[a] This table does not purport to be comprehensive. It contains entries encountered in researching this book, and is meant to serve as an entry into the literature on polymorphic colourants designated by the Colour Index. There are no doubt many additional instances that have not been noted (e.g. Whitaker 1995).
[b] Refers to location in Herbst and Hunger (1997).

8-I

Quinacridone itself is trimorphic[3] (Manger and Struve 1958; Struve 1959), the three forms being designated as α, β, and γ, distinguished by their X-ray powder patterns. The α and γ are red, while β is violet. On the basis of some subtle variations in the powder pattern of the γ form there was also an earlier claim of the existence of an additional, γ' phase (Whitaker 1977a). The data from the full crystal structure analysis of the γ form were used to calculate the powder pattern, and it was demonstrated that those subtle variations could be simulated by varying the average crystallite size included in the calculation (Potts *et al.* 1994), thus quite conclusively proving the identity of the γ and γ' forms. As noted above, the properties of pigments are functions of a variety of factors, one of them of course being the polymorphic form. As discussed in Chapter 4, the existence and characterization of a new polymorphic form should be determined by the use of a number of analytical techniques, including preferably X-ray diffraction as applied in this case.

[3] Actually, up to at least seven polymorphic forms have been claimed, mostly in the patent literature. See Labana and Labana (1967) and references 15–21 in Potts *et al.* (1994). Polymorphic behaviour has also been claimed for the 2,9-dichloro- (Bohler and Kehrer 1963; Deuschel *et al.* 1964; Nagai *et al.* 1965) and 2,9-dimethoxy derivatives (Ciba 1965).

The structures of the α and β forms remain somewhat enigmatic (Paulus *et al.* 1989), at least in the open literature. Leusen (1996) has claimed that computational methods for generating possible structures and subsequent comparison with X-ray powder diffraction patterns have been used to produce all three of the quinacridone structures, in the correct stability order. However, details are still to be published (Leusen *et al.* 1996). Lincke and Finzel (1996) published a proposed structure of the α-form from a powder pattern that was computed by generating the structure by perturbing the known γ structure (Potts *et al.* 1994). However, the rather high R-factor (20.6 per cent) and the presence of quite a few peaks on the final difference plot, raise some doubts as to the correctness of this proposed structure.

8.3.2 *Perylenes*

Perylene pigments have the basic structure **8-II**. The anhydride (X = O) (Pigment Red 224) is an important intermediate in the preparation of other derivatives, but has also been used as a pigment in a number of applications (Herbst and Hunger 1997). The crystal structures of two polymorphs reported earlier (Möbus *et al.* 1992; Lovinger *et al.* 1984) have recently been carried out using electron crystallography on microcrystalline thin films (Ogawa *et al.* 1999). The two forms have similar packing modes, with experimentally significant differences in cell constants and mutual orientation of neighbouring molecules.

8-II

Perhaps the most widely used of the perylene pigments is Pigment Red 179 **8-II** (X = N–CH$_3$), also known as perylene red. In spite of the intensive investigation and widespread use of this compound, it can join the ranks of sucrose and naphthalene as very commonly crystallized compounds that as yet have exhibited no evidence of polymorphism.

Pigment Red 149 (X = (3′, 5′ dimethyl) phenyimido) is also widely used, especially in textile applications. Three polymorphs are known (Mitsubishi 1976; Bäbler 1983; Mizuguchi 2001), exhibiting different shades of red. The β form has a more yellow tint than the thermodynamically stable α form that is the polymorph used commercially.

Following the initial synthetic (Graser and Hädicke 1980, 1984) and structural (Hädicke and Graser 1986a,b) investigations, Klebe *et al.* (1989) reviewed the *crystallochromy*—effect of solid state on the colour properties—in a series of perylene **8-II** (X = N–R) derivatives for a variety of R-substituents. The latter group found 24 different modes of overlap between neighbouring perylene moieties, and correlated the packing with the absorption properties of the solid pigments. Kazmeier and

Hoffmann (1994) also investigated this problem theoretically, and proposed the concept of a quantum interference effect to account for spectral shifts, a model which contrasts with the exciton explanation given by Mizuguchi (1997). Klebe *et al.* found that packing patterns could be correlated with molecular conformation and steric requirements, but this information could not be used for reliable predictions of crystal structure from the structure of an isolated molecule. Attempts to computationally model and predict some polymorphic structures of perylene pigments were only partially successful (McKerrow *et al.* 1993), which is not surprising, considering the present state of the art of crystal structure prediction even for considerably smaller molecules (Section 5.10) (Lommerse *et al.* 2000; Erk 2001*b*; Erk 2000; Mizuguchi 2001; Motherwell 2001).

Mizuguchi (1997, 1998*a*) carried out combined structural and spectroscopic studies on evaporated films of two perylene pigments **8-III** and **8-II** (X = NCH₂CH₂Phenyl) [PDC] that indicate that the films can undergo polymorphic changes upon exposure to solvent vapours. In the case of **8-III** the original evaporated film is violet, while exposure to acetone leads to a colour change to reddish-purple. The X-ray diffraction patterns of the two films indicate that a structural change has indeed taken place.

8-III

The single crystal structure determination of PDC (Pigment Black 31) was reported by Hädicke and Graser (1986*a*), but Mizuguchi found that vapour phase growth initially led to a brilliant amorphous red film, rather than the black colour usually assigned to this pigment. Exposure to acetone vapour or to temperatures above 100 °C led to a new crystalline phase that differs from that of Hädicke and Graser (Fig. 8.1). The first structure is characterized by a parallel arrangement of neighbouring molecules in $C2/c$, while that determined by Mizuguchi from vapour-phase grown crystals (Mizuguchi 1998*b*) exhibits a herring-bone bond arrangement in $P2_1/c$. Mizuguchi also presents solid state polarized reflection spectra (Section 6.3.2) and discusses how the amorphous (red) to crystalline Form II (black) transition (and vice versa), involving an optical absorption at about 635 nm, can be utilized in optical disk technology applications. Mizuguchi's interpretation of these effects on the basis of exciton theory differs from that given by Kazmaier and Hoffmann (1994).

8.3.3 *Phthalocyanines*

Phthalocyanines **8-IV** (Pc's) comprise one of the most important and widely used single class of pigments, used in printing inks, paints, plastics, and automotive finishes. Annual production of all Pc's currently amounts to about 80 000 tonnes/year with

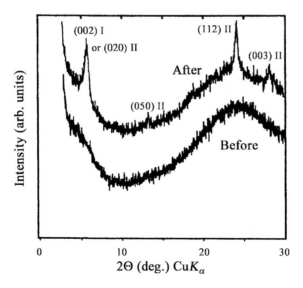

Fig. 8.1 X-ray diffraction pattern of films of PDC (Pigment Black 31) **8-II** (X = NCH₂CH₂Phenyl) [PDC] (Pigment Black 31) before and after exposure to acetone vapours. The material is clearly amorphous prior to exposure, and crystallinity is readily apparent following treatment. The diffraction peaks have been assigned Miller indices by the original author based on correspondence with one of the two known crystal structures. Modification II dominates the transition to the crystalline phase. (From Mizuguchi (1998a), with permission.)

copper phthalocyanine (M = Cu) CuPc accounting for a little over 60 per cent of the total. Activity in the field has spawned a number of books (Moser and Thomas 1963, 1983; Woehrle 1989; Leznoff and Lever 1996), reviews (Booth 1971; Fryer *et al.* 1981; Löbbert 2000) and a specialist journal entitled *Journal of Porphyrins and Phthalocyanines.*

8-IV

Historically, the first mention of Pc's appears to have been by Braun and Tscherniak who in 1907 described a greenish residue on their filter, but provided no further characterization of the material (Herbst and Hunger 1997). Twenty years later de Diesbach and von der Weid (1927) noted that the same preparation lead to a blue product which was recognized and characterized at Scottish Dyes (1929*a*,*b*). Robertson carried out the first crystal structure determination of the metal-free (1936) and Ni derivatives (Robertson and Woodward 1937), at that time the largest molecules to be determined by single crystal methods, and also reported (1935) the cell constants of a number of other Pc derivatives. Other early reports on the polymorphic behaviour, include the X-ray powder diffraction patterns by Susich (1950; FIAT 1948) and the X-ray diffraction patterns and IR spectra by Ebert and Gottlieb (1952).

8.3.3.1 *Copper Phthalocyanine*

For commercial reasons copper phthalocyanine (CuPc) has received the most attention. In most cases new polymorphic forms resulted from changes in crystallization process conditions or in changes in synthetic procedures. At least five modifications—designated α, β, γ, δ, ε—are known (Horn and Honigman 1974; Löbbert 2000, and references therein), and at least four more have been claimed in the literature (Gieren and Hoppe 1971); 'R': Pfeiffer (1962); 'π': Brach and Six (1975); 'X': Miller *et al.* 1972; Moser and Thomas (1983); 'ρ': Komai *et al.* (1977). Additional patents (Byrne and Kurz 1967; Brach and Lardon 1973; Brach and Six 1973; Sharp *et al.* 1972) have led to some confusion, indeed controversy (Assour 1965), as to the existence and identity of various forms. As in other cases in which there are clearly multiple polymorphic modifications, part of the problem lies in nomenclature (see Section 1.2.3), as well in the determination of clear distinguishing characteristics for each of the claimed forms. For instance, in a 1956 patent, Eastes claimed the β form, which all subsequent workers have referred to as the γ form (e.g. Brand 1964).

Herbst and Hunger (1997) have presented the diffraction patterns for five of the CuPc modifications (Fig. 8.2). In addition, Whitaker (1995) carried out a survey and critical evaluation of the X-ray powder diffraction patterns of the various forms of CuPc. He notes that in an earlier review (Whitaker 1977*b*), he had compiled claims for nine different crystalline modifications of CuPc. On the basis of the more recent review, in which he cites the sources of many of the powder patterns, he suggests the following sources as the preferred references for the X-ray powder diffraction patterns of the various forms:

α: preferred listing is given in the Powder Diffraction File (PDF, see Section 1.3.4), (PDF 36-1883);

β: preferred listing (PDF – 37-1846) has been indexed from the single crystal structure determination (Brown 1968*a*);

γ: Whitaker (1977*b*); see also Komai *et al.* (1977) and Wheeler (1979).

δ: the powder patterns published subsequent to the Whitaker (1977*b*) review (Komai *et al.* 1977; Enokida and Hirohashi 1991) differ significantly from that in the review. Whitaker's comparison of the patterns indicates that one of the Komai patterns claimed for the so-called δ_K form matches that for the ε form, while a second, purported new pattern for this δ_K form actually corresponds to the X form. The pattern given by Enokida and Hirohashi matches that in the Whitaker review

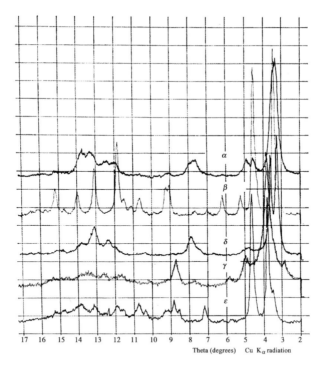

Fig. 8.2   X-ray diffraction patterns of five modifications of CuPc. (From Herbst and Hunger 1997, with permission.)

save an apparent typographical error for one d-spacing which should be 8.75 Å in the review rather than 7.85 Å.

ρ: this form has been reported by Reznichenko et al. (1984) and in a series of patents by Ninomiya et al. (1979). As Whitaker points out, most of these are from the same source, raising the question of reproducibility, but the pattern does seem to be unique.

σ: according to Whitaker, the pattern claimed by Enokida and Hirohashi (1991) appears to be unique.

In addition, Whitaker notes that the new pattern claimed by Wheeler (1979) is identical to that of the α form, the one claimed by Suzuki et al. (1989a) is an improved crystalline specimen of the π form, and that claimed by Suzuki et al. (1989b) can be interpreted as a poorly crystalline sample of the α form.

On the basis of solubilities in benzene, Horn and Honigman (1974) estimated the relative thermodynamic stability of the five most commonly recognized polymorphs as $\alpha = \gamma < \delta < \varepsilon < \beta$. Benyon and Humphreys (1955) determined the enthalpy difference between the two commercially most significant α and β forms as $10.75\,\mathrm{kJ\,mol^{-1}}$, which is consistent with the results of the solvent-mediated $\alpha \to \beta$

phase transformation reported by Cardew and Davey (1985). Other modifications also transform to the β form upon heating in inert, high-boiling solvents (Löbbert 2000). Synthesis of CuPc generally leads to the β form. Dissolution of this modification in concentrated H_2SO_4 or treatment with 55–90 per cent H_2SO_4 leads to the greenish yellow sulphate (Honigman 1964), which upon hydrolysis gives α CuPc. Grinding in the presence or the absence of additives also induces the $\beta \rightarrow \alpha$ transformation, which has also been studied by a variety of methods (Suito and Uyeda 1963, 1965, 1974). Investigations of crystal growth have been carried out using both electron microscopy (Ashida *et al.* 1971) and XRD (Haman and Wagner 1971), and many other aspects of the transitions between the α and β phases have been reviewed (Fryer *et al.* 1981).

The crystal structure of the β form has been reported by Brown (1968a); that of the α form has not yet been reported due to the difficulty of obtaining suitable single crystals (Honigmann *et al.* 1965), but some confusion has been perpetrated in the literature concerning the nature of this structure. Horn and Honigman (1974) appear to have been the first to publish the schematic drawing of the presumed structure of α-CuPc. The same figure, appearing in a number of subsequent publications (e.g. Fryer *et al.* 1981; Law 1993; Herbst and Hunger 1997; Hunger 1999; Löbbert 2000), does not correctly represent the structure reported by Brown, which in fact is the structure of α-PtPc (Brown 1968b). Brown (1968b) noted that the cell constants of α-CuPc determined from X-ray powder diffraction photographs of α-CuPc bore a strong resemblance to the cell constants of PtPc 'which suggested that the two crystal structures might be similar'.[4] The structure of α-PtPc is reported in a non-conventional space group $C2/n$, with the following cell constants: $a = 26.29$, $b = 3.818$, $c = 23.92$ Å, $\beta = 94.6°$. Brown also presented these in the more conventional setting as space group $C2/c$: $a = 26.29$, $b = 3.818$, $c = 34.09$ Å, $\beta = 135.6°$, which we employ here. Since $Z' = 0.5$ centrosymmetric molecules lie on inversion centres at the unit cell origin (with coordinates 0, 0, 0 for the metal atom) and because the cell is C-centred there is also a molecule with metal atom at the coordinates 0.5, 0.5, 0. If α-CuPc and α-PtPc are indeed isostructural (an assumption for which we do not have sufficient data to verify or deny), then at the very least the schematic representation for the structure should be as presented in Fig. 8.3. Note that the nature of the stacking along the short b axis is not altered by correcting the view of this structure, but that neighbouring parallel stacks are offset by one half the unit cell translation along the b axis (i.e. by 1.9 Å).

The IR, UV-VIS, and XPS, dark current and photocurrent have been compared for a number of the polymorphs of CuPc (Knudsen 1966; Enokida and Hirohashi 1991). The most important characteristic of a pigment is its colour; the variation in spectral response five CuPc polymorphs, presented in Fig. 6.9, shows considerable

---

[4] The reference to the determination of the cell constants of α-CuPc is given as 'C. J. Brown, to be published'. The present author could find no evidence that those cell constants have been published.

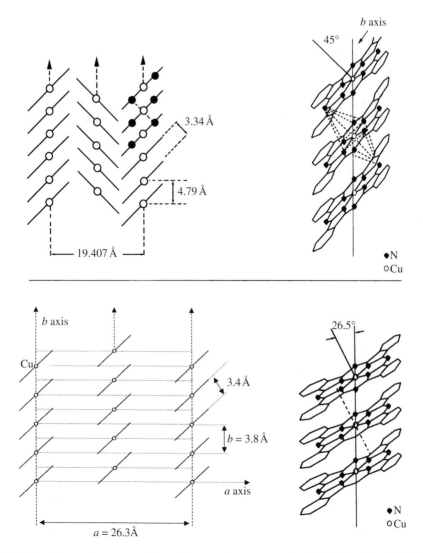

Fig. 8.3   Schematic representation of the crystal structures of α-CuPc (lower) and β-CuPc (upper). This diagram is based on Horn and Honigman (1974) but the schematic representation of the packing of α-CuPc has been modified to correspond with the discussion in the text. The translation distance along the b axis for α-CuPc has been increased somewhat out of proportion and the ab face has been outlined to facilitate illustrating the C-centring.

variability, which is manifested in different shades of blue. Herbst and Hunger (1997) have listed six commercially available CuPc pigments, based on the α, β and ε crystal modifications (including two with 0.5–1 Cl atom per molecule, but retaining the α structure) that vary from greenish to reddish blue.

8.3.3.2   *Metal-free phthalocyanine*

Metal-free phthalocyanine (MfPc) is another important class of polymorphic phthalo-cyanines.[5] Many of the aspects of the polymorphic behaviour were covered in the previous section, but because of its historic importance (Robertson 1936, 1953), and current potential use as a photoconductor for laser printers (Loutfy *et al.* 1988), additional information is provided here.

The number of known and characterized polymorphs of MfPc appears to be still a matter of some question. Herbst and Hunger (1997) indicate that there are five different crystal modifications, designated α, β, γ, κ, and τ. Whitaker (1995) notes that in his earlier (1977*b*) review there were X-ray powder patterns for three (designated α, β, and χ) forms, but that in the intervening years the situation had become more complicated, with the claimed discovery of several more forms, named ε, η, τ, modified η and γ. He also notes the difficulties that may arise in obtaining polymorph-defining powder diffraction patterns due to uncertainty about polymorphic purity in the samples.[6] The original crystal structure of the β form carried out in two-dimensional projection (Robertson 1936) has recently been redone (Matsumoto *et al.* 1999). On the basis of the cell constants and (equivalent) space group, this structure determination is identical with that reported by Kubiak and Janzcak (1992).[7] They have also determined the structure of α-MfPc. The optical properties of the α, β, and χ forms have been studied by Loutfy (1981). Only the α form is marketed, as Pigment Blue 16.

8.3.4   *Some other pigments—old and new*

Indanthrone **8-V** (Pigment Blue 60) is one of the oldest vat dyes known, having been originally synthesized by René Bohn at BASF in 1901. The polymorphic behaviour was well characterized by Susich before WW II (Susich 1950; FIAT 1948). The most stable of the four known polymorphs is α, appearing as greenish needles, and for which the crystal structure is known (Bailey 1955), while the δ form appears as plates. The β, γ, and δ forms convert to α upon heating to 250 °C. It is also possible to interconvert the various forms by dissolving them in sulphuric acid or by transforming them to

---

[5] The second most important Pc's are $Cl_{16}CuPc$ (Pigment Green 7) and $Cl_{12}Br_4CuPc$ (Pigment Green 36), which have not yet been reported to be polymorphic.

[6] From his own studies and a careful evaluation of the available literature Erk (2001*b*) has concluded that there are seven modifications of MfPc, for which α, β, and γ correspond to the respective congeners for CuPc, and four additional forms, which he designates η, τ, τ', and τ''.

[7] However, although Kubiak and Janzcak located and reported the hydrogen atoms on the periphery of the molecule in both of these structures, they did not report any information regarding the two hydrogens required in the core for the molecule to be aromatic. Indeed they indicate that the two hydrogens are absent in the β-MfPc. Matsumoto *et al.* (1999) did locate the hydrogens in the core, thereby confirming their presence. In their determination of the crystal structure of α-MfPc (Jansczak and Kubiak 1992), which does appear to be isostructural with α-PtPc (itself used as the model for α-CuPc—see text), the hydrogen atoms were treated in a similar way as in their β-MfPc, although some residual electron density in the core was viewed as partial occupancy (0.0096 per site) of bismuth, rather than interpreted as disordered hydrogen. While there is some question about the chemistry of these reports regarding the treatment of the core hydrogens, it appears that the molecular frameworks and crystal structures are essentially correct.

the leuco compound (FIAT 1948). The β form has a reddish-blue shade, the γ form has a red shade (Herbst and Hunger 1997), and the δ form has a greenish-grey colour (FIAT 1948). In the same publication, Susich noted the difficulty of distinguishing among the forms, although Jelinek *et al.* (1964) later showed that the δ form could be distinguished from the α using the electron microscope.

8-V   8-VI

On the other hand diketopyrrolo-pyrrol, prototypically **8-VI** (R, R$'$ = H, DPP), is a relatively new family of red pigments (Herbst and Hunger 1997), having been discovered serendipitously by Farnum *et al.* (1974). The most widely used is P. R. 254 (R = 4-Cl, R$'$ = H), which is dimorphic (Hao *et al.* 1999*a*). The *m*-chloro (R = 3-Cl, R$'$ = H) and *m*-methyl (R = 3-CH$_3$, R$'$ = H) derivatives have also recently been shown to be polymorphic (Hao *et al.* 1999*b,c*). The parent compound (R, R$'$ = H) has been shown to be trimorphic.

MacLean *et al.* (2000) have recently studied the dimorphic behaviour of the pigment precursor ('latent' pigment) derivative of **8-VI** (R = COO*t*-but, R$'$ = H) (abbreviated DPP-Boc). The 'latency' is due to the thermal decomposition reaction of both polymorphs resulting in the commercially important pigment DPP. The α form of DPP-Boc contains three half molecules in the asymmetric unit (see also Ellern *et al.* 1994) while the β form contains one half molecule per asymmetric unit. Hence, they are easily distinguishable by solid state NMR as well as by X-ray powder diffraction. The crystal structure solution from powder data and Rietveld refinement of both polymorphs is an exemplary study demonstrating the potential of these methods in determining the detailed crystal structure of these compounds which are often difficult to crystallize.

8.4   Isomorphism of pigments

Little reference has been made in this volume to the subject of *isomorphism*, the appearance of essentially identical crystal structures of different compounds, a phenomenon also recognized by Mitscherlich (1820). The two structures might be more correctly described as isostructural (see Section 6.2.1), but the former term has gained wide acceptance and use in the literature (e.g. Kálmán *et al.* 1993*a,b*), including the development of a numeric index to measure the similarity of structures (Fábián and Kálmán 1999). The phenomenon is not uncommon among pigments, by virtue of the nature of the way compounds for potential use as pigments have been developed.

As implicitly noted above, pigments are usually classified by families, and synthetic efforts at preparing new pigments and modifying the properties of known compounds have been directed at changing substituents in a rather systematic manner: for example, H, $-CH_3$, $-OCH_3$, Cl, Br, etc. The 'core' of many of the prototypical dye families (or chromophore class) is quite large and rigid and essentially determines the limited number of possibilities for the molecules to pack in a crystal structure, even if polymorphism is possible. That, combined with the similarity of size (i.e. van der Waals radius) and polarity of some of these substituents can lead to very similar structures for different compounds, which is manifested in X-ray powder diffraction patterns that are also very similar. These observations are consistent with the "chloro-methyl exchange rule" (Desiraju and Sarma 1986), and the tendency for chloro- and bromo-substituted analogues to form isostructural crystals (Csöregh *et al.* 2001), although some exceptions to the former have been noted (e.g. Csöregh *et al.* 2001).

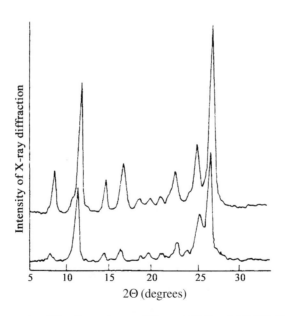

Fig. 8.4 X-ray powder diffraction patterns of Pigment Yellow 14 **8-VII**, R = CH$_3$ (upper) and Pigment Yellow 63 **8-VII**, R = Cl (lower). (From Shenmin *et al.* 1992, with permission.)

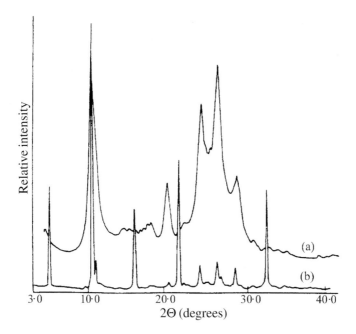

Fig. 8.5 X-ray powder diffraction patterns for the two polymorphs of Pigment Yellow 17 **8-VII**, R = OCH₃ (a) modification obtained by heating sample for 90 °C for one hour after synthesis; (b) modification obtained by recrystallization from nitrobenzene. The lower pattern shows evidence of virtually every peak that appears in the upper pattern, indicating that it may not be a pure modification. Nevertheless, there are diffraction maxima that do appear to be unique to a second form. (From Shenmin *et al.* 1992, with permission.)

For example the monobromo and dibromo analogues of the monoazo C.I. Pigment Yellow 3 are isomorphous (Chapman and Whitaker 1971). In the isomorphous C.I. Pigment Yellow 14 **8-VII**, R = CH₃, and C.I. Pigment Yellow 63 **8-VII**, R = Cl two methyl groups in the former are replaced by Cl in the latter (Shenmin *et al.* 1992). The powder patterns (Fig. 8.4) clearly indicate the isostructurality. Substitution at the same position by the larger and more polar methoxy group **8-VII**, R = OCH₃ is sufficient to overcome the tendency for isostructurality, and in fact apparently leads to a dimorphic system, for which both powder patterns are shown in Fig. 8.5. Similarly, the 2,9-dimethyl- and 2,9-dichloroacridones are isomorphous (Paulus *et al.* 1989). In another variation on this theme, a 50:50 mixture of **8-VIII** (R = n–C₃H₇ and

R = n–C_4H_9) has been reported to crystallize in four polymorphic forms (Brandt *et al.* 1982). In the copper phthalocyanines Pigment Blue 15, containing no Cl, is isomorphous with Pigment Blue 15 : 1, which on an average contains 0.5–1.0 atoms of chlorine per molecule (Hao and Iqbal 1997). Recognition of this phenomenon can be useful, for instance, in attempting to work out the crystal structures of unknown polymorphic forms.

9

Polymorphism of high energy materials

Although personally I am quite content with existing explosives, I feel we must not stand in the path of improvement . . . (Churchill 1950)

9.1 Introduction

High energy materials include a variety of substances that react rapidly to produce light, heat and gas. Some of these materials are used for nonexplosive purposes, while those that are explosive (i.e. cause a noisy, large-scale rapid expansion of matter into a volume much greater than its original volume) can be classified into three general categories: pyrotechnics, propellants and high explosives. The last, in turn, are subdivided into primary, or initiating explosives and secondary explosives (Cooper and Kurowski 1996). Some of factors that influence the performance of explosives include sensitivity to detonation (by impact, friction, shock or electrostatic charge), the rate of deflagration to detonation transition (i.e. the rate at which a burning reaction transforms to a much faster reaction), the detonation velocity, the detonation pressure, the crystal density, thermal and shock stability and crystal morphology. Many of these factors ultimately depend on the solid state structure of the solid high energy material (e.g. Rudel *et al.* 1990; Dick 1995), and hence on the polymorphic modification, which is often ignored (e.g. Zeman 1980). Since it is crucial to assess the effectiveness and safety of these materials in their ultimate end form a great deal of work is normally carried out to fully characterize them. Moreover, a fundamental requirement of such a material is the reproducibility of its performance, and that reproducibility can be compromised by polymorphic variations and transformations. Hence, it is not surprising that polymorphism has been discovered among many compounds used as high energy materials.

For obvious reasons many of these compounds have been developed by government laboratories and military research establishments or programmes funded externally but with broad restrictions on publication. Some of the reports emanating from such research are eventually declassified, but even then often they are not widely abstracted, referenced or readily accessible. There are excellent sources of general information on high energy materials (e.g. Meyer 1987; Cooper and Kurowski 1996 and general references compiled in both of these monographs) and the study of their properties (e.g. Brill 1992; Brill and James 1993, references therein, a continuing series of papers by Brill and coworkers (e.g. Beal and Brill 2000) and a recent review by Agrawal (1998)). We have attempted here to compile some of the information relating to the polymorphism of highly energetic molecular materials in the hope that some structural, physical and chemical data will now be more readily accessible in one source. Where space considerations have limited the amount of information to be included we have attempted to provide details on the primary sources of that information. There are of course many high energy compounds for which polymorphism has not yet been reported; these are not covered here. Moreover, in keeping with the general theme of this monograph, we have not included here discussions of some well-known polymorphic inorganic high energy materials, such as ammonium nitrate and lead (and other) azides.

9.2 The 'alphabet' of high energy molecular materials

Just as pigments are often specified by their Colour Index designation (Chapter 8), so high energy materials are usually designated by acronyms. These are defined here for reference throughout the chapter.

DATB   1,3-diamino-2,4,5-trinitrobenzene
DATH   1,7-diazido-2,4,6-trinitro-2,4,6-triazaheptane
DiPEHN   dipentaerythritol hexanitrate
DPT   3,7-dinitro-1,3,5,7-tetrazabicyclo[3.3.1]nonane
HMX   cyclotetramethylene tetranitramine
HNAB   hexanitroazobenzene
HND   hexanitrophenylamine
HNIW   hexanitrohexa-azaisowurtzitane
NTO   5-nitro-2,4-dihydro-3H-1,2,4-triazol-3-one
OHMX   1,7-dimethyl-1,3,5,7-tetranitrotrimethylenetetramine
PETN   pentaerythritol tetranitrate
PTTN   1,2,3-propanetriol trinitrate
RDX   cyclo-1,3,5-trimethylene-2,4,6-trinitramine
TAGN   triaminoguanidinium nitrate
TNDBN   1,3,5,7-tetranitro-3,7-diazabicyclo[3.3.1]nonane
TNT   (2,4,6-) trinitrotoluene
triPEON   tripentaerythritol octanitrate
2-methyl-1-(2′,4′,6′-trinitrophenyl)-4,6-dinitrobenzimidazole
1,1′-dinitro-3,3′-azo-1,2,4-triazole
1,8-dinitronaphthalene (McCrone book, p. 139)
hexammonium dinitramide
lead styphnate (2,4,6-trinitroresorcinol)
pentanitrophenylazide
2,6-dinitrotoluene

Lead styphnate   2-methyl-1-(2′,4′,6′-trinitrophenyl)-4,6-dinitrobenzimidazole

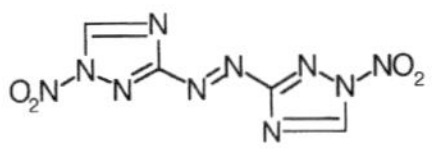

1,1'-dinitro-3,3'-azo-1,2,4-triazole

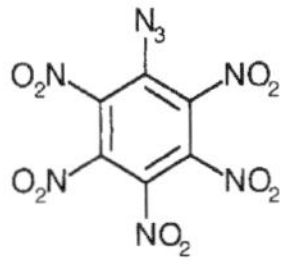

pentanitrophenylazide

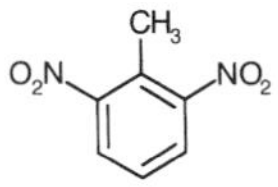

2,6-dinitrotoluene

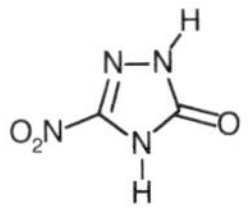

NTO

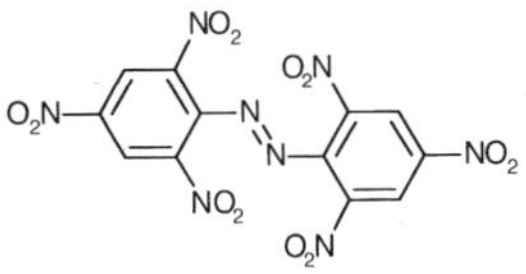

HNAB

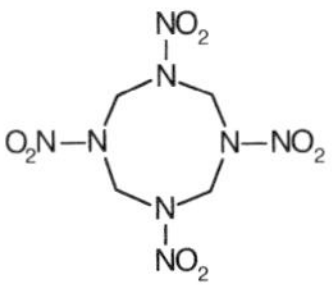

HMX

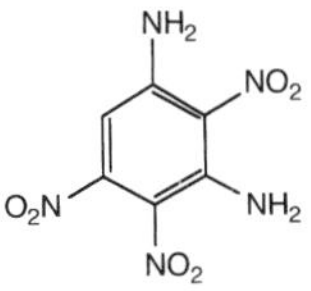

DATB

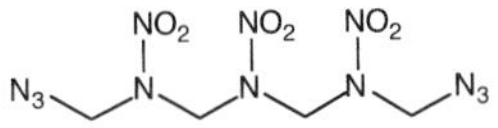

DATH

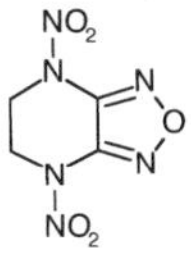

DNFP

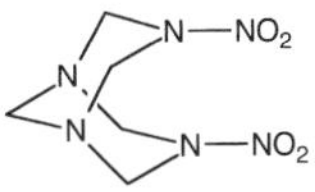

DPT

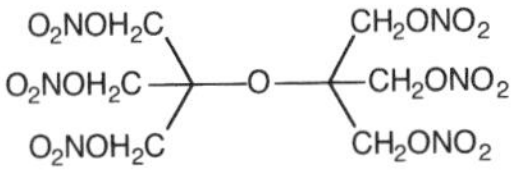

DiPEHN

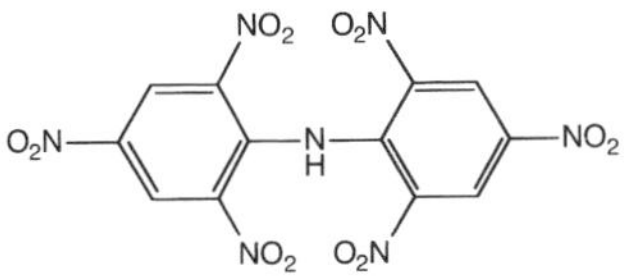

HND

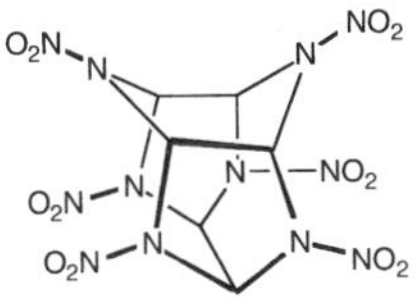

HNIW

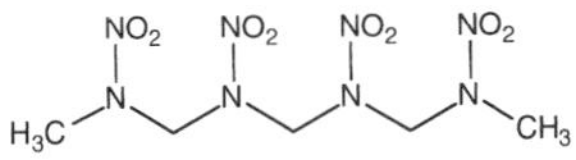

OHMX

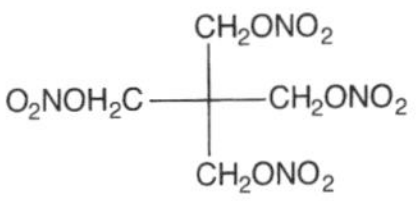

PETN

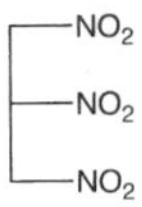

PTTN

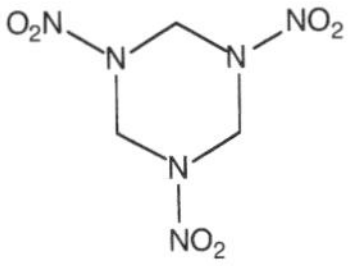

RDX

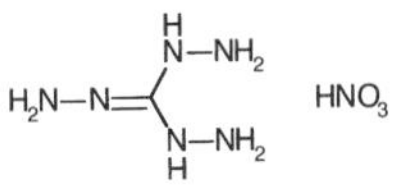

TAGN

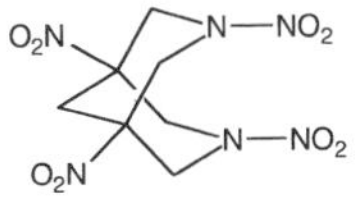

TNDBN

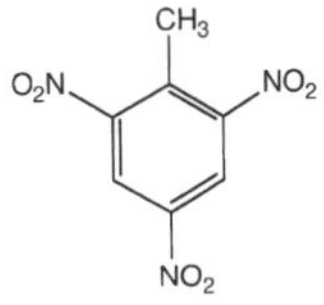

TNT

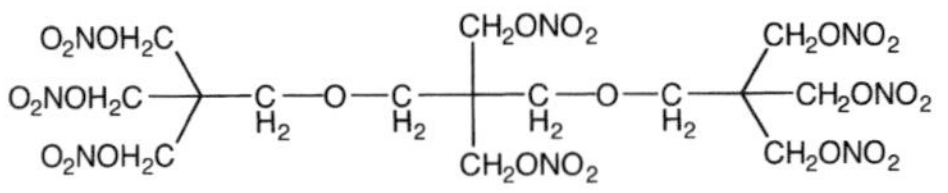

triPEON

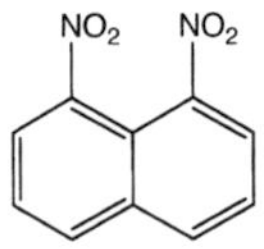

1,8-dinitronaphthalene

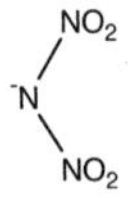

dinitramide anion

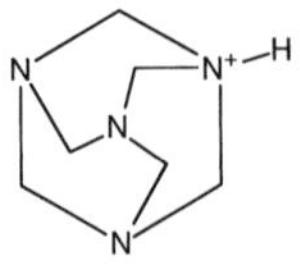

hexammonium cation

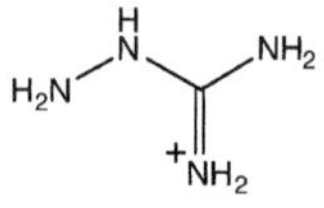

aminoguanidinium cation

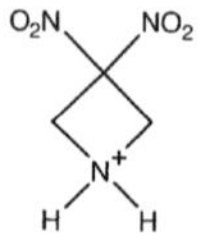

azetidinium cation

9.3   Individual systems

In the following descriptive sections, we have attempted to collect the structural data, or when readily accessible, references to structural data on the polymorphic system. If a crystal structure has been reported and is entered in the Cambridge Structural Database (see Section 1.3.3), then the appropriate REFCODE(s) are given for direct access to that information. Less accessible information is quoted here, together with the appropriate references. Other descriptions of studies of polymorphic systems by a variety of increasingly sophisticated techniques are meant to be representative, rather than comprehensive. Two of the most widely studied systems are **HMX** and **HNIW**, the former having been developed in the USA during the Second World War (Blomquist and Ryan 1944), and the latter more recently developed as 'the densest and most energetic explosive known' (Miller 1995), as quoted by Sorescu *et al.* (1998*a*). The reports on these systems provide excellent examples of the nature of investigations on polymorphic systems and hence will be covered in more detail than the others. Following a common practice, high energy materials are broadly classified as aliphatic or aromatic.

9.3.1   *Aliphatic materials*

9.3.1.1   *HMX*

Four polymorphs of **HMX**, currently designated α, β, γ, δ have been identified and quite well characterized in terms of many physical properties (Cady and Smith 1962; Holston Defense Corp. 1962). McCrone (1950*a*) also published a thorough optical and thermal study of the four polymorphs (including diagrams indicating morphology, form and habit, and interfacial angles) albeit with a different labelling system for the four modifications than that given here. A summary of some of the comparative data on **HMX** polymorphs is given in Table 9.1. Cyclotetramethylene tetranitramine has also been prepared as stoichiometric solids with over 100 organic molecules (George *et al.* 1965; Selig 1982) and among them the solvate with dimethylformamide has

Table 9.1 Data on polymorphs of **HMX**[a]

	β (I)	α (II)	δ (IV)	γ (III)
Crystal structure, CSD REFCODE	OCHTET01, OCHTET04, OCHTET12	OCHTET	OCHTET03	DEDBUJ[c]
Calculated density[b,d]	1.893, 1.902 (1.96)	1.839 (1.87)	1.759 (1.78)	1.78, 1.82 (1.82)
Melting point (°C)[e]	246–247	256–257	279–280	280–281.5
Transition temperatures (°C)[f]	β → δ 167–183 β → γ 154	α → δ 193–201 α → β 116		γ → δ 167–182
Transition temperatures (°C)[g]	β → δ 102–104	α → γ metastable α → δ 160–164		
Impact sensitivity (cm)[h]	31–32	5–50	6–25	6–12
Heat of sublimation (kcal mol^{-1})[i]	41.9	39.3		38.0
Heat of solution (kcal mol^{-1})[h]	4.4.	3.8	3.7	1.5

[a] The polymorphic designation in parentheses is that of McCrone (1950a).
[b] Calculated from X-ray data as reported in crystal structure.
[c] The cell constants reported by Cady and Smith (1962) are actually those of a hemihydrate of HMX (Main et al. 1985) and not those of the γ form (as recognized also by a different REFCODE in the CSD). This misnomer has also been perpetuated in Kohno et al. (1996). Cady and Smith did cite a report by Krc (1955) that gave the following cell constants for the γ form (no space group indicated): a = 16.80, b = 7.95 c = 10.97 Å, β = 130°, Z = 4, and indicated that these could not be transformed to the cell that later turned out to be that of the monohydrate.
[d] Value in parenthesis is given by Meyer (1987).
[e] Teetsov and McCrone (1965).
[f] Meyer (1987).
[g] Gibbs and Popolato (1980).
[h] Holston Defense Corporation (1962).
[i] Taylor and Crookes (1976).
OCHTET, OCHTET01—Cady et al. (1963); OCHTET03—Cobbledick and Small (1974); OCHTET04—Zhitomirskaya et al. (1987) and OCHTET12—Choi and Boutin (1970).

been claimed to be polymorphic (Cobbledick and Small 1975; Haller et al. 1983), although Marsh (1984) has shown that the latter structure (reported in space group C2/c) is most likely identical to the earlier one (reported in space group $R\bar{3}c$).

Some of the crystallization conditions reported by McCrone (1950a), reflect the relative stabilities and the need to use kinetic conditions to obtain the less stable forms.

β is the form stable at room temperature, and can be prepared, for instance, by very slow cooling from solutions in acetic acid, acetone, nitric acid, or nitromethane. The α modification can be prepared from the same solvents, but with more rapid cooling, and even more rapid cooling can yield the γ form. δ may be obtained from solvents in which it is only slightly soluble, and even then by rapidly chilling, even to pouring the solution over ice. Teetsov and McCrone (1965) quantitatively determined the stability ranges and transformation temperatures using the method of eutectics later followed by Yu *et al.* (2000) (Section 5.9); the results on **HMX** are also summarized in Table 9.1. Teetsov and McCrone (1965) have noted that some of these transition temperatures may be quite variable, in particular $\alpha \leftrightarrow \beta$, depending on the amount of strain in the β form. Also, a number of authors (e.g. Cady and Smith 1962) reported that trace amounts of residual **RDX** can affect the results from studies on **HMX** transformations; the two substances may be easily distinguished on the polarizing microscope by the difference in their indices of refraction (McCrone 1957).

The polymorphs of **HMX** have been studied by a wide variety of techniques. One of the crucial parameters for high energy materials is the impact sensitivity, which in this case was determined by determining the height from which a dropped 5 kg weight would initiate reaction. Low values therefore indicate increased sensitivity, which can also be influenced by the presence of grit and to some degree by crystal size and shape. In particular, the authors noted that crystal habit had a significant effect in the cases of α- and γ-**HMX**, but very little on the δ form. A rather thorough study of this property was carried out by Cady and Smith (1962), who concluded that (a) the impact sensitivity of β is reproducible and independent of particle size, and (b) in spite of variations in the results of different tests, the order of sensitiveness of **HMX** polymorphs is $\delta > \gamma > \alpha > \beta$.

Structurally, **HMX** exhibits conformational polymorphism, as demonstrated in Fig. 9.1. Brill and Reese (1980) analysed the relative stabilities of the α, β and δ forms in terms of coulombic forces to interpret the thermophysical behaviour of the three forms. They concluded that the chair conformation found in the β modification is more stable than the chair–chair conformation in the other two forms. The relative stability to pyrolysis could also be accounted for by the analysis of the coulombic attractions and repulsions around the molecule in each of the modifications. More recently, Henson *et al.* (1999) have followed the change in second harmonic generation response during the $\beta \rightarrow \delta$ phase change, which has long been implicated in the thermal decomposition of **HMX** (Karpowicz and Brill 1982).

The IR spectra of all four crystal modifications were reported by Cady and Smith (1962) and by Holston (1962). The latter did point out some distinguishing features among the polymorphs, but Cady and Smith noted that problems with sample preparation and conversions among forms indicated that the optical properties described by McCrone (1950*a*) were the basis for the best rapid qualitative and even rough quantitative analysis. Raman spectroscopy, which requires less potentially destructive sample preparation, has been used to distinguish the polymorphs (Goetz and Brill 1979). The low resolution ^{1}H NMR spectra of the four crystal modifications were reported by Landers *et al.* (1985). The ^{14}N nuclear quadrupole resonance spectrum

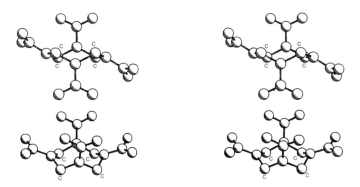

Fig. 9.1   Steroviews of the **HMX** molecule in the β (top), α (bottom) forms. In both structures the view is on the plane of C−N−C of the ring. Labels for nitrogen and oxygen atoms have been omitted. (From Bernstein 1987, with permission.)

of the β form has been determined, and has been used to suggest a mechanism for the $\beta \rightarrow \delta$ phase transition (Landers *et al.* 1981).

In the last few years, there has been a spate of computational studies on the structures of polymorphs of high energy materials, in particular the nitramines, including **HMX**. The molecular energetics were studied by Smith and Bharadwaj (1999), using high level (B3LYP/6-311G**) quantum mechanical geometry optimizations, with subsequent single-point energy calculations at the MP2/6-311G** level. Of the four low-energy conformers, two corresponded to the conformers found in the crystal structures, but the two lowest energy conformers have not been observed. This led the authors to suggest that the conformations in the observed structures are influenced by intermolecular interactions. Lattice energy calculations, based on the atom–atom potential method (see Section 5.3), including MNDO-derived partial charges led to sublimation energies similar in magnitude to those listed in Table 9.1, but did not correspond to the observed order of stability. While the authors attributed the discrepancy to the overestimation of the coulombic energy, this study illustrates some of the difficulties and limitations of computing lattice energetics of polymorphs as discussed in Chapter 5. In a subsequent study employing a different computational strategy, the molecular structures and energetic stabilities of the α, β and δ modifications were computed using a first-principle electronic structure method (Lewis *et al.* 2000). The computed results were compared with experimental molecular volumes, the bulk modulus and its pressure derivative 'in reasonable accord with experiment'; the predicted energetic ordering of the three polymorphic phases does correspond with experiment.

Isothermal–isobaric molecular dynamics simulations of the α, β and δ modifications have been carried out over the temperature range 4.2 - 553 K, using a force field developed for **RDX**, together with charges derived from *ab initio* calculations (Sorescu *et al.* 1998b). These gave results in close agreement with the experimentally determined crystal structures. Another molecular dynamics study (Kohno *et al.*

1996) indicated the importance of compressed N−N bonds in the initial decomposition process of nitramines in general, and the polymorphs of **HMX** in particular.

9.3.1.2 *HNIW*

Just as **HMX** is one of the oldest and widely used high energy materials, **HNIW** (also sometimes referred to as CL-20) is one of the newest cyclic nitramines and has generated considerable interest, although the early work is in references of limited accessibility (see for instance, references 1–4 in both Russell *et al.* 1992, 1993, and Sorescu *et al.* 1998*a*, and references 1–7 therein). These same authors agree that five polymorphs are known, although Nedelko *et al.* (2000) indicate the existence of six crystal modifications.

Nielsen *et al.* (1998) have described the preparation of modifications α, β, γ, and ε, in considerable detail. The α modification crystallizes as a hemihydrate, from which the water may be removed to yield an anhydrate. The other three modifications are unsolvated. The same authors and another group (Jacob *et al.* 1999) have reported the crystal structures of all four modifications (under REFCODES PUBMIInn (α) and PUBMUUnn (β, γ, ε)).

In a study of the crystallization behaviour at room temperature, Kim *et al.* (1998) found that the initially appearing β form gradually converted to the ε form in a solution-mediated process, indicating that the latter is more stable in accord with Ostwald's Rule (see Section 2.3). These authors also present comparative XRD patterns and qualitative and quantitative FTIR analyses of mixtures of these two forms. Another study, ostensibly to determine the polymorphic purity of the material, in fact used a qualitative visual examination to estimate the maximum amount of the β form as an impurity in samples of the ε form from three different suppliers (Bunte *et al.* 1999). FTIR, in combination with thermal and microscopic methods has also been used to determine the thermally generated polymorphic forms among the modifications (Foltz *et al.* 1994*b*), with a specific application to a formulation based on the ε form as the active ingredient (Foltz 1994). Spectroscopic characterization of the various forms, including FT-Raman, NMR, CIMS and UV are also cited by Nielsen *et al.* (1998); SEM images of the various phases have been presented by Foltz *et al.* (1994*a,b*).

At room temperature and atmospheric pressure, the relative thermal stability of the four modifications has been determined by solvent-mediated transformations and thermal analysis to be $\varepsilon > \gamma > \alpha$-hemihydrate $> \beta$ (Foltz *et al.* 1994*a,b*; Foltz 1994). Of the anhydrate materials ε is indeed the most dense as expected from this ordering (see Section 2.3), but in contradiction to the expected correlation between density and relative stability β is more dense than γ (Nielsen *et al.* 1998).

In a rather complete study Russell *et al.* (1993) used a specially designed high-temperature/high-pressure diamond anvil cell to determine the phase diagram and stability fields between −125 and 340 °C and atmospheric pressure to 14.0 GPa. They characterized the ζ phase which exists at room temperature above 0.7 GPa pressure, and obtained FTIR spectra for all five polymorphs in various regions of

$P–T$ space. The data were used to characterize nine observed interphase transitions, among which four are reversible and five are unidirectional. The pressure and thermal ranges of decomposition were determined for the α, γ and ζ polymorphs.

The kinetics of thermal decomposition of three of the modifications were studied by thermogravimetry, IR spectroscopy and optical and electron microscopy (Nedelko *et al.* 2000), with the conclusion that the rate increases in the series $\alpha > \gamma > \varepsilon$. However, it was found that the results for a particular polymorph also depend upon the morphological features of the crystals as well as their size distribution and mean size.

The molecular conformations found in the crystal structures of the β, ε and γ modifications differ (Sorescu *et al.* 1998*a,b*), indicating that these are conformational polymorphs (Chapter 5). The conformation found in the α hemihydrate is similar to that in the γ form. The computed lattice energies, employing an intermolecular potential developed for **RDX** (Sorescu *et al.* 1997) including charges calculated at the HF/6-31G** level, gave an ordering of the relative stability of the ε (-50.35 kcal mol^{-1}), β (-49.62 kcal mol^{-1}) and γ (-48.19 kcal mol^{-1}) forms compatible with the experimental results of Russell *et al.* (1993); the differences in lattice energies between polymorphs are also in the range expected.

There have been a number of studies of the mechanism of thermal decomposition of **HNIW** (Patil and Brill 1993) with a particular emphasis on the role of free radicals (Pace 1991, 1992). Ryzhkov and McBride (1996) compared the reactions at low temperature in the α and β modifications, and found that the same cavities that contain water in the hemihydrate play an important role in differences in the solid state chemistry between the two modifications.

9.3.1.3 *RDX*

Cyclo-1,3,5-trimethylene-2,4,6-trinitramine is one of the most important and widely used high energy materials and the stable α form is well characterized (Meyer 1987), including its crystal structure (Choi and Prince 1972). McCrone (1950*b*, 1957) prepared the unstable β form on the microscope, where he determined some optical properties, and noted the difficulties in obtaining it in quantities suitable for further characterization. However, visually the β form is easily distinguishable from the α form, although apparently not detectable by thermal methods (Hall 1971). Subsequent efforts to obtain the β form (Sergio 1978) eventually led to the preparation, by crystallization from high-boiling solvents, of quantities sufficient to determine the FTIR spectra (Karpowicz *et al.* 1983). The study of the spectra of both solid forms of **RDX** with those in the vapour and solution phases and a comparison with those of **HMX** suggested that the former also exhibits conformational polymorphism (Karpowicz and Brill 1984). Miller *et al.* (1991) subsequently carried out a $P–T$ study similar to that described above for **HNIW**, and using FTIR, energy dispersive X-ray powder diffraction and optical microscopy in a diamond anvil cell, further characterized these two phases in addition to a high pressure γ phase, which transforms to either α or β rather than decomposing.

9.3.1.4 *PETN*

The polymorphism of the secondary explosive **PETN** was first reported by Blomquist and Ryan (1944), and the crystal structure of the common Form I was originally reported by Booth and Llewellyn (1947) and later by Trotter (1963). It crystallizes in a space group ($P\bar{4}2_1c$) that is quite rare for molecular substances. Also noteworthy is the fact that the $\bar{4}$ crystallographic site available in this space group corresponds to the molecular point symmetry, since Kitaigorodskii (1961) indicated that the exigencies of packing efficiency generally lead to situations in which the crystallographic site symmetry for a molecule is lower than the full molecular point symmetry.

The early work on **PETN** suggested that Form I was stable to its melting point of 142.9 °C. Cady and Larson (1975) found that careful measurement of the melting points of the two polymorphs indicated that Form II actually melts 0.2 °C *higher* than Form I, suggesting that at that temperature Form II is the more stable form. Form II also forms spontaneously on a face of Form I growing from a supercooled melt. At lower temperatures (i.e. <130 °C) Form II transforms rapidly to Form I. The details of these transformations were studied in considerable detail, but unfortunately are available in a document that is difficult to access (Cady 1974). Cady and Larson also determined the crystal structure of Form II in space group *Pcnb*, in which obviously there is no longer a crystallographic site symmetry of $\bar{4}$, but the molecular symmetry is nearly retained.

Sorescu *et al.* (1999) have successfully modelled Form I up to pressures of about 5 GPa.

9.3.1.5 *TAGN*

The crystal structure of the room temperature phase of **TAGN** was determined simultaneously by two groups (Bracuti 1979; Choi and Prince 1979). Oyumi and Brill (1985) reported endotherms in the DTA traces at 258, 270 and 407 K. In the course of a study of the rigid body motion of the nitrate ion in this material, a low temperature (−10 °C) polymorph was discovered (Bracuti 1988), as anticipated in the data presented by Oyumi and Brill. The room temperature structure crystallizes in space group *Pbcm*, while that at low temperature is *Pbca*. The *b* and *c* crystallographic axes are essentially identical for the two structures, but the *a* axis is quadrupled (from 8.366(2) to 33.47(1) Å on going to the low temperature modification.

9.3.1.6 *PTTN (Nitroglycerine)*

Nitroglycerine is a liquid at room temperature with a solidification point of 13.2 °C for the stable modification. The compound has been known since 1914 to be polymorphic (Hibbert 1914), with a less stable modification solidifying at 2.2 °C. The structure of the stable orthorhombic modification has been reported twice (Espenbetov *et al.* 1984; Litvinov *et al.* 1985), while that of the triclinic metastable form has not yet been reported.

9.3.1.7 *DATH*

The crystallization behaviour has been described by Oyumi *et al.* (1987*a*). An amorphous phase, unstable at room temperature, can be prepared under kinetic conditions by rapid removal of solvent (acetone or acetonitrile) or rapid cooling of the melt. It initially appears as a transparent waxy material that transforms into a crystalline material over about an hour. It can also be prepared on a DTA instrument by cooling from the melting point of the crystalline material (406 K) to the temperature range 333–290 K. On the other hand, crystals suitable for single crystal structure determination (carried out by the same authors) can be grown by slow evaporation from the same solvents.

9.3.1.8 *OHMX*

1,7-dimethyl-1,3,5,7-tetranitrotrimethylenetetramine is the acyclic analogue of **HMX**. Thermal analysis and IR spectroscopy were used to identify five polymorphs (Oyumi *et al.* 1987*b*). Rapid evaporation of solutions of a number of different solvents led to polycrystalline Form IV. Forms II and III were obtained by slow evaporation, the first from DMF, and the second from acetone. Form I could be obtained by cooling warm Form II to room temperature or by applying mechanical friction to Form II or Form IV on an NaCl IR sample cell. All four can be maintained at room temperature. Form V was described as a 'pre-melt phase (which) is nearly thermally neutral with respect to Form IV...' The only crystals suitable for single crystal structure determination were of Form II; those of Form III exhibit twinning.

Differences and similarities in the IR spectra (that of Form V is indistinguishable from Form IV) were used to interpret structural changes and constancies resulting from transitions between the various modifications.

9.3.1.9 *NTO*

As an insensitive high explosive, **NTO** has been considered as the high energy material in the activation of auto air bags (Wardle *et al.* 1990) as well as replacement material for **RDX** in bomb fill (Lee and Gilardi 1993). The compound was reported to be dimorphic (which may have led to some of the confusion in the thermal decomposition data (Williams and Brill 1995; Botcher *et al.* 1996), both forms being obtained from aqueous solutions. The more stable α form apparently crystallizes under thermodynamic conditions, while the β form is obtained (with some difficulty and in small amounts) under kinetic conditions, as well as by recrystallization from methanol or ethanol/methylene chloride solvents. They can be distinguished by morphology and IR spectra. The crystal structures of both have been determined (Lee and Gilardi 1993) with α crystallizing in $P\bar{1}$ and β in $P2_1/c$. Both are characterized by (different) chains of hydrogen-bonded molecules along the a crystallographic axis. In the α modification this leads to polymer-type behaviour with crystals shattering parallel to that axis and crystal ends fraying much like rope. They can even be bent without breaking (see also Yakobson *et al.* 1989).

9.3.1.10   *TNDBN*

1,3,5,7-tertranitro-3,7-diazabicyclo[3.3.1]nonane is another compound that is structurally and compositionally similar to **HMX**; it also exhibits conformational polymorphism. Oyumi *et al.* (1986*a*) identified five polymorphic modifications of **TNDBN**
by DTA and variable temperature (above and below room temperature) mid-IR spectroscopy. Phases III and IV could be obtained at room temperature by crystallization
from acetone and accetonitrile with IV appearing more often than III. The latter, crystallizing as orthorhombic crystals in space group *Pbca* with $Z' = 2$, is the only crystal
structure reported to date. The two molecules in the asymmetric unit adopt different
conformations. On the basis of this observation the authors considered the possibility that transitions among other phases might involve conformational changes; the
analysis of the IR spectra indicated that the nitro groups are similarly disposed in the
other four polymorphs, suggesting that the conformation is also constant.

9.3.1.11   *DPT*

3,7-dinitro-1,3,5,7-tetrazabicyclo[3.3.1]nonane is yet another analogue of **HMX**, for
which the crystal structure of one form was reported by Choi and Bulusu (1974).
Although it was the subject of thermal analysis (Hall 1971), prior to a study by Oyumi
et al. (1986*a*) no polymorphism had been reported. By following the mid-IR spectrum
as a function of temperature the latter authors detected a gradual, reversible phase
transition beginning at 343 K and being completed at 440 K. A careful redetermination
of the DSC indicated a phase transition at 339 K with $\Delta H_t = 1.2\,\text{kcal mol}^{-1}$. The
changes in the IR spectra are consistent with a possible inversion in the amine lone
pairs accompanied by some increase in the rotational freedom of the molecule.

9.3.1.12   *Dinitramide salts*

The dinitramide anion is a relatively new entry in the field of energetic materials
(Bottaro *et al.* 1991; Russell *et al.* 1997 and references therein) and a number of salts
have been prepared and structurally characterized (Gilardi and Butcher 1998*b* and
references 7–12 therein; Sitzmann *et al.* 2000, and references 2b–i therein), including
some polymorphic ones, described briefly here.

9.3.1.12.1 Hexammonium dinitramide: The material is dimorphic: crystallization
from water yields the monoclinic $P2_1/c$ modification, while crystallization from
polar organic solvents leads to a triclinic $P\bar{1}$ form (Gilardi and Butcher 1998*b*). Both
structures contain $\mathbf{R}_2^2(9)$ hydrogen-bonded rings, with two nitro oxygens as acceptors
and one −NH and one −CH as donors. The overall packing is quite similar with
alternating layers of cations and anions.

9.3.1.12.2 Aminoguanidinium dinitramide: The material was originally prepared
from an aqueous medium as white crystals, with a melting point of 70–79 °C. Several
recrystallizations from ethyl acetate led to a melting point of 91–94 °C. The crystal

structures of a triclinic $P\bar{1}$ modification (density $= 1.639\,\mathrm{g\,cm^{-3}}$) and a noncen-trosymmetric monoclinic Pc modification (density $= 1.650\,\mathrm{g\,cm^{-3}}$) are reported, but it is not clear if the melting points are indeed different. The anions differ in conformation by about $11.5°$ in one torsion angle, suggesting that the pair of structures are conformational polymorphs.

9.3.1.12.3 *Ammonium dinitramide and dinitro azetidinium dinitramide:* For both of these materials the pressure/temperature and reaction phase diagram have been determined using a high-temperature–high-pressure diamond anvil cell with FTIR spectroscopy, Raman spectroscopy and optical microscopy. For ammoninm dinitramide energy dispersive X-ray diffraction was also employed (Russell *et al.* 1996, 1997).

At ambient conditions ammonium dinitramide has not been found to be polymorphic; the crystal structure of the orthorhombic α form under those conditions has been reported by Gilardi *et al.* (1997). The pressure/temperature studies led to the discovery of a second monoclinic phase (β) formed from the α phase at 2.0 ± 0.2 GPa. As the pressure is further raised, a solid state rearrangement, melting, and thermal decomposition take place.

The structure of the one known crystal (α) modification of dinitro azetidinium dinitramide at ambient conditions has been reported by Gilardi and Butcher (1998*a*). It crystallizes in space group $Cm21$. The pressure studies led to a rapid and reversible phase transition at 1.05 ± 0.05 GPa to the β modification, which could not be maintained to ambient conditions. Additional studies led to the determination of the stability fields of both of these polymorphs and the liquidus phase, as well as a determination of the temperature/pressure conditions for decomposition (Russell *et al.* 1997). The FTIR spectra of both solid phases were also analysed in detail.

In a number of these studies, Russell *et al.* have shown that the exploration of temperature/pressure phase space with a relatively simple apparatus (Block and Piermarini 1976), combined with a number of rapidly developing analytical techniques, can lead to detailed understanding of polymorphic systems, the relationships among crystal modifications and the characterization of their properties. More studies of this type, especially to explore the pressure domain of polymorphic systems, should be encouraged.

9.3.2 *Aromatic materials*

9.3.2.1 *TNT*

TNT proved to be a very difficult material to prepare in single crystalline form ... there is still a great deal of confusion regarding the structure of the crystalline products and the conditions under which they are obtained. (Sherwood and Gallagher 1984)

How can a material so well known be so poorly understood? (Lowe-Ma 2000, quotation from 1990)

The history (perhaps one should say 'saga') of the crystal chemistry and poly-morphism of this compound is summarized in these two statements by two prominent workers in the chemical crystallography of high energy materials. 2,4,6-Trinitrotoluene is the most common aromatic explosive (Cooper and Kurowski 1996). In 1949 McCrone reported that there was 'no evidence of polymorphism for TNT'; only a few years later, he twice mentioned an unstable polymorph of the mate-rial (McCrone 1957), although Grabar et al. (1969) subsequently doubted whether McCrone had observed what later became recognized as the monoclinic modification. Yet the confusion on the polymorphism of **TNT** had begun well before.

The first chemical crystallographic (goniometric) studies of **TNT** were apparently reported by Friedländer (1879), who claimed that the crystals exhibited rhombic sym-metry. The next report by Artini (1915) indicated monoclinic symmetry, although Ito (1950) noted the similarity in the axial ratios in these two descriptions. The orthorhom-bic/monoclinic dilemma continued with the initial X-ray investigations: Hertel and Römer (1930; ZZZMUC02) reported a monoclinic cell with a β angle close to 90° (89°29′29″), whereas Hultgren (1936; ZZZMUC03) reported a rhombic cell with unit cell axes sufficiently similar to those of the monoclinic cell to raise doubts about the significance of the difference.

Ito (1950) carried out his own study, including an analysis of the twinning that he recognized from Weissenberg photographs, and concluded that **TNT** is trimorphic, with one rhombic form and two monoclinic forms. The three forms reported by Ito have identical b and c axial lengths, with the length of the a axis doubling and quadrupling from the rhombic to the two monoclinic forms.

Burkhardt and Bryden (1954) then reported for crystals grown by sublimation at 78 °C cell constants with e.s.d.s for a $P2_1/c$ monoclinic cell with $\beta = 111.2$° that could be transformed to a pseudo-orthorhombic cell with $\beta = 90.5$°. The trans-formed cell again had b and c axes identical to an orthorhombic form (grown from ether at -70 °C) in space group $Pmca$ or $P2ca$, within experimental error, with the a axis length (20.07 Å) slightly less than half that of the transformed monoclinic cell (39.81 Å). However, they also reported three other monoclinic and two other orthorhombic modifications obtained under different crystallization conditions and some transformations that took place with ageing and grinding.

J.R.C. Duke (Lowe-Ma 2000; also reference 10 (dated 1981) in Gallagher and Sherwood 1996) determined the crystal structures of both modifications. He prepared the monoclinic form with $P2_1/c$ (Z = 8) (called Form A) by annealing cast **TNT**. Orthorhombic crystals were prepared by quenching the melt and were reported as $Pb2_1a$ (Z = 8) (called Form B). Again the two have very similar, but nevertheless statistically significant different cell constants. The latter appears to be identical to the orthorhombic structure reported (in equivalent, but transformed space group $Pca2_1$) by Carper et al. (1982; ZZZMUC01), wherein the authors again noted the propensity for twinning of **TNT** single crystals. In both structures Z′ = 2, the two independent molecules in the asymmetric unit exhibiting different degrees of twist of the nitro groups out of the plane of the benzene ring.

Table 9.2 Comparison of some of the crystallographic constants reported for **TNT**[a]

	Carper *et al.* (1982)	Golovina *et al.* (1994)	Duke (1981)[b]
Orthorhombic form			
a (Å)	14.991 (1)	20.041 (20)	15.005 (2)
b (Å)	6.077 (1)	15.013 (8)	20.024 (4)
c (Å)	20.017 (2)	6.084 (5)	6.107 (3)
Space group	$Pca2_1$	$P2_1ab$	$Pb2_1a$
R-factor	0.057	0.055	0.043
Monoclinic form			
a (Å)		21.407 (20)	21.275 (2)
b (Å)		15.019 (8)	6.093 (3)
c (Å)		6.0932 (5)	15.025 (1)
Monoclinic angle (°)		(γ)111.00 (2)	(β) 110.14
Space group		$P2_1/b$	$P2_1/c$
R-factor		0.061	0.049

[a] The axis assignments of equivalent space groups are given here as reported by the authors. Appropriate transformation should be made for the purpose of comparison.
[b] Cited by Gallagher and Sherwood (1996).

Much of the confusion surrounding the crystal chemistry of **TNT** as noted in the quotations at the start of this section has now been resolved, due to a large extent, to the availability of the crystal structures of the monoclinic and orthorhombic forms (Carper *et al.* 1982; Golovina *et al.* 1994) and the work of Gallagher and Sherwood (1996), who actually made use of the unpublished crystal structures of Duke (1981 *vide supra*). There is general agreement among the crystallographic constants reported by these various authors (Table 9.2), but some do differ by more than 3 e.s.d.s of the reported values. Gallagher and Sherwood found that the orthorhombic form could be obtained from cyclohexanol and ethanol; the monoclinic form, stable from room temperature to the melting point of 82 °C, crystallizes from a number of other solvents, including methanol, toluene, acetone (almost always with twinning) and ethyl acetate (apparently free from twinning).

Some of the past confusion regarding the two polymorphs is obvious from the similarities in the cell constants and may be understood by examining the crystal structures, Fig. 9.2. By visual inspection, the packing is very similar. Gallagher and Sherwood have carefully analysed the symmetry relations between the two independent molecules (A and B) in the asymmetric unit for both structures. In the monoclinic modification, centrosymmetrically related pairs of molecules are arranged in layers almost parallel to the *b–c* plane, with alternate layers related by a pseudo-glide plane. This leads to a molecular stacking sequence of . . . AA BB AA BB . . . For the orthorhombic structure, there is a true glide plane parallel to the *b* axis, so that the molecular stacking sequence is . . . AB AB AB AB . . . The sheet structure appears almost identical, but subsequent sheets are displaced relative to one another by *c*/2 with respect to the monoclinic axial system.

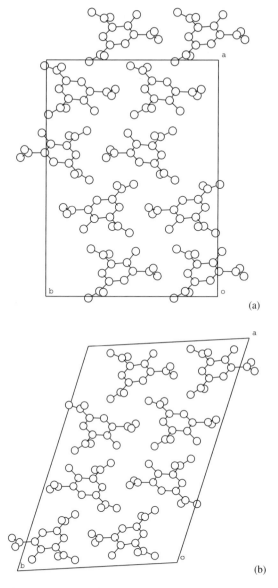

Fig. 9.2   Packing diagrams of (a) the orthorhombic, and (b) the monoclinic structures of **TNT**, including the orientation on the unit cell. Starting from the top, a row by row comparison, which is facilitated by noting the position and orientation of the methyl group, indicates the great similarity in the packing arrangement, although the molecules are located differently with respect to the unit cell axes. In the monoclinic structure the γ angle (between the a and b axes) is $\sim 111°$, since the authors chose the unconventional unit cell setting with the c axis unique, rather than the conventional choice of b as the unique axis. (After Golovina *et al.* 1994.)

The similarity of the two crystal structures leads to very similar X-ray powder diffraction patterns (Fig. 9.3), reminiscent of the situation in terephthalic acid (Section 4.4). Careful inspection reveals that the orthorhombic modification can be distinguished by a peak (the 511 reflection in the Golovina *et al.* (1994) cell) at $2\theta = 27.24°$, while the monoclinic modification can be distinguished peak (from 211) at $2\theta = 19.26°$.

Thermal (DSC) studies clearly confirm that the monoclinic form is stable from room temperature up to its melting point, while the orthorhombic modification goes through a small endotherm at $70\,°C$, corresponding to a solid–solid phase transition with a heat of transition of $0.22\,kcal\,mol^{-1}$ (Connick *et al.* 1969; Gallagher and Sherwood 1996). Over a two month period, the orthorhombic material transforms to the monoclinic.

Gallagher *et al.* (1997) have also carried out lattice energy calculations on both forms. They obtained a value of $-28.83\,kcal/mol^{-1}$ for the monoclinic form and $-28.24\,kcal\,mol^{-1}$ for the orthorhombic. The difference is that expected for polymorphs (see Chapter 5); moreover, the absolute values compare favourably with experimental sublimation enthalpies reported by Edwards (1950) ($28.3 \pm 1.0\,kcal\,mol^{-1}$, albeit with no specification of the polymorphic form), although values as low as $23.7\pm0.5\,kcal\,mol^{-1}$ (Pella 1976, 1977) have been reported (Chickos 1987). The Gallagher *et al.* (1997) calculations were also used to investigate and compare the specific interactions between molecules in the twinned monoclinic phase and the two untwinned phases.

TNT also provides a good historical example of Ostwald's Rule and some other aspects of crystal growth and habit. Groth's entry (Groth 1917, Vol. IV, p. 364) for the compound (taken from Friedländer 1879), shown in Fig. 9.4, is clearly that of the orthorhombic form (also portrayed by McCrone 1949), now known to be less stable. A supplementary note on p. 766 of the same volume of Groth refers to the Rendic 1915 work describing the subsequently discovered more stable monoclinic form. In their studies of **TNT**, Gallagher *et al.* (1997) determined the habit and computationally modelled the expected morphology for the orthorhombic, monoclinic and twinned monoclinic phases, using attachment energies. For the orthorhombic case, there is similarity in the prominent crystal faces to be expected, but a lack of correspondence between theory and experiment on the relative sizes of those faces.

These recent studies have transformed the previously enigmatic 2,4,6-**TNT** system into one which is now quite well understood and over which considerable control can be exercised. It should be noted here that 2,4,5-**TNT**, one of the main impurities in 2,4,6-**TNT**, has also been shown to be polymorphic (Chick and Thorpe 1970, 1971), and has been characterized by thermal and microscopic methods as well as by IR spectroscopy (Section 9.3.2.7).

9.3.2.2 *HNAB*

Hexanitroazobenzene **HNAB** is known to crystallize in five polymorphic forms (McCrone 1967). The structures of Form I, stable from room temperature to $185\,°C$

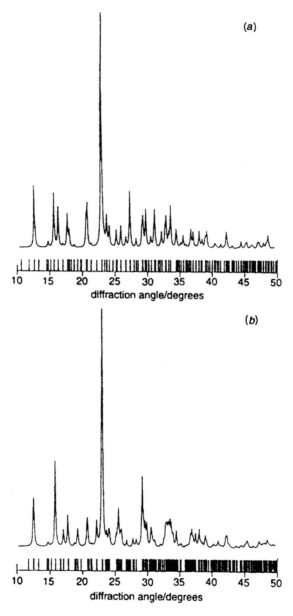

Fig. 9.3 Calculated X-ray powder diffraction patterns for the (a) orthorhombic and (b) mon-oclinic polymorphs of **TNT** (from Gallagher and Sherwood 1996, with permission). The experimental powder patterns reported by Connick *et al*. (1969) are also listed in the PDF.

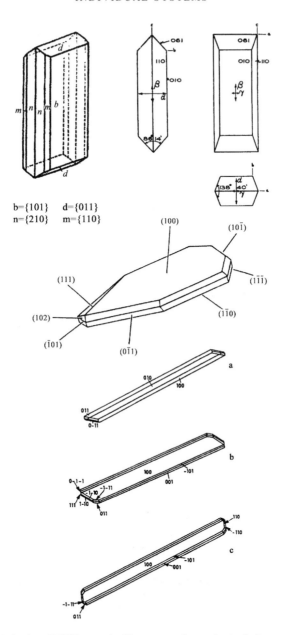

b={101}   d={011}
n={210}   m={110}

Fig. 9.4 Morphologies of **TNT** crystals. Upper: experimental morphology of the orthorhombic form, from Groth (1917) (left, in perspective) and McCrone (1949) (three views in projection), with permission; middle: experimental morphology of the monoclinic form grown by slow cooling from ethyl acetate (from Gallagher and Sherwood 1996, with permission); lower: computationally predicted morphology of the (a) orthorhombic, (b) monoclinic; (c) twinned monoclinic phases. (From Gallagher et al. 1997, with permission.)

and Form II, stable from room temperature to 205 °C have been published (Graeber and Morosin 1974), and are examples of conformational polymorphism (Fig. 5.2) (Bernstein 1987). Form III has a stability similar to Form I and apparently has a strong tendency to form twinned crystals, while Forms IV and V have been obtained only upon supercooling of the melt. Thermal gradient sublimation has been used to grow clear single crystals of Form III (Firsich 1984), but the structure has not yet been determined.

9.3.2.3   Lead styphnate (lead trinitroresorcinate)

The so-called 'normal' material contains the dibasic resorcinate dianion. This material is used mainly as a primary, or initiating explosive together with lead azide. Neither McCrone and Adams (1955) nor Meyer (1987) make any mention of polymorphic behaviour; the trimorphic polymorphic behaviour of monobasic lead styphnate has been known for quite some time (Brün 1934; Böttger and Will 1846; Tausen 1935; Hitchens and Garfield 1941).

The two normal lead styphnate structures are polymorphic monohydrates whose structures have been published by Pierce-Butler (1982, 1984).

9.3.2.4   DNFP

Oyumi et al. (1986b) have carried out a rather thorough study on the tetramorphic system. Forms I and II may be obtained selectively from specific solvents; Form III is obtained by cooling Form IV; Form IV is obtained by heating any of the other three modifications. They have been characterized by thermal analysis, IR spectroscopy and analysis of decomposition products. The crystal structure of Form I has been reported by two different groups (Oyumi et al. 1986b; Lowe-Ma et al. 1990).

9.3.2.5   1,1'-Dinitro-3,3'-azo-1,2,4-triazole

This material has been considered as a potential candidate for high-energy propellant applications. A yellow form may be obtained from ethanol, while a pale-orange modification crystallizes from acetone. They crystallize in (equivalent) $P2_1/c$ and $P2_1/a$ space groups respectively, both with $Z' = 0.5$, requiring that the molecule lies on a crystallographic inversion centre and exhibits an essentially planar molecular conformation (Cromer et al. 1988).

9.3.2.6   2-Methyl-1-(2',4',6'-trinitrophenyl)-4,6-dinitrobenzimidazole

This material crystallized from an ethanol/acetone solvent as easily separable concomitant polymorphs (Section 3.5): Form I is orange–brown triangular tabular platelets, while Form II appears as cream–yellow rods with hexagonal cross-section (Lowe-Ma et al. 1989; Freyer et al. 1992). The densities of the two forms differ significantly (1.658 and 1.712 g cm^{-3}) and the full crystal structure analysis indicated that they may also be considered conformational polymorphs, differing in the rotation of the phenyl ring about the exocyclic bond and in the rotations of the nitro groups on that phenyl ring. However, both forms have 'unexceptional explosive properties . . . (and) . . . uninspiring detonation velocities . . . and detonation pressures . . . .' (Fryer *et al.* 1992).

9.3.2.7 *Also worthy of note*

In this section, we list a number of other compounds, sometimes included among energetic materials, that have exhibited polymorphism.

1,8-dinitronaphthalene is used in some explosive mixtures with ammonium nitrate (Meyer 1987). Detailed information on the preparation and characterization of the two dimorphs is given by McCrone (1951), who also studied the solution phase transformation (in thymol) between two polymorphs (McCrone 1957). The structure of the orthorhombic form was initially published by Akopyan *et al.* (1965), and subsequently refined at room temperature and 97 °C by Ciechanowicz-Rutkowska (1977). The CSD contains an entry (DNTNAP03) for the monoclinic form (Kozin 1964), but no coordinates are available.

2,6-dinitrotoluene is an important product in the manufacture of both powdery and gelatinous commercial explosives. It has been shown to be trimorphic (McCrone 1954). Well-formed modifications I (stable below 40 °C) and II (stable above 40 °C) can both be obtained from thymol; Form III is obtained as feathery dendrites only from the melt and is unstable at all temperatures. No crystal structures have been reported.

2,4,5-Trinitrotoluene (also known as γ-**TNT**) is one of the main impurities in military and commercial grades of **TNT**. Chick and Thorpe (1971) characterized two polymorphs. Form I (mp 376.2 K) may be obtained by recrystallization from alcohol or solidification of the melt. Form II (mp 347.2 K) is produced in small quantities with difficulty from an undercooled melt. It readily converts to Form I by mechanical perturbation or even spontaneously. Chick and Thorpe also determined latent heats of fusion, entropies of fusion, specific heats, IR spectra. Due to the conversion induced by grinding no X-ray data were presented for either form. No crystal structures have been reported.

Pentanitrophenylazide is apparently at least dimorphic, the two forms exhibiting different impact sensitivity, H_{50} (α, 53 cm; β, 17 cm) (Lowe-Ma 2000).

Hexanitrophenylamine **HND**, although toxic, has been employed in underwater explosives together with **TNT** and aluminium powder (Meyer 1987). Two poly-morphs were characterized by McCrone (1952), including a listing of the X-ray powder diffraction lines for Form I, which he determined to be orthorhombic, with cell constants essentially matching those reported later by Dickinson and Holden (1977). The latter also apparently carried out the structure analysis (CSD Refcode HNIDPA) but no coordinates are given.

Dipentaerythritol hexanitrate **DiPEHN** is formed as a by product in the synthesis of **PETN**, and hence can influence its performance. It is dimorphic, the stable modifi-cation is Form I (mp 76.0 °C microscope hot stage) (Cady 1974), which can be grown as dendrites from a supercooled melt. Form II starts growing from the supercooled melt, but large crystals can be grown from the melt above 70 °C. The melting point of Form II, also determined on a hot stage microscope is 74.4 °C. The similarity of the melting points and additional thermal data indicate that Form I is more stable relative to Form II at room temperature than in the vicinity of the melting point.

Tripentaerythritol octanitrate **triPEON** is an additional impurity in the production of **PETN**. Using hot stage microscopic methods Cady (1974) identified four polymorphs, with melting points 83.3, 72.1, 74.6 and 69.0 °C and heats of fusion (cal g^{-1}) of 18, 13, 13.1 and ~10.5) for Forms I–IV, respectively. This report also contains a rather detailed evaluation and discussion of previous work on the characterization of the polymorphic forms of this compound.

10

Polymorphism and patents

Modern science is not a solitary undertaking . . . Litigation is. Real science is the study of facts that are regular, but a courtroom trial is quintessentially singular. Science depends on placing facts in an orderly context, but a trial frames facts in isolation. Good science transcends the here and now, the individual and the idiosyncratic, the single laboratory, the single nations, the single planet, even the single galaxy, but a trial typically examines the single datum, and demands that scientific truths be rediscovered anew every time. Scientific facts emerge from many isolated observations, as data are accumulated, vetted for error, tested for significance, correlated, regressed, and reanalyzed, but trials are conducted retail. Good science is open, collegial, and cumulative, but the courtroom setting is discrete, insular, and closed-a one shot decision. (Huber 1991)

Scientists do not deal in certainties, only in likelihoods. In mathematical terms, we deal in 'probabilities'. Unlike most lawyers, we are not in the *absolute* business of proving something is 'true' and another thing is 'false'. Applied to our technological studies, we believe that this sort of 'lawyers' language' would take us into the realm of metaphysics because such categorical statements are not relatable to opinions derived from observations. (Smith 1993)

Courts of law are not the optimal fora for trying questions of scientific truth . . . (Zenith Laboratories v. Bristol-Myers Squibb 1994)

10.1 Introduction

A patent is a social contract, known since the Middle Ages. According to the Oxford English Dictionary definition, it is

a license to manufacture, sell, or deal in an article or commodity, to the exclusion of other persons; in modern times, a grant from the government to a person or persons conferring for a certain definite time the exclusive privilege of making, using, or selling some new invention.

Originally patents granted the right to sell, specifically tobacco, and were expanded to other commodities. Today's patents only give the right to exclude. As noted by Maynard and Peters (1991),

patent systems reward the competitive, creative drive with a temporary, limited, exclusive right, in return for the cooperation of an inventor in teaching the rest of society how to use his or her findings for all time thereafter (see also Grubb 1986).

Polymorphism presents interesting issues to patent systems. Since crystal modifications of a substance represent different crystal structures with potentially different properties, the discovery or preparation of a new crystal modification represents an opportunity to claim an invention that potentially can be recognized in the awarding of a patent. The rules and regulations of such patents differ from country to country, although some degree of standardization has resulted from the World Trade Agreement and subsequent legislation, but there are still important differences in the nuances of the granting and enforcement of patents in various venues. For instance, in the United Kingdom, a new crystal modification is not prima facia patentable; the inventor must demonstrate that it is an unobvious variant of the previously known material. As seen in earlier chapters, a particular crystalline modification can possess considerable chemical, physical or biological advantages over its congeners, and the granting and maintenance of patent exclusivity over the rights to particular polymorphic form(s) may have considerable economic consequences. As a result, it is not surprising that the past 30 years or so have witnessed a number of patent litigations essentially involving different crystal modifications, many concerning the definitions, descriptions and analytical techniques presented in earlier chapters. Partly because of the size and economic impact of the pharmaceutical industry, some of the most visible cases have involved some widely used drugs. In addition, in the United States there are special patent provisions for pharmaceuticals (Engelberg 1999) which also may contribute to the frequency and nature of these legal battles (Barton 2000).

In this chapter we will review some of the cases that serve as the meeting ground between the worlds of science and the law alluded to in the preambulary quotations of this chapter (see also Faigman 1999; Foster and Huber 1999). The intention is to demonstrate the scientific issues that were raised in the course of these litigations, not to provide legal opinions or precedents. Many of these cases also involved principles and nuances of patent law, which also will not be covered here. Litigations involving patent issues can generate hundreds of thousands of pages of documents and testimony with the discovery of many facts and the expression of many (often opposing) scientific opinions on both sides of the issue. We will limit the descriptions here to what is given in the official records of the cases considerd, mainly from the patents, judicial decisions and the reports of them. Of course, even these are subject to controversy, since court rulings can be and are reversed, thus perhaps altering the way a scientific issue is viewed by a court of law and the society that is guided by that law. Moreover, both science and the law are dynamic, and the interpretations and ramifications of any particular case can and do change with time.

10.2   Ranitidine hydrochloride

Ranitidine was developed in the 1970s by Allen & Hanburys Ltd. of the Glaxo Group (later Glaxo Wellcome and now Glaxo SmithKline) in the flurry of activity following the identification of the histamine H_2 receptor (Black *et al.* 1972) and other H_2 antagonists (Bradshaw 1993) for the treatment of peptic ulcers. In June 1977, David Collin, a Glaxo chemist, first prepared ranitidine hydrochloride **(RHCl)**, and within a month Glaxo filed a U.S. patent application, which resulted in the issue of U.S. Patent No. 4,128,658 (Price *et al.* 1978) (the '658 patent). Example 32 of this patent gives the procedure for the preparation of the hydrochloride (Fig. 10.1) from ranitidine base.

United States Patent [19]

Price et al.

[11] **4,128,658**

[45] **Dec. 5, 1978**

[54] **AMINOALKYL FURAN DERIVATIVES**

[75] Inventors: **Barry J. Price**, Hertford; **John W. Clitherow**, Sawbridgeworth; **John Bradshaw**, Ware, all of England

[73] Assignee: **Allen & Hanburys Limited**, London, England

[21] Appl. No.: **818,762**

[22] Filed: **Jul. 25, 1977**

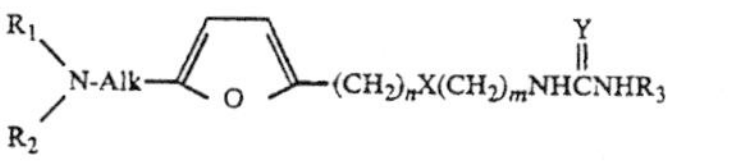

and physiologically acceptable salts thereof and N-oxides and hydrates, in which R$_1$ and R$_2$ which may be the same or different represent hydrogen, lower alkyl, cycloalkyl, lower alkenyl, aralkyl or lower alkyl interrupted by an oxygen atom or a group

EXAMPLE 32

N-[2-[[[5-(Dimethylamino)methyl-2-furanyl]methyl]thio]ethyl]-N'-methyl-2-nitro-1,1-ethenediamine hydrochloride

N-[2-[[[5-(Dimethylamino)methyl-2-furanyl]methyl]thio]ethyl]-N'-methyl-2-nitro-1,1-ethenediamine (50 g, 0.16 mole) was dissolved in industrial methylated spirit 74° o.p. (200 ml) containing 0.16 of an equivalent of hydrogen chloride. Ethyl acetate (200 ml) was added slowly to the solution. The hydrochloride crystallised and was filtered off, washed with a mixture of industrial methylated spirit 74° o.p. (50 ml) and ethyl acetate (50 ml) and was dried at 50°. The product (50 g) was obtained as an off-white solid m.p. 133°–134°.

Fig. 10.1 Portion of the title page and Example 32 from the '658 patent, giving the procedure for the preparation of **RHCl**, which subsequent to the discovery of a second polymorphic form became known as the procedure for the preparation of Form I.

In the course of subsequent scale up, Glaxo developed a pilot plant process called 3A, and then one called 3B. On 15 April 1980, for unknown reasons,[1] the thirteenth batch of **RHCl** prepared using the latter process produced crystals that gave different IR spectra and X-ray powder diffraction patterns from previous batches. Glaxo concluded that a new polymorph, designated Form 2, had been produced, and the earlier form, described in the '658 patent, was designated Form 1.[2] Glaxo subsequently developed a process, referred to as 3C, to manufacture all the **RHCl** it has sold commercially as the active ingredient in Zantac.

In October 1981, Glaxo filed a patent application on Form 2, from which two patents were eventually granted in June of 1985 as U.S. Patent No. 4,521,431 (the '431 patent) and June 1987 as U.S. Patent No. 4,672,133. The abstract of the '431 patent states simply, 'A novel form of ranitidine . . . hydrochloride, designated Form 2, and having favourable filtration and drying characteristics, is characterized by its infra-red

---

[1] In fact, many of the discoveries of new crystal modifications have been made serendipitously (e.g. Silvestri v. Grant 1974), as have many other important scientific discoveries (Roberts 1989).

[2] The following description is taken from the court's opinion in the case of Glaxo Inc. and Glaxo Group Limited v. Novopharm Ltd. No. 91-759-CIV-5-BO.

spectrum and/or by its X-ray powder diffraction pattern.' These advantageous filtering and drying characteristics are due, in part, at least, to the fact that Form 2 tends to crystallize as more needle-like crystals than the plate-like Form 1 (see Fig. 4.41).

By 1991, Zantac sales had reached nearly $3.5 billion, nearly twice the sales of the next best selling drug. A number of generic drug firms undertook efforts (under the provisions of the Waxman-Hatch Law (Engelberg 1999)) to prepare to go on the market with Form 1 in 1995, upon anticipated expiration of the '658 patent.[3] One of these generic companies was Novopharm Ltd., the defendant in the first **RHCl** litigation that went to trial. Novopharm scientists unsuccessfully attempted to prepare Form 1 faithfully following the procedure of example 32 of the '658 patent and therefore sought approval to market Form 2, claiming that the product is, and always has been, Form 2 **RHCl**. In November 1991, Novopharm filed an abbreviated new drug application (ANDA) at the FDA to market Form 2 beginning in 1995. As required by the Waxman-Hatch Act, Novopharm notified Glaxo of its contention that the '431 patent was unvalid. Glaxo sued Novopharm for infringement of the '431 patent. Novopharm admitted infringement of the '431 patent, but contended that it was invalid, claiming that Form 2 was inherent in the '658 patent. Novopharm claimed that Glaxo *never* performed Example 32 precisely as written either before or after including it in the '658 patent (emphasis in the Court's original). Novopharm theorized that if Glaxo had performed Example 32, the result would have been Form 2, not Form 1. Therefore the Form 2 patent was invalid.

Glaxo argued, *inter alia*, that Novopharm's experiments were contaminated with 'seed' crystals of Form 2, and therefore, were not faithful replications of Example 32 (quotation marks in Court's original). To make its point on the inherency argument, Glaxo proved that Example 32 does not invariably lead to Form 2, and in fact had led to Form 1. In support of its position Glaxo compared David Collin's original experiments as described in his notebooks in minutiae with Example 32, and also presented evidence that Example 32 as performed in 1993 at Oxford University had led to Form 1. This was sufficient evidence to convince the court that Example 32 does lead to the Form 1 product and that the '431 patent was valid.

Novopharm then examined the possibility of marketing Form 1, and developed a stable, reproducible process for the manufacture of Form 1. In April 1994,[4] Novopharm filed a new ANDA, this time seeking approval to market Form 1 **RHCl** upon the expiration of the '658 patent. Shortly thereafter, Glaxo sued Novopharm again, alleging that Novopharm had sought permission to manufacture and market a product which would contain not pure Form 1, but rather a mixture of Forms 1 and 2, thereby infringing upon Glaxo's Form 2 patents. Novopharm's ANDA, as

---

[3] At the time a U.S. patent was valid for 17 years from the date of issue. Subsequent international trade agreements have led to changes in the period of enforcement, and during the changeover period there have been some extensions to the terms of some existing patents.

[4] Much of the following description is taken from the original district court opinion in the case of Glaxo Inc. and Glaxo Group Limited v. Novopharm Ltd. No. 5:94-CV-527-BO(1), 931 F.Supp. 1280 and the appeals court's opinion in the case of Glaxo Inc. and Glaxo Group Limited v. Novopharm Ltd. 96-1466, DCT. 94-CV-527.

initially filed, specified that the marketed product be approximately 99 per cent pure Form 1 **RHCl** (with impurities that may include Form 2 **RHCl**) as determined by IR. Amended ANDAs filed by Novopharm would have permitted the marketed product to have a Form 1 **RHCl** of purity as low as 90 per cent. Novopharm, however, submitted X-ray evidence at trial that demonstrated that its actual samples of **RHCl** did not contain detectable Form 2.

The court found that Novopharm had established that its product would not contain Form 2, and that if the product did contain Form 2, then it would be present as an independent component or impurity, not as the basis for some improvement or equivalent. The court thus allowed Novopharm to market mixtures of Forms 1 and 2.

The appeals court upheld the district court's decision holding that based on all the available evidence, Glaxo had not proven that Novopharm was likely to market a product that contained Form 2. Notably, the appeals court pointed to Glaxo's failure to test Novopharm's samples. In reviewing a later case (Glaxo v. Torpharm 1998), the same appeals court noted it had explicitly declined to address the question of whether small amounts of Form 2 **RHCl** in a mixture containing primarily Form 1 would infringe the '431 patent (Glaxo v. Torpharm 1998).

The two Glaxo v. Novopharm cases involved many aspects of the study and analysis of polymorphic materials described in earlier chapters: the (often serendipitous) discovery and recognition of polymorphic forms, the role of solvent, heating, stirring and other experimental techniques in attempts to control the polymorph obtained, the use and development of analytical methods for the characterization of polymorphic forms, the relative stability of polymorphic forms, the phenomenon of disappearing polymorphs, the role of seeding, both intentional and unintentional, and the distinction between polymorphic identity and polymorphic purity. Many of these issues also arose in the other litigations involving **RHCl** (for instance: Glaxo Inc. v. Geneva 1994, 1996; Pharmaceuticals *et al*. Nos. 94-1921, 94-4589 and 96-3489, D. NJ, Glaxo Inc. v. Boehringer-Ingelheim 1996; Glaxo Inc. *et al*. v. Torpharm Inc. *et al*. 1997).

10.3   Cefadroxil

Cefadroxil (also sometimes spelled cephadroxil), an antibiotic originally developed by Bristol-Myers (BM) and marketed under the trade names Ultracef and Duricef, had combined sales in the US of $100 million in 1988. A crystalline monohydrate (apparently discovered serendipitously), claimed in U.S. Patent No. 4,504,657 (the '657 patent) issued in March 1985, is described in the patent in terms of its X-ray diffraction pattern (Fig. 10.2). The improvement of this substance over the earlier known material included greater stability and high bulk density, properties which enabled the production of smaller pills.

Prosecution of this patent was in fact quite lengthy, the original application dating from August 1979, and involved questions of similarities and differences in crystalline modifications. (See Kalipharma v. Bristol-Myers 1989 for details of the history of the prosecution of the patent application.) The prior art included a 1973 patent (U.S.

United States Patent [19]

Bouzard et al.

[11] Patent Number: 4,504,657

[45] Date of Patent: Mar. 12, 1985

[54] CEPHADROXIL MONOHYDRATE

[75] Inventors: Daniel Bouzard, Franconville;
                   Abraham Weber; Jacques Stemer,
                   both of Paris, all of France

[73] Assignee: Bristol-Myers Company, New York,
                   N.Y.

[21] Appl. No.: 358,567

[22] Filed: Mar. 16, 1982

           Related U.S. Application Data

[60] Continuation of Ser. No. 931,800, Aug. 7, 1978, aban-
      doned, which is a continuation of Ser. No. 874,457,
      Feb. 2. 1978, Pat. No. 4,160,863, which is a division of
      Ser. No. 785,392, Apr. 7, 1977, abandoned.

[30]   Foreign Application Priority Data

  Apr. 7, 1977 [GB] United Kingdom 17028/76

[51] Int. Cl.³ C07D 501/22; C07D 501/12
[52] U.S. Cl. ....,.................. 544/30
[58] Field of Search 544/30

[56]   References Cited
              U.S. PATENT DOCUMENTS

  3,489,752 1/1970 Crast, Jr. 260/243

  3,655,656   4/1972 Van Heyningen 260/243
  3,781,282 12/1973 Garbrecht 260/243
  3,957,773   5/1976 Burton 260/243
  3,985,741 10/1976 Crast, Jr. 260/243
  4,091,215   5/1978 Bouzard 544/30
  4,160,863   7/1979 Bouzard et al. 544/30
  4,162,314   7/1979 Gottschlich 544/30

          FOREIGN PATENT DOCUMENTS

   829758 12/1977 Belgium .
  1240687   7/1971 United Kingdom .

             OTHER PUBLICATIONS

Dunn et al., The Journal of Antibiotics, 29: 65–80, (Jan.
1976).

Primary Examiner—Mark L. Berch
Attorney, Agent, or Firm—Robert E. Carnahan; David
M. Morse

[57]   ABSTRACT

A novel crystalline monohydrate of 7-[D-α-amino-α-(p-
hydroxyphenyl)acetamido]-3-methyl-3-cephem-4-car-
boxylic acid is prepared and found to be a stable useful
form of the cephalosporin antibiotic especially advantageous for pharmaceutical formulations.

1 Claim, 1 Drawing Figure

We claim:

65   1. Crystalline   7-[D-α-amino-α-(p-hydroxyphenyl-
)acetamido]-3-methyl-3-cephem-4-carboxylic   acid
monohydrate exhibiting essentially the following x-ray
diffraction properties:

Line	Spacing d(A)	Relative Intensity
1	8.84	100
2	7.88	40
3	7.27	42
4	6.89	15
5	6.08	70
6	5.56	5
7	5.35	63
8	4.98	38
9	4.73	26
10	4.43	18
11	4.10	61
12	3.95	5
13	3.79	70
14	3.66	5
15	3.55	12
16	3.45	74
17	3.30	11
18	3.18	14

-continued

Line	Spacing d(A)	Relative Intensity
19	3.09	16
20	3.03	29
21	2.93	8
22	2.85	26
23	2.76	19
24	2.67	9
25	2.59	28
26	2.51	12
27	2.46	13
28	2.41	2
29	2.35	12
30	2.30	2
31	2.20	15
32	2.17	11
33	2.12	7
34	2.05	4
35	1.99	4
36	1.95	14
37	1.90	10

. . . . .

Fig. 10.2 Upper, the title page of U.S. patent 4,504,657 for the 'Bouzard form' of cephadroxil
monohydrate; lower, the claim for the Bouzard from, giving the powder diffraction pattern in
tabular form.

Patent No. 3,781,282) (the '282 patent) in which a modification of cefadroxil had also been prepared. This was termed the 'Micetich form' after one of the inventors. One of the questions surrounding the '657 patent was whether the crystalline material described therein (known as the 'Bouzard form' or 'Bouzard monohydrate') was the same as or different from the 'Micetich form'. In the end, the '657 patent was granted, in essence recognizing the difference between the two.

In anticipation of the expiration of '282 patent, there were attempts to make the Micetich form according to Example 19 in that patent. Those attempts invariably led to the Bouzard form, leading the companies involved in carrying out the experiments to claim inherency of the Bouzard form in the '282 patent. BMs' explanation of these results included the role of unintentional seeding that tended to favour the formation of the Bouzard form rather than the Micetich form. A number of litigations ensued, which involved these issues (e.g. Bristol-Myers v. United States International Trade Commission *et al.* 1989; Kalipharma v. Bristol-Myers 1989). In the former case, there was also a question of the identity of the various crystal modifications, determined mainly by powder X-ray diffraction, with conflicting opinions by various experts. In the latter case, the court found that the '. . . plaintiff had established a prima facie case of invalidity (of the '657 patent)', and that '. . . its scientists have replicated the claimed cefadroxil monohydrate according to . . . Gambrecht Example 7. . .' The court also did not accept the defendant's '. . . only challenge, that seeds from its cefadroxil monohydrate acted as a template such that any attempt to repeat Garbrecht Example 7 anywhere in the world would yield BM cefadroxil monohydrate. . .'.

There have been many subsequent litigations and appeals. Many experiments have been done and many issues arose in those cases. Some of them are noted here:

1 The X-ray powder pattern of the Bouzard form was indexed to determine that every line in the pattern could be accounted for by that particular crystal modification.
2 Attempts were made to prepare the Micetich form in an unseeded environment with the measure of success being a matter of contention.
3 Experiments were run in a clean room equipped with filters that would eliminate particles the size of bacteria (which are several microns in size), but could not eliminate seeds that can be hundreds of times smaller (See Section 3.2).
4 A series of side-by-side experiments were run with one sample sealed from seed crystals and the other not sealed—the seed-free produced the Micetich form and the open flasks produced the Bouzard form (Tarling 2001).

The series of litigations continued over other issues, including the nature of crystal modifications (Zenith v. Bristol-Myers Squibb 1992). Zenith prepared and formulated a hemihydrate of cefadroxil, for which it submitted an ANDA to the FDA. Bristol-Myers (now Bristol-Myers Squibb) contended that the FDA should require a much more extensive new drug application (NDA) for the hemihydrate, rather than approve it within the framework of the FDA monograph for the monohydrate. BM also alleged that the Zenith product converted to the monohydrate, thereby infringing the '657 patent. Bristol-Myers' theory of infringement was that the Zenith product converts to the Bouzard monohydrate after ingestion in the gullet and the stomach, as it is mixed with liquid. A federal court found in favour of BM. Upon appeal, Zenith contended that the '657 patent does not cover Bouzard form crystals which might form momentarily in a patient's stomach, and there were both legal and scientific grounds for the basis of that contention. One of the main points of contention was whether the X-ray pattern of the material found by BM in a patient's stomach matched the claim of the Bouzard patent, as shown in Fig. 10.2. Bristol-Myers had identified 15 of the 37 diffraction lines cited in the claim. The court found that, 'Although the term "essentially" recited in the claim permits some leeway in the exactness of the comparison with the specified 37 lines of the claim, it does not permit ignoring a substantial number of lines altogether', and that '. . . there was a failure of proof as to whether any crystals, assumed to form in the stomach from ingested cefadroxil . . . , literally infringe the '657 claim'. On this basis, the appeals court reversed the lower court's decision. One result of this decision is that many subsequent patent applications including claims based on substances characterized by X-ray diffraction patterns and/or IR spectra were framed in language that differs from that used in the '657 patent.

10.4   Terazosin hydrochloride

Terazosin hydrochloride is a drug developed by Abbott Laboratories for the treatment of hypertension and benign prostatic hyperplasia, and marketed as the dihydrate under the trade name Hytrin since 1987. The patent of interest here is U.S. Patent No. 5,504,207 (the '207 patent), filed in October, 1994 for which the fourth claim is a particular anhydrous crystalline form of terazosin hydrochloride, designated by Abbott as 'Form IV',[5] a form not marketed by Abbott. In the '207 patent, Form IV is defined by reference to the X-ray powder diffraction pattern, listing the peak positions of several of the peaks. To be within the scope of claim 4 of the '207 patent, a product must be anhydrous terazosin hydrochloride and must exhibit a powder X-ray diffraction pattern having each of the principal peaks identified in claim 4.

The defendants were planning to market anhydrous terazosin hydrochloride rather than the dihydrate found in Abbott's Hytrin tablets, the latter being protected by a patent that expired in 2000. Plaintiffs sued defendants for infringing the '207 patent. Geneva argued that the fourth claim in the '207 patent was invalid because Form IV terazosin hydrochloride was 'on sale' in the United States more than one year before the filing date for the '207 patent. One of the defendant's experts in X-ray crystallography analysed two samples of anhydrous terazosin hydrochloride purchased prior to the 1994 filing date and found that the patterns for both contained all of the 'principal peaks' of claim 4 of the '207 patent (note the difference from the cefadroxil case, *vide ante*). One batch was pure Form IV terazosin hydrochloride, while the second was a mixture of Forms IV and II. Abbott argued that 'it (was) conceivable

---

[5] Much of the following is taken essentially verbatim from the opinion of the court in Abbott v. Geneva (1998).

that the substances (in those samples) were initially manufactured as a less stable crystal form' of terazosin hydrochloride, and that these 'substances converted over time to Form IV'. The court viewed this as 'little more than speculation', adding that, 'Despite Abbott's extensive work with terazosin hydrochloride, Abbott was unable to produce any expert testimony that such a transformation occurs or is likely to occur with terazosin hydrochloride'. The court therefore granted summary judgement in favour of the three defendants, holding claim 4 invalid because Form IV terazosin hydrochloride was on sale more than one year before Abbott filed its application for the '207 patent.

10.5 Aspartame

Aspartame, the artificial sweetener marketed as NutraSweet®, is a dipeptide, which was discovered in 1965 by accident to be 100–200 times sweeter than sucrose. The discovery was originally made at G.D. Searle, which was later acquired by Monsanto.

In the original conventional manufacturing process, the crystallization from aqueous solutions tended to produce needles with diameter of 10 μm or less. The crystals were very fine with large specific volume. These characteristics led to many problems in the filtration and drying processes, the formation of scale on reactor surfaces, and the high dustability and hygroscopicity of the final product, which made it difficult to handle and unsuitable for use as a direct (i.e. table top) sweetener (Ajinomoto 1983; Kishimoto et al. 1989).

Five crystal modifications of Aspartame are known (Ajinomoto 1983; Hatada et al. 1985; Nagashima et al. 1987; Kishimoto and Naruse 1988; Tsuboi et al. 1991; Furedi-Milhofer et al. 1999). A number of analytical methods have been used recently to characterize three of them: two hemihydrate polymorphs and a dihemihydrate (Leung et al. 1998a,b; Zell et al. 1999). The crystal structure of one of the hemihydrate forms, known as Form I, was published earlier (Hatada et al. 1985). More recently, synchroton radiation was used to determine the structure of the 'low humidity' form (denoted Form Ib by the authors), with an asymmetric unit comprised of three aspartame molecules and two water molecules (Meguro et al. 2000).

Many attempts were made to improve the qualities of the crystals obtained from the conventional manufacturing process. The Japanese Ajinomoto Co. had a license to make aspartame and, as a result of 'intensive investigations to improve the workability {of the crystallization} step in the production . . .', discovered that cooling aqueous solutions of aspartame without stirring led to 'bundle-like' crystal aggregates of crystals (Kishimoto and Naruse 1987). These crystals had significantly improved handling characteristics compared to the conventional crystals. This process was subsequently refined and led to an Ajinomoto European patent, granted in 1985.

In April 1992, the Opposition Division of the European Patent Office concluded that the bundle-like crystals were the same as the conventional crystals. In response to that decision, Ajinomoto presented a detailed study of the correlation of the differences between the two (Tarling 1992). The two types of crystals were different habits rather than polymorphs. The bundle-like crystals, produced without stirring, had a higher degree of crystal perfection than the conventional crystals, produced with agitated stirring. The Tarling report identified some 'significant and reproducible differences' in a number of properties determined with a variety of analytical techniques. The difference formed the basis of the Ajinomoto response to the Opposition, and included the following:

- specific volumes: the bundle-like crystals have smaller volumes and hence higher bulk density;
- rate of dissolution: because of their smaller surface to volume ration the larger bundle-like crystals would be expected to dissolve more slowly than the conventional crystals; however, the opposite is true;
- water content: bundle-like crystals contain significantly less water than conventional crystals;
- density: a small but significant difference in the density was determined, with the conventional crystals having a lower density, consistent with the higher strain noted for those crystals;
- solid state NMR spectra: consistent shifts in position and changes in the widths of peaks could be detected between the two modifications, indicating slightly different molecular environments;
- polarizing light microscopy: bundle-like crystals show very sharp extinction indicating a high degree of structural perfection, which is not exhibited by conventional crystals;

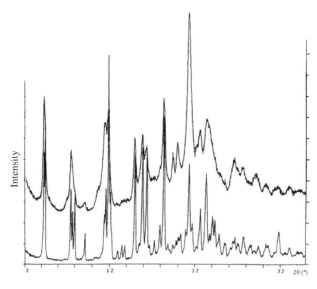

Fig. 10.3 Comparison of the measured X-ray diffraction patterns of aspartame. Upper, conventional crystals; lower, bundle-like crystals. (From Tarling 2001, with permission.)

- X-ray powder diffraction: the powder patterns for the bundle-like and conventional crystals are shown in Fig. 10.3. They are quite similar, as expected for materials that have essentially the same crystal structure. However, there are some slight differences in the peak positions, which was attributed to the difference in water content and degree of internal crystal strain. There are some more obvious differences in relative intensities, including the position of the strongest peak. There are clear differences in the peak widths, which also reflects differences in the degree of crystallinity;
- scanning electron microscopy, single crystal X-ray diffraction (Hatada *et al.* 1985; Kishimoto *et al.* 1989), and crystal habit modelling were used to determine the preferred orientation and to simulate the X-ray powder patterns of both forms to show a good comparison with the experimental data.

On 27 May 1997, the Boards of Appeal of the European Patent office set aside the 1992 opposition to the Ajinomoto patent (European Patent Office 1997), and a New European Patent Specification was issued (Ajinomoto 1998).

10.6   Concluding remarks

We have presented here but a sampling of a number of litigations that involved controversies over the nature, identity, quantity and uniqueness of crystalline modifications. As in other aspects of the law each case has its own character and idiosyncrasies. As evidenced in numerous examples throughout this book, the same variability exists in the nature of the polymorphic behaviour of a substance. That behaviour differs because the substances are different, and to be properly understood each substance has to be investigated on its own. Nevertheless, the polymorphism of molecular systems can be understood and utilized on fundamental principles, in light of Buerger and Bloom's comments quoted at the opening of Chapter 1, we hope that this book serves to remove some of the mystery that they suggested surrounds the subject.

References

Following each reference are the page numbers on which that reference is cited. A following 'f' denotes that the reference is in a figure caption. A following 't' denotes that it appears in a table.

Aakeröy, C. B., Nieuwenhuyzen, M. and Price, S. L. (1998). Three Polymorphs of 2-amino-5-nitropyrimidine: experimental structures and theoretical predictions. *J. Am. Chem. Soc.*, **120**, 8986–3. [184]

Abbott Laboratories v. Geneva Pharmaceuticals, Inc., Abbott Laboratories v. Novopharm Limited, Abbott Laboratories v. Invamed, Inc. Nos 96 C 3331, 96 C 5868, 97 C 7587; 1998 WL 566884 (N. D. Ill.), Sept. 1, 1998. [304]

Addadi, L., Berkovitch-Yellin, Z., Weissbuch, I., van Mil, J., Shimon, L. J. W., Lahav, M. and Leiserowitz, L. (1985). Growth and dissolution of organic crystals with 'tailor–made' inhibitors—implication in stereochemistry and materials science. *Angew. Chem. Int. Ed. Engl.*, **24**, 466–85. [47]

Adyeeye, C. M., Rowley, J., Madu, D., Javadi, M. and Sabnis, S. S. (1995). Evaluation of crystallinity and drug-release stability of directly compressed theophylline hydrophilic matrix tablets stored under varied moisture conditions. *Int. J. Pharm.*, **116**, 65–75. [5]

Aerts, J. (1996). Polymer crystal silverware: a fast method for the prediction of polymer crystal structures. *Polym. Bull.*, **36**, 645–52. [182]

Agatonovic-Kustrin, S., Wu, V., Rades, T., Saville, D. and Tucker, I. G. (1999). Powder diffractometric assay of two polymorphic forms of rantidine hydrochloride. *Int. J. Pharmacol.*, **184**, 107–14. [125]

Agatonovic-Kustrin, S., Wu, V., Rades, T., Saville, D. and Tucker, I. G. (2000). Ranitidine hydrochloride X-ray assay using a neural network. *J. Pharm. Biomed. Sci.*, **22**, 985–92. [125]

Agrawal, J. P. (1998). Recent trends in high energy materials. *Prog. Energy Combust. Sci.*, **24**, 1–30. [275]

Aguiar, A. J., Krc, J., Kinkel, A. W. and Samyn, J. C. (1967). Effect of polymorphism on the absorption of chloramphenicol from chloramphenicol palmitate. *J. Pharm. Sci.*, **56**, 847–53. [244, 247f]

Ahr, G., Voith, B. and Kuhlmann, J. (2000). Guidances related to bioavailability and bioequivalence: European industry perspective. *Eur. J. Drug Metab. Pharmacokinet.*, **25**, 25–7. [244]

Ajinomoto Co. Inc. (1983). Process for crystallizing alpha-L-asparatyl-L-phenylalanin-methyl ester. European Patent Application 0 091 787 A1. Issued as European Patent Specification 0 091 787 B1 (04.09.85). [305]

Ajinomoto Co. Inc. (1998). Process for crystallizing alpha-L-aspartyl-L-phenylalanine-methyl ester. New European Patent Specification EP 0 091 787 B2. [307]

Akopyan, Z. A., Kitaigorodskij, A. I. and Struchkov, Yu. T. (1965). Steric hindrance and molecular conformation. XII. The crystal and molecular structure of 1,8-dinitronaphthalene. *Zh. Strukt. Khim.*, **6**, 729–44. [295]

Aldoshin, S. M., Kozina, O. A., Gutsev, G. L., Atovmyan, E. G., Atovmyan, L. O. and Nedvetskii, V. S. (1988). *Izvestia Akademii Nauk SSSR Seriya Khimicheskaya.* 2301. [208]

Aldridge, P. K., Evans, C. L., Ward, H. W. II, Cogan, S. T., Boyer, N. and Gemperline, P. J. (1996). Near-IR detection of polymorphism and process-related substances. *Anal. Chem.*, **68**, 997–1002. [129, 130]

Ali, A. A. and Farouk, A. (1981). Comparative studies on the bioavailabilities of ampicillin anhydrate and trihydrate. *Int. J. Pharm.*, **9**, 239–43. [245]

Allemand, P. M., Fite, C., Srdanov, G., Keder, N., Wudl, F. and Canfield, P. (1991). On the complexities of short-range ferromagnetic exchange in a nitronyl nitroxide. *Synthetic Metals*, **41–43**, 3291–5. [201]

Allen, F. A., Davies, J. E., Galloy, J. J., Johnson, O., Kennard, O., Macrae, C. F., Mitchell, E. M., Mitchell, G. F., Smith, J. M. and Watson, D. G. (1991). The development of Versions 3 and 4 of the Cambridge Structural Database system. *J. Chem. Inf. Comput. Sci.*, **31**, 187–204. [16]

Allen, F. H. and Kennard, O. (1993). 3D search and research using the Cambridge Structural Database. *Chem. Des. Autom. News*, **8**, 31–7. [16, 183]

Allen, F. H., Kennard, O., Watson, D. G., Brammer, L., Orpen, A. G. and Taylor, R. (1987). Tables of bond lengths determined by X-ray and neutron diffraction. Part 1. Bond lengths in organic compounds. *J. Chem. Soc., Perkin Trans.* **2**, S1–19. [151]

Allen, F. H., Kennard, O. and Watson, D. G. (1994). Crystallographic databases: Search and retrieval of information from the Cambridge Structural Database. In *Structure correlation* (ed. H.-B. Bürgi, and J. D. Dunitz), pp. 71–110. VCH, Weinheim. [16]

Allen, F. H., Harris, S. E. and Taylor, R. (1996). Comparison of conformer distributions in the crystalline state with conformational energies calculated by *ab initio* techniques. *J. Comput. Aided Mol. Des.*, **10**, 247–54. [155]

Almeida, M. and Henriques, R. T. (1997). Perylene based conductors. In *Handbook of organic conductive molecules and polymers*, Vol. 1, pp. 87–149. John Wiley & Sons, Chichester. [191]

Alper, A. S. (1999). The Gibbs phase rule revisited; interrelationships between components and phases. *J. Chem. Educ.* **74**, 1167–1569. [30]

Amici, G. B. (1844). Note sur un appareil de polarisation. *Ann. Chim. Phys., Ser. 3*, **12**, 114–20. [21]

Amorós, J. L. (1959). Notas sobre la historia de la cristalografia I. La controversia Haüy–Mitscherlich. *Bol. Real Soc. Espan. Hist. Nat.*, **57**, 5–30. [20]

Amorós, J. L. (1978). *Le gran aventura del cristal. Naturaleza y evolucion de la ciencia de los cristales.*, pp. 205–9. Editorial de la Universidad Complutense, Madrid. [20]

Andersgaard, H., Finholt, P., Gjermundsen, R. and Hoyland, T. (1974). Rate studies on dissolution and enzymatic hydrolysis of chloramphenicol palmitate. *Acta Pharm. Suec.*, **11**, 239–48. [247, 248f]

Andreev, G. A. and Hartmanová, M. (1989). Floatation method of precise density measurements. *Phys. Stat. Solidi. A: Appl. Res.*, **116**, 457–68. [147]

Andreev, Y. G., Lightfoot, P. and Bruce, P. G. (1997). A general Monte Carlo approach to structure solution from powder-diffraction data: application to poly(ethylene oxide)3:LiN(SO_2CF_3)$_2$. *J. Appl. Cryst.*, **30**, 294–305. [111]

Andronis, V. and Zografi, G. (2000). Crystal nucleation and growth of indomethacin polymorphs from the amorphous state. *J. Noncryst. Solids*, **271**, 236–48. [252]

Anex, B. G. (1966). Optical properties of highly absorbing crystals. *Mol. Cryst.*, **1**, 1–36. [226]

Anex, B. G. and Fratini, A. V. (1964). Polarized single crystal reflection spectra of auramine perchlorate. *J. Mol. Spectr.*, **14**, 1–26. [226, 228]

Anex, B. G. and Simpson, W. T. (1960). Metallic reflection from molecular crystals. *Rev. Mod. Phys.*, **32**, 466–76. [226, 231]

Anonymous. (1949). Crystallographic data 21. *o*-aminobenzoic acid (anthranilic acid). *Anal. Chem.*, **21**, 1016–7. [59]

Antipin, M. Y., Timofeeva, T. V., Clark, R. D., Nesterov, V. N., Dolgushin, F. M., Wu, J. and Leyderman, A. (2001). Crystal structures and molecular mechanics calculation of nonlinear optical compounds: 2-cyclooctylamino-5-nitropyridine (COANP) and 2-adamantylamino-5-nitropyridine (AANP). New Polymorphic modifcation of AANP and electrooptic effects. *J. Mater. Chem.*, **11**, 351–8. [213]

Apperley, D. C., Fletton, R. A., Harris, R. K., Lancaster, R. W., Tavener, S. and Threlfall, T. L. (1999). Sulfathiozole polymorphism studied by magic–angle spinning NMR. *J. Pharm. Sci.*, **88**, 1275–80. [135, 139f]

Arnautova, E. A., Zakharova, M. V., Pivinia, T. S., Smolenskii, E. A., Sukhachev, D. V. and Shcherbukhin, V. V. (1996). Methods for calculating the enthalpies of sublimation of organic molecular crystals. *Russ. Chem. Bull.*, **45**, 2723–32. [168]

Arnold, P. R. and Jones, F. (1972). Polymorphism in anthranilic acid crystals. Examination by DTA and DSC [differential scanning calorimetry]. *Mol. Crys. Liq. Cryst.*, **19**, 133–40. [59]

Artini, E. (1915). *Rend. Ac. Lincei.*, **21**, 274. [288]

Ashida, M. Uyeda, N. and Suito, E. (1971). Thermal transformation of vacuum condensed thin films of copper phthalocyanines. *J. Cryst. Growth*, **8**, 45–56. [268]

Ashizawa, K. (1989). Polymorphism and crystal structure of 2R,4S,6-fluoro-2methyl-spiro[chroman-4,4'-imidazoline]2'-5-dione (M79175). *J. Pharm. Sci.*, **78**, 256–60. [129]

Ashwell, G. J., Bahra, G. S., Brown, C. R., Hamilton, D. G., Kennard, C. H. L. and Lynch, D. E. (1996). 2,4-bis[4-(N,N-dibutylamino)phenyl]squaraine: X-ray crystal structure of a centrosymmetric dye and the second-order non-linear optical properties of its non-centrosymmetric Langmuir-Blodgett films. *J. Mater. Chem.*, **6**, 23–6. [87, 205]

Ashwell, G. J., Wong, G. M. S., Bucknall, D. G., Bahra, G. S. and Brown, C. R. (1997). Neutron reflection from 2,4-bis(4-(N-methyl-N-octylamino)phenyl)squaraine at the air-water interface and the linear and nonlinear optical properties of its Langmuir-Blodgett and spin–coated films. *Langmuir*, **13**, 1629–33. [205]

Assour, J. M. (1965). On the polymorphic modifications of phthalocyanines. *J. Phys. Chem.*, **69**, 2295–9. [266]

Azaroff, L. V. and Burger, M. J. (1958). *The powder method in x-ray crystallography*. McGraw-Hill Book Company, New York, U. S. A. [112]

Bäbler, F. (1978). Stable gamma modification of an isoindolinone pigment—coloring PVC, lacquers, cellulose ether(s) and ester(s) etc., in greenish-yellow shades. Ciba-Geigy AG Patent DE 2 804 062; GB 1 568 198. [261t]

Bäbler, F. (1983). Perylene tetracarboxylic bisdimethylphenylimide—in gamma modification is useful for pigmenting plastics, paint, lacquer and ink. Ciba–Geigy AG Patent EU 0023191 B1. [261t, 263]

Baczynski, W. L. and von Niementowski, S. (1919). Structure of hydroxyquinacridone. *Berichte*, **52B**, 461–84. [259]

Bailey, M. (1955). The crystal structure of indanthrone. *Acta Crystallogr.*, **8**, 182–5. [270]

Bailey, M. and Brown, C. J. (1967). The crystal structure of terephthalic acid. *Acta Crystallogr.*, **22**, 387–91. [52f, 113, 115t]

Bailey, M. and Brown, C. J. (1984). The crystal structure of terephthalic acid: errata. *Acta Crystallogr. C*, **40**, 1762. [52f, 113]

Ballard, B. E. and Biles, J. A. (1964). Effect of crystallizing solvent absorption rates of steroid implants. *Steroids*, **4**, 273–8. [246]

Ballester, L., Gil, A. M., Gutierrez, A., Perpinan, M. F., Azcondo, M. T., Sanchez, A. E., Amador, U., Campo, J. and Palacio, F. (1997). Polymorphism in [Cu(cyclam)(TCNQ)$_2$](TCNQ) stacked systems (cyclam=1,4,8,11-tetraazacyclotetradecane, TCNQ=7,7,8,8-tetracyanoquinodimethane). *Inorg. Chem.*, **36**, 5291–8. [202]

Bally, T., Haselbach, E., Lanyiova, S., Mardcher, F. and Rossi, M. (1976). Electronic structure and physico-chemical properties of azo-compounds. Part XIX. Concerning the conformation of isolated benzylideneaniline. *Helv. Chim. Acta*, **59**, 486–98. [228]

Banister, A. J., Bricklebank, N., Clegg, W., Elsegood, M. R. J., Gregory, C. I., Lavender, I., Rawson, J. M. and Tanner, B. K. (1995). The first solid-state paramagnetic 1,2,3,5-dithidiazolyl radical; x-ray crystal structure of [p-NCC$_6$F$_4$NSSN]. *J. Chem. Soc. Chem. Commun.*, 679–80. [200, 202]

Banister, A. J., Bricklebank, N., Lavender, I., Rawson, J. M., Gregory, C. I., Tanner, B. K., Clegg, W., Elsegood, M. R. J. and Palacio, F. (1996). Spontaneus magnetization in a sulfur-nitrogen radical at 36K. *Angew. Chem., Int. Ed. Engl.*, **35**, 2533–5. [200, 202, 203]

Bar, I. and Bernstein, J. (1985). Conformational polymorphism. 6. The crystal and molecular structures of Form II, Form III and Form V of N′-2-pyridilsulfonamide (sulfapyridine). *J. Pharm. Sci.*, **74**, 255–63. [8, 120f, 251]

Bar, I. and Bernstein, J. (1987). Modification of crystal packing and molecular conformation via systematic substitution. *Tetrahedron*, **43**, 1299–305. [169]

Barker, T. V. (1908). XII. Krystallographische Untersuchung der Dinitrobenzole und Nitrophenol. *Z. Kristallogr.*, **44**, 154–61. [212]

Barnes, A. F., Hardy, M. J. and Lever, T. J. (1993). A review of the applications of thermal methods within the pharmaceutical industry. *J. Therm. Anal.*, **40**, 499–509. [108]

Barton, J. H. (2000). Reforming the patent system. *Science*, **287**, 1933–4. [298]

Bassett, W. A. and Takahushi, T. (1974). X-ray diffraction studies up to 300 kbar. *Adv. High Pressure Res.* **4**, 164–247. [238]

Bauer, J., Spanton, S., Henry, R., Quick, J., Dizki, W., Porter, W. and Morris, J. (2001). Ritonavir: An extraordinary example of conformational polymorphism. *Pharm. Res.*, **18**, 859–66. [90, 149, 170, 252]

Bauer, M., Harris, R. K., Rao, R. C., Apperley, D. C. and Rodger, C. A. (1998). NMR study of desmotropy in Irbesartan, a tetrazole-containing pharmaceutical compound. *J. Chem. Soc. Perkin Trans.* **2**, 475–81. [139, 140f]

Bauer, N. and Lewin, S. Z. (1972). Determination of density. In *Physical methods of organic chemistry*, Vol. 1., pt. I (ed. A. Weissberger, B. W. Rossiter), pp. 131–90. John Wiley & Sons, New York, U. S. A. [147]

Bauer, W. H. and Kassner, D. (1992). The perils of Cc: comparing the frequencies of falsely assigned space groups with their general population. *Acta Crystallogr., B*, **48**, 356–69. [178, 183]

Baur, W. H. and Tillmanns, E. (1986). How to avoid unnecessarily low symmmetry in crystal structure determinations. *Acta Crystallogr. B*, **42**, 95–111. [115]

Bavin, M. (1989). Polymorphism in process development. *Chem. Ind.* (August), **21**, 527–9. [27, 240]

Bavin, P. G. M., Sly, J. C. P., Tovey, G. D. and Ward, R. J. (1979). Polymorph of cimetidine. *GB Patent* 1,543,238. [73]

Bayard, F., Decoret, C. and Royer, J. (1990). Structural aspects of polymorphism and phase transition in organic molecular crystals. In *Structure and properties of molecular crystals—studies in physical and theoretical chemistry*, Vol. 69, pp. 211–34. Elsevier, Amsterdam. [188]

Beach, S., Latham, D., Sidgwick, C., Hanna, M. and York, P. (1999). Control of the physical form of salmeterol xinofoate. *Org. Process Res. Develop.*, **3**, 370–6. [256]

Beal, R. W. and Brill, T. B. (2000). Thermal decomposition of energetic materials 77. Behavior of N−N bridged bifurazan compounds on slow and fast heating. *Propell. Explos. Pyrot.*, **25**, 241–6. [275]

Bechgaard, K., Jacobsen, C. S., Mortensen, K., Pedersen, H. J. and Thorup, N. (1980). The properties of five highly conducting salts: $(TMTSF)_2X$, $X = PF_6^-$, AsF_6^-, SbF_6^-, BF_4^- and NO_3^-. *Solid State Commun.*, **33**, 1119–25. [191]

Bechgaard, K., Kistenmacher, T. J., Bloch, A. N. and Cowan, D. O. (1977). The crystal and molecular structure of an organic conductor from 4,4′,5,5′-tetramethyl-$\Delta^{2,2'}$-bis-1,3-diselenole and 7,7,8,8-tetracyano-p-quinodimethane. *Acta Crystallogr. B*, 417–22. [190]

Becker, H.-D., Hall, S. R., Skelton, B. W. and White, A. H. (1984). Structural studies in the lepidopterene system. *Aust. J. Chem.*, **37**, 1313–27. [164]

Becker, R. and Döring, W. (1935). The kinetic treatment of nuclear formation in supersaturated vapors. *Ann. Phys.*, **24**, 719–52. [44]

Beckmann, W. (2000). Seeding the desired polymorph: background, possibilities, imitations and case studies. *Org. Process Res. Dev.*, **4**, 372–83. [256]

Beckmann, W., Nickisch K. and Budde, U. (1998). Development of a seeding technique for the crystallization of the metastable A modification of abecarnil. *Org. Process Res. Dev.*, **2**, 298–304. [252]

Begley, M. J., Crombie, L., Griffiths, G. L. and Jone, R. C. F. (1981). Charge-transfer and non-charge-transfer crystal forms of (**E**)–(5,5′-dimesitylbifuranylidenediones: and X-ray structural investigation. *J. Chem. Soc., Chem. Commun.*, 823–5. [216]

Beilstein Institute (1978). *How to use Beilstein. Beilstein handbook of organic chemistry*. Beilstein Institute, Frankfurt.

Bell, S. E. J., Burns, D. T., Dennis, A. C. and Speers, J. S. (2000). Rapid analysis of ecstasy and related phenethylamines in seized tablets by Raman spectroscopy. *Analyst*, **125**, 541–4. [132]

Bennema, P. and Hartman, P. (1980). The attachment energy as a habit controlling factor. *J. Cryst. Growth*, **49**, 145–56. [47]

Benyon, J. H. and Humphreys, A. R. (1955). The enthalpy difference between α- and β-copper phthalocyanine measured with an isothermal calorimeter. *Trans. Faraday Soc.*, **51**, 1065–71. [267]

Bergerhoff, G., Hundt, R., Sievers, R. and Brown, I. D. (1983). Inorganic Crystal Structure Database. *J. Chem. Inf. Comput. Sci.*, **23**, 66–9. [18]

Berkovitch-Yellin, Z. and Leiserowitz, L. (1982). Atom-atom potential analysis of the packing characteristics of carboxylic acids. A study based on experimental electron density distributions. *J. Am. Chem. Soc.*, **104**, 4052–64. [52f, 113]

Berman, H. M., Jeffrey, G. A. and Rosenstein, R. D. (1968). The crystal structures of the α and β forms of D-mannitol. *Acta Crystallogr. B*, **24**, 442–9. [73]

Berman, H. M., Westbrook, J., Feng, Z., Gilliland, G., Bhat, T. N., Weissig, H., Shindyalov, I. N. and Bourne, P. E. (2000). The protein Data bank. *Nucleic Acids Res.*, **28**, 235–42. (http://www.rcsb.org/pdb/). [18]

Bernstein, F. C., Koetzle, T. F., Williams, G. J. B., Meyer, A., Brice, M. D., Rodgers, J. R., Kennard, O., Shimanouchi, T. and Tasumi, M. (1977). The protein data bank: a computer-based archival file for macromolecular structures. *J. Mol. Biol.*, **112**, 535–42. [9, 18]

Bernstein, H. I. and Quimby, W. C. (1943). The photochemical dimerization of *trans*-cinnamic acid. *J. Am. Chem. Soc.*, **65**, 1845–6. [234]

Bernstein, J. (1979). Conformational Polymorphism. III. The crystal and molecular structures of the second and third forms of iminodiacetic acid. *Acta Crystallogr. B*, **35**, 360–6. [55, 158]

Bernstein, J. (1984). Crystal forces and molecular conformation. In *X-ray crystallography and drug action: current perspectives* (ed. A. S. Horn and C. J. DeRanter), pp. 23–44. Oxford University Press, New York, U. S. A. [151]

Bernstein, J. (1987). Conformational polymorphism. In *Organic Solid State Chemistry*, Vol. 32: *Studies in organic chemistry* (ed. G. R. Desiraju). Elsevier, Amsterdam. [2, 83, 152, 155f, 157, 159f, 160f, 161f, 162f, 163f, 163, 169, 281f, 294]

Bernstein, J. (1988). Polymorph IV of 4-Amino-N-2-pyridinyl benzene sulfonamide, sulfapyridine. *Acta Crystallogr. C*, **44**, 900–2. [120f]

Bernstein, J. (1991*a*). Polymorphism of L-glutamic acid: decoding the α-β phase relationship vis graph-set analysis. *Acta Crystallogr. B*, **47**, 1004–10. [56, 58]

Bernstein, J. (1991*b*). Polymorphism and the investigation of structure–property relations in organic solids. In *Organic crystal chemistry* (ed. J. B. Garbarczyk and D. W. Jones), pp. 6–26. International Union of Crystallography Book Series, Oxford University Press, Oxford. [56, 190f, 191f]

Bernstein, J. (1992). Effect of crystal environment on molecular structure. In *Accurate molecular structures* (ed. A. Domenicano and I. Hargittai), pp. 469–97. Oxford University Press, Oxford. [151]

Bernstein, J. (1993). Crystal growth, polymorphism and structure–property relationships in organic crystals. *J. Phys. D*, **26**, B66–76. [189]

Bernstein, J. (1999). Structural and crystallographic aspects of supramolecular engineering. In *Supramolecular engineering of synthetic metallic materials* (ed. J. Veciana, C. Rovira and D. B. Amabilino), pp. 23-40. Kluwer, Dordrecht, The Netherlands. [194f, 200f]

Bernstein, J. and Davis, R. E. (1999) Graph set analysis of hydrogen bond motifs. In *Implications of molecular and materials structure for new technologies*, NATO Science Series E: Applied Science (ed. J. A. K. Howard and F. H. Allen and G. P. Shields), Kluwer Academic Publishers, Dordrecht, Vol. 360 275–290 [57f, 58f, 81].

Bernstein, J. and Goldstein, E. (1988). The polymorphic structures of a squarylium dye. The monoclinic (green) and triclinic (violet) forms of 2,4-bis(2-hydroxy-4-diethylaminophenyl)-1,3-cyclobutadienediylium 1,3-diolate. *Mol. Cryst. Liq. Cryst.*, **164**, 213–29. [87, 88f, 89f, 205]

Bernstein, J. and Hagler, A. T. (1978). Conformational polymorphism. The influence of crystal forces on molecular conformation. *J. Am. Chem. Soc.*, **100**, 673–81. [2, 155, 157, 169, 180]

Bernstein, J. and Henck, J.-O. (1998). Disappearing and reappearing polymorphs—an anathema to crystal engineering? *Cryst. Eng.*, **1**, 119–28. [92, 95, 255]

Bernstein, J. and Izak, I. (1976). Molecular conformation and electronic structure. III. The crystal and molecular structure of the stable form of N-(p-chlorobenzylidene)-p-chloroaniline. *J. Chem. Soc., Perkin Trans. 2*, 429–34. [72, 227]

Bernstein, J. and Schmidt, G. M. J. (1972). Conformational studies. Part IV. The crystal and molecular structure of the metastable form of N-(p-Chlorobenzylidene)-p -chloroaniline, a planar anil. *J. Chem. Soc., Perkin Trans. 2*, 951–5. [72, 227]

Bernstein, J., Anderson, T. and Eckhardt, C. J. (1979). Conformational influence on electronic spectra and structure. *J. Am. Chem. Soc.*, **101**, 541–5. [227f, 227, 228f, 229f, 230f]

Bernstein, J., Hagler, A. T. and Engel, M. (1981). An *ab initio* study of the conformational energetics of N-benzylideneaniline. *J. Chem. Phys.*, **75**, 2346–53. [227]

Bernstein, J., Etter, M. C. and MacDonald, J. C. (1990). Decoding hydrogen-bond patterns. The case of iminodiacetic acid. *J. Chem. Soc., Perkin Trans. 2*, 695–8. [56, 58]

Bernstein, J., Davis, R. E., Shimoni, L. and Chang, N.-L. (1995). Patterns in hydrogen bonding: functionality and graph set analysis in crystals. *Angew. Chem., Int. Ed. Engl.*, **34**, 1555–73. [55, 57, 58, 60f]

Bernstein, J., Davey, R. J. and Henck, J.-O. (1999). Concomitant polymorphs. *Angew. Chem. Int. Ed.*, **38**, 3440–61. [42f, 43, 45f, 75, 82f]

Berzelius, J. (1844). Verbindungen des phosphors mit schwefel. *Jahresbericht*, **23**, 44–55. [4, 17, 21]

Beyer, T., Day, G. M. and Price, S. L. (2001). The prediction, morphology and mechanical properties of the polymorphs of paracetamol. *J. Am. Chem. Soc.*, **123**, 5086–94. [241]

Bingham, A. L., Hughes, D. S., Hursthouse, M. B., Lancaster, R. W., Travener, S. and Threlfall, T. L. (2001). Over one hundred solvates of sulfathiazole. *Chem. Commun.*, 603–4. [81]

Biradha, K. and Zaworotko, M. (1998). Supramolecular isomerism and polymorphism in dianion salts of pyromellitic acid. *Cryst. Eng.*, **1**, 67–78. [81]

Bish, D. L. and Reynolds, R. C. (1989). Sample preparation for x-ray diffraction. In *Reviews in mineralogy, modern powder diffraction*, Vol. 20 (ed. D. L. Bish and J. E. Post), pp. 73–99. Mineralogy Society of America, Washington, D. C. [112]

Black, J. W., Duncan, W. A. M., Durant, C. J., Ganellin, C. R. and Parsons, E. M. (1972). Definition and antagonism of histamine H_2-receptors. *Nature*, **236**, 385–90. [298]

Black, S. N. and Davey, R. J. (1988). Crystallization of amino acids. *J. Cryst. Growth*, **90**, 136–44. [70]

Black, S. N., Williams, L. J., Davey, R. J., Moffat, F., McEwan, D. M., Sadler, D. E., Docherty, R. and Williams, D. J. (1990). Crystal chemistry of 1-(4-chlorophenyl)-4,4-dimethyl-2-(1-*H*-1,2,4-triazol-1-yl)pentan-3-one, a Paclobutrazol Intermediate. *J. Phys. Chem.*, **94**, 3223–6. [47]

Black, S. N., Davey, R. J., Morley, P. R., Halfpenny, P., Shepherd, E. E. A. and Sherwood, J. N. (1993). Crystal-growth and characterization of the electrooptic material 3-(2,2-dicyanoethenyl)-1-phenyl-4,5-dihydro-1H-pyrazole. *J. Mater. Chem.*, **3**, 129–32. [212]

Blagden, N., Davey, R. J., Lieberman, H., Williams, L., Paynem, R., Roberts, R., Rowe, R. and Docherty, R. (1998*a*). Crystal chemistry and solvent effects in polymorphic systems: sulphathiazole. *J. Chem. Soc., Faraday Trans.*, **98**, 1035–45. [81, 93, 253]

Blagden, N., Davey, R. J., Rowe, R. and Roberts, R. (1998*b*). Disappearing poly-morphs and the role of reaction by-products: the case of sulphathiazole. *Int. J. Pharm.*, **172**, 169–77. [81, 253]

Blake, A. J., Gould, R. O., Halcrow, M. A. and Schröder, M. (1993). Confor-mational studies on [16]aneS₄. Structures of α- and β-[16]aneS4 ([16]aneS4 = 1,5,9,13-tetrathiacyclohexadecane). *Acta Crystallogr. B*, **49**, 773–9. [80]

Blanco, M., Coello, J., Iturriaga, H., Maspoch, S. and Perex-Maseda, C. (2000). Determination of polymorphic purity by near infrared spectrometry. *Anal. Chim. Acta*, **407**, 247–54. [129]

Block, S. and Piermarini, G. (1976). The diamond cell stimulates high pressure research. *Phys. Today*, **29**, 44–7. [287]

Block, S., Weir, C. E. and Piermarini, G. J. (1970). Polymorphism in benzene, naphthalene and anthracene at high pressure. *Science*, **169**, 586–7. [238]

Blomquist, A. T. and Ryan, J. F. Jr. (1944). Studies related to the stability of PETN. *OSRD Report NDRC-B-3566*. [278, 284]

Bock, H., Rauschenbach, A., Näther, C., Havlas, Z., Gavezzotti, A. and Flippini, G. (1995). Orthorhombic and monoclinic 2,3,7,8-tetramethoxythianthrene—small structural difference—large lattice change. *Angew. Chem., Int. Ed. Engl.*, **34**, 76–8. [178]

Bock, H., Schödel, H., Näther, C. and Butenschein, F. (1997). Interactions in crystals: the dimorphism of (2-pyridyl)(2-pyrimidyl)amine. *Helv. Chim. Acta*, **80**, 593–605. [73]

Bock, M., Depke, G., Egner, U., Muller-Fahrnow, A. and Winter, G. (1994). Detection and characterization of polymorphic modifications of the anxiolytic drug Abecarnil. *Tetrahedron*, **50**, 13125–34. [160]

Boese, R, Polk, M. and Bläser, D. (1987). Cooperative effects in the phase transfor-mation of triethylcyclotriboroxane. *Angew. Chem, Int. Ed. Engl.*, **26**, 245–7. [237]

Bohler, H. and Kehrer, F. (1963). Pure crystal forms of 5,12-dihydroquino[2,3-b]acridine-7,14-diones. Sandoz Ltd. Belgian Patent 611, 271; *Chem. Abstr.*, **58**, 4675. [262]

Boldyreva, E. (1999). Interplay between intra- and intermolecular interactions in solid-state reactions: general overview. In *Reactivity of molecular solids, the molecular solid state*, Vol. 3 (ed. E. Boldyreva and V. Boldyrev). pp. 1–50. [239]

Boldyreva, E. V., Shakhtshieder, T. P., Vasilchenko, M. A., Ahsbahs, H. and Uchtmann, H. (2000). Anisotropic crystal structure distortion of the monoclinic polymorph of acetaminophen at high hydrostatic pressures. *Acta Crystallogr. B*, **56**, 299–309. [239]

Boman, C.-E., Herbertsson, H. and Oskarsson, A. (1974). Crystal and molecular structure of a monoclinic phase of iminodiacetic acid. *Acta Crystallogr. A*, **30**, 378–82. [158]

Bonafede, S. J. and Ward, M. D. (1995). Selective nucleation and growth of an organic polymorph by ledge-directed epitaxy on a molecular crystal substrate. *J. Am. Chem. Soc.*, **117**, 7853–61. [93]

Bondi, A. (1963). Heat of sublimation of molecular crystals. A catalog of molecular structure increments. *J. Chem. Eng. Data*, **8**, 371–81. [168]

Boon, W. and Vangerven, L. (1992) Magnetization of electron-spin pairs in a free-radical at low-temperatures in a high magnetic field. *Physica B*, **177**, 527–30. [203]

Boone, C. D. G, Derissen J. L. and Schoone, J. C. (1977). Anthranilic acid II (*o*-aminobenzoic acid). *Acta Crystallogr. B*, **33**, 3205–6. [59]

Booth, A. D. and Llewellyn, F. J. (1947). The crystal structure of pentaerythritol tetranitrate. *J. Chem. Soc.*, 837–46. [284]

Booth, G. (1971). Phthalocyanines. In *The chemistry of synthetic dyes*, Vol. V (ed. K. Venkataraman). Academic Press, New York. [265]

Borka, L. (1991). Review on crystal polymorphism of substances in the European Pharmacopoeia. *Pharma. Acta Helv.*, **66**, 16–22. [95, 240, 241]

Borka, L. (1995). Crystal polymorphism and related phenomena of substances in the European Pharmacopoia. An updated review for fasciculae 13 to 19. *Pharmeuropa*, **7**, 586–93. [241]

Borka, L. and Backe-Hansen, K. (1968). IR spectroscopy of chloramphenicol palmitate. Polymorph alteration caused by the potassium bromide disk technique. *Acta Pharm. Suec.*, **5**, 271–8. [246]

Borka, L. and Haleblian, J. K. (1990). Crystal polymorphism of pharmaceuticals. *Acta Pharm. Jugoslav.*, **40**, 71–94. [241]

Borsenberger, P. M. and Weiss, D. S. (1993). *Organic photoreceptors for xerography*. Marcel Dekker, New York. [204]

Bos, M. and Weber, H. T. (1991). Comparison of the training of neural networks for quantitative X-ray fluorescence spectrometry by genetic algorithm, and backward error propagation. *Anal. Chim. Acta*, **247**, 97–105. [124]

Botcher, T. R., Berdall, D. J., Wight, C. A., Fan, L. M. and Burkey, T. J. (1996). Thermal decomposition mechanism of NTO. *J. Phys. Chem.*, **100**, 8802–6. [285]

Botha, S. A., Guillory, J. K. and Lotter, A. P. (1986). Physical characterization of solid forms of urapidil. *J. Pharm. Biomed. Anal.*, **4**, 573–87. [111]

Botsaris, G. D. (1976). Secondary nucleation: a review. In *Industrial crystallization* (6th Symposium, Usto nad Labem) (ed. J. W. Mullin), pp. 3–22. Plenum Press, New York, U. S. A. [68]

Bottaro, J. C., Penwell, P. E. Ross, D. S. and Schmidt, R. J. (1991). Novel N,N-dinitroamide salts—useful as oxidizers in rocket fuels, exhibiting high temperature stability, high energy density and an absence of smoke generating halogen(s). World Intellectual Property Organization Application Number PCT.US91/04268. [286]

Böttger, R. and Will, H. (1846). Über eine neue, der Pikrinsaüer nahesteheude Saüer. *Annalen*, **58**, 275–300. [294]

Bottom, R. (1999). The role of modulated temperature differential scanning calorimetry in the characterisation of a drug molecule exhibiting polymorphic and glass forming tendencies. *Int. J. Pharm.*, **192**, 47–53. [251]

Bouas-Laurent, H. and Durr, H. (2001). Organic photochromism. *Pure Appl. Chem.* **73**, 639–65. [216]

Bouche, R. and Draguet-Brughmans, M. (1977). Polymorphism of organic drug substances. *J. Pharm. Belgique*, **32**, 23–51. [240]

Boucherle, J. X., Gillon, B., Maruani, J. and Schweizer, J. (1987). Crystal structure determination neutron diffraction of 2,2-diphenyl-1-picrylhydrazyl (DPPH) benzene solvate (1/1). *Acta Crystallogr C*, **43**, 1769–73. [203]

Boyd, P., Mitra, S., Raston, C. L., Rowbottom, G. L. and White, A. H. (1981). Magnetic and structural studies on copper(II) dialkyldithiocarbamates. *J. Chem. Soc., Dalton Trans.*, 13–22. [198]

Boyd, R. H. (1994). Prediction of polymer crystal-structures and properties. *Atomistic Model. Phys. Prop. Adv. Polym. Sci.*, **116**, 1–25. [182]

Boyle, R. (1661). *Essay on the unsuccessfulness of experiments.* Henry Herrington, London. [66]

Brach, P. J. and Lardon, M. A. (1973). π-form metal-free phthalocyanine. Xerox Corporation Patent US 3 708 293; DE-Os 2 218 788 (1972); GB 1 395 769 (1972); FR 2 138 865 (1972). [266]

Brach, P. J. and Six, H. A. (1973). π-form metal phthalocyanine. Xerox Corporation Patent US 3 708 292; DE-Os 2 218 767 (1972); GB 1 395 615 (1972); FR 2 138 730 (1972). [266]

Brach, P. J. and Six, H. A. (1975). Process of making x-form metal phthalocyanine. Xerox Corporation Patent US 3 927 026; DE-Os 2 026 057 (1970); GB 1 312 946 (1975). [266]

Bracuti, A. J. (1979). 1,2,3-triaminoguanidinium nitrate. *Acta Crystallogr.*, **35**, 760–1. [284]

Bracuti, A. J. (1988). Discovery of a low temperature form of TAGN. U. S. Army Armament Research & Development Center, Picatinny Arsenal, NJ. *ARAED-TR-88009.* Access No. E780-1782. [284]

Bradshaw, J. (1993). Ranitidine. In *Annals of drug discovery* (ed. D. Lednicer). Vol. 3, pp. 45–81. American Chemical Society, Washington, DC. [298]

Braga, D. and Grepioni, F. (1991). Effect of molecular shape on crystal building and dynamic behavior in the solid state: from crystalline arenes to crystalline metal arene complexes. *Organometallics*, **10**, 2563–9. [85]

Braga, D. and Grepioni, F. (1992). Crystal structure and molecular interplay in solid ferrocene, nickelocene, and ruthenocene. *Organometallics*, **11**, 711–8. [153, 163, 183]

Braga, D., Cojazzi, G., Abati, A., Maini, L., Polito, M., Scaccianoce, L. and Grepioni, F. (2000). Making and converting organometallic pseudo-polymorphs via non-solution methods. *J. Chem. Soc., Dalton Trans.*, 3969–75. [92]

Braga, D., Grepioni, F. and Sabatino, P. (1990). On the factors controlling the crystal packing of first row transition metal binary carbonyls. *J. Chem. Soc., Dalton Trans.*, 3137–42. [153]

Braga, D., Grepioni, F., Dyson, P. J., Johnson, B. F. G., Frediani, P., Bianchi, M. and Piacenti, F. (1992). $Ru_6(CO)_{17}$—A case of organometallic crystal polymorphism. *J. Chem. Soc., Dalton Trans.*, 2565–71. [153, 163]

Brahadeeswaran, S., Venktaramanan, V. and Baht, H. L. (1999). Non-linear activity of anhydrous and hydrated sodium *p*-nitrophenolate. *J. Cryst. Growth*, **205**, 548–53. [213]

Brand, B. P. (1964). Preparation of delta copper phthalocyanine. Imperial Chemical Industries Patent US 3 150 150. [266]

Brand, J. C. D. and Speakman, J. C. (1960). *Molecular structure, the physical approach*, pp. 248–50. E. Arnold, London. [152]

Brandstätter, M. (1947). The isomorphous replacability of H, OH, NH$_2$, CH$_3$ and Cl in the *m*-dinitrobenzene series. *Monatschefte Chem.*, **77**, 7–17. [98]

Brandt, H., Hörnle, R. and Leverenz, K. (1982). Color stable modification of monoazo dyestuff—viz butyl cyano methyl nitrophenyl azo pyridone, used for dyeing polyester fibers. Bayer AG Patent DE 3 046 587. [274]

Brehmer, T. H., Weber, E. and Cano, F. H. (2000). Balance of forces between contacts in crystal structure of two isomeric benzo-condensed dibromidihydroxy-containing compounds. *J. Phys. Org. Chem.*, **13**, 63–74. [153]

Brill, T. B. (1992). Connecting the chemical-composition of a material to its combustion characteristics. *Prog. Energy Combust. Sci.*, **18**, 91–116. [275]

Brill, T. B. and James, K. J. (1993). Kinetics and mechanisms of thermal-decomposition of nitroaromatic explosives. *Chem. Rev.*, **93**, 2667–92. [275]

Brill, T. B. and Reese, C. O. (1980). Analysis of intra- and intermolecular interactions relating to the thermophysical behavior of α, β, and δ-ocahydro-1,3,5,7-tetranitro-1,3,5,7-tetraazocine. *J. Phys. Chem.*, **84**, 1376–80. [280]

Bristol-Meyers Company v. United States International Trade Commission, Gema, S. A., Kalipharma, Inc., Purepac Pharmaceutical Co., Istituto Biochimico Italiano Industria Giovanni Lorenzini, Institut Biochimique, S. A., and Biocraft Laboratories, Inc. (Dec. 8, 1989) No. 89–1530 [See also United States International Trade Commission, Investigation No. 337-TA-293]. [303]

Brittain, H. G. (1997). Spectral methods for the characterization of polymorphs and solvates. *J. Pharm. Sci.*, **86**, 405–12. [125, 133]

Brittain, H. G. (1999*a*). Methods for the characterization of polymorphs and solvates. In *Polymorphism in pharmaceutical solids*, Vol. 95 of Drugs and the Pharmaceutical Sciences, Marcel Dekker, Inc. New York, pp. 227–78. [94]

Brittain, H. G. (ed.) (1999*b*). Polymorphism in Pharmaceutical Solids. In *Drugs and the pharmaceutical sciences* (ed. J. Swarbrick), Vol. 95. Marcel Dekker, New York. [240, 249]

Brittain, H. G. (1999*c*). Application of the phase rule to the characterization of polymorphic systems. In *Drugs in the pharmaceutical sciences* (ed. J. Sarbrick), Vol. 95: *Polymorphism in pharmaceutical solids* (ed. H. G. Brittain), pp. 35–72. Marcek Dekker, New York. [30]

Brittain, H. G. and Fiese, E. F. (1999). Effects of pharmaceutical processing on drug polymorphs and solvates, in Brittain, H. G. (ed.) *Polymorphism in pharmaceutical solids*, vol. 95 of Drugs and the Pharmaceutical Sciences, Marcel Dekker, Inc. New York, pp. 331–61. [256]

Brittain, H. G. and Grant, D. J. W. (1999). Effects of polymorphism and solid-state solvation on solubility and dissolution rate. In *Polymorphism in pharmaceutical solids* (ed. H. G. Brittain), pp. 279–330, Vol. 95 of *Drugs and the pharmaceutical sciences* (series ed. J. Swarbrick), Marcel Dekker, Inc. New York. [111, 244, 245f, 246]

Brittain, H. G., Ranadive, S. A. and Serajuddin, A. T. M. (1995). Effect of humidity-dependent changes in crystal structure on the solid-state fluorescene properties of a new HMG-COA reductase inhibitor. *Pharm. Res.*, **12**, 556–9. [5]

Brock, C. P. and Dunitz, J. D. (1994). Towards a grammar of crystal packing. *Chem. Mater.*, **6**, 1118–27. [319]

Broderick, W. E., Eichorn, D. M., Liu, X., Toscano, P. J., Owens, S. M. and Hoffman, B. M. (1995). Three phases of [Fe(C$_5$Me$_5$)$_2$]$^+$[TCNQ]$^-$: ferromagnetism in a new structural phase. *J. Am. Chem. Soc.*, **117**, 3641–2. [199]

Brown, C. J. (1968*a*). Crystal structure of β-copper phthalocyanine. *J. Chem. Soc.*, 2488–93. [266, 268]

Brown, C. J. (1968*b*). Crystal structure of platinum phthalocyanine. A reinvestigation. *J. Chem. Soc. (A)*, 2494–8. [268]

Brown, C. J. (1968*c*). Crystal structure of anthranilic acid. *Proc. R. Soc. London. Ser. A: Math. Phys. Sci.*, **302**, 185–9. [59]

Brown, C. J. and Ehrenberg, M. (1985). Anthranilic acid I, C$_7$H$_7$NO$_2$, by neutron diffraction. *Acta Crystallogr. C*, **41**, 441–3. [59]

Bruce, A. D. and Cowley, R. A. (1981). *Structural phase transitions*, Taylor & Francis Ltd., London. [223]

Brückner, S. (1982). An unusual example of packing among molecular layers: the structures of two crystalline forms of 2,2-aziridinecarboxamide, C$_4$H$_7$N$_3$O$_2$. *Acta Crystallogr. B*, **38**, 2405–8. [63f, 65]

Brückner, S., Calligaris, M., Nardin, G. and Randacio, L. (1971). The crystal structure of the form of N-N'- ethylenebis(salicylaldehydeiminato)cobalt(II) inactive towards oxygenation. *Acta Crystallogr. B*, **25**, 1671–4. [238]

Brün, W. (1934). Priming mixture. Remington Arms U. S. Patent 1 942 274. [294]

Bruno, I. J., Cole, J. C., Lommerse, P. M., Rowland, R. S., Taylor, R. and Verdonk, M. L. (1997). IsoStar: a library of information about nonbonded interactions. *J. Comput. Aided Mol. Des.*, **11**, 525–37. [186]

Buckley, H. E. (1951). *Crystal growth*. Wiley, New York, U. S. A. [47]

Buckton, G. and Darcy, P. (1999). Assessment of disorder in crystalline powders—a review of analytical techniques and their application. *Int. J. Pharm.*, **179**, 141–58. [254]

Buerger, M. J. (1951). Crystallographic aspects of phase transformations. In *Phase transformations in solids* (ed. R. Smoluchowski, J. E. Mayer and W. A. Weyl), pp. 183–211. John Wiley & Sons, New York, U. S. A. [31, 32f, 33f, 34]

Buerger, M. J. (1956). *Elementary crystallography: an introduction to the fundamental geometrical features of crystals*. Wiley, New York, U. S. A. [46]

Buerger, M. J. and Bloom, M. C. (1937). Crystal polymorphism. *Z. Kristallogr.*, **96**, 182–200. [1, 3, 9, 26, 240, 307]

Bugay, D. E. (1993). Solid-state nuclear magnetic resonance spectroscopy: theory and pharmaceutical applications. *Pharm. Res.*, **10**, 317–27. [134f, 142]

Bugay, D. E. (1999). Quantitative and regulatory aspects of polymorphism, presented at the 1st International Symposium on Aspects of Polymorphism and Crystallization—Chemical Development Issues, Hinckley, Leicestershire, UK, June, 1999. Organized by Scientific Update, Wyvern Cottage, High Street, Mayfield, East Sussex, TN20 6AE, UK, http://www.scientificupdate.co.uk [130, 131f, 132f, 143f]

Bugay, D. E. (2001). Characterization of the solid state: spectroscopic techniques. *Adv. Drug Deliv. Revs.*, **48**, 43–65. [94, 128, 130, 133]

Bugay, D. E., Newman, A. W. and Findlay, W. P. (1996). Quantitation of cefepime.2HCl dihydrate in cefepime.2HCl monohydrate by diffuse reflectance IR and powder x-ray diffraction techniques. *J. Pharm. Biomed. Anal.*, **15**, 49–61. [130, 132]

Bunte, G., Pontius, H. and Kaiser, M. (1999). Analytical characterization of impurities in new energetic materials. *Propell. Explos. Pyrot.*, **24**, 149–55. [282]

Burg, J., Grobet, P., Van den Bosch, A. and Vansummeren, J. (1982). Magnetic interaction and spin diffusion in solvent-free DPPH. *Solid State Commun.*, **43**, 785–7. [203]

Burger, A. (1975). Zur polymorphie oraler antidiabetika 2 Mitteilung: tolbutamide. *Sci. Pharm.*, **43**, 161–8.

Burger, A. (1982a). Thermodynamic and other aspects of the polymorphism of drug substances. *Pharm. Int. [The Netherlands]*, **3**, 158–63. [40, 106]

Burger, A. (1982b). Interpretation of polymorphism studies. *Acta Pharm. Technol.*, **28**, 1–20. [40, 106]

Burger, A. (1983). The relevance of polymorphism. In *Topics in pharmaceutical science* (ed. D. D. Breimer and P. Speiser), pp. 347–58. Elsevier, Lausanne. [3, 240]

Burger, A. and Ramberger, R. (1979a). On the polymorphism of pharmaceuticals and other molecular crystals. I. Theory of thermodynamic rules. *Mikrochim. Acta* **II**, 259–272. [25, 34, 38, 41, 128, 147]

Burger, A. and Ramberger, R. (1979b). On the polymorphism of pharmaceuticals and other molecular crystals. *Mikrochim. Acta*, **II**, 273–316. [34, 38, 40, 41, 128, 147]

Burger, A., Schulte, K. and Ramberger, R. (1980). Aufkläruing thermodynamischer beziehungen zwischen fünf polymorphen modifikationen von sulfapyridin mittels DSC. *J. Therm. Anal.*, **19**, 475–84. [251]

Burger, A., Ratz, A. W., and Brox, Werner. (1986). Polymorphic pharmaceuticals in the European pharmacopoeia: Oxytetracycline hydrochloride. *Pharm. Acta Helvet.*, **61**, 98–106. [249]

Burger, A., Henck, J.-O. and Dunser, M. N. (1996). On the polymorphism of dicarboxylic acids: I pimelic acid. *Mikrochim. Acta*, **122**, 247–57. [97]

Burger, A., Rollinger, J. M. and Bruggeller, P. (1997). Binary system of (R)- and (S)-nitrendipine—Polymorphism and structure. *J. Pharm. Sci.*, **86**, 674–9. [249]

Burger, A., Henck, J.-O., Hetz, S., Rollinger, J. M., Weissnicht, A. A. and Stottner, H. (2000). Energy/temperature diagram and compression behavior of the polymorphs of D-mannitol. *J. Pharm. Sci.*, **89**, 457–68. [81, 95, 141, 242]

Bürgi, H.-B. and Dunitz, J. D. (1994). Structure correlation; the chemical point of view. In *Structure correlation*, Vol. 1 (ed. H.-B. Bürgi and J. D. Dunitz), pp. 163–204. VCH, Weinheim. [157]

Bürgi, H.-B., Hulliger, J. and Langley, P. J. (1998). Crystallization of supramolecular materials. *Curr. Opin. Solid State Mater. Sci.*, **3**, 425–30. [92]

Burkert, W. B. and Allinger, N. L. (1982). *Molecular Mechanics*, ACS Monographs No 177, American Chemical Society, Washington, D.C. [165]

Burkhard, H., Muller, C. and Senn, O. (1968). Process for the production of finely divided, heat stable 1-phenylamino-2-nitrobenzene-4-sulphonic acid, its production and uses. Sandoz Patents Ltd. GB patent 1 040 607. [260t]

Burkhardt, L. A. and Bryden, J. H. (1954). X-ray studies of 2,4,6-trinitrotoluene. *Acta Crystallogr.*, **7**, 135–6. [288]

Burns, G. R., Cunningham, C. W. and McKee, V. (1988). Photochromic formazans: Raman specta, X-ray crystal structure and ^{13}C magnetic resonance spectra of the orange and red isomers of 3-ethyl-1,5-diphenylformazan *J. Chem. Soc., Perkin Trans. II*, 1275–80. [216]

Burzlaff, H., Zimmermann, H. and de Wolff, P. M. (1983). Crystal lattices. In *International tables for crystallography*, Vol. A (ed. T. Hann), pp. 734–44. D. Reidel Publishing Company, Dordrecht, The Netherlands. [115]

Busing, D., Jenau, M., Reuter, J., Wurflinger, A. and Tamarit, J. L. (1995). Differential thermal-analysis and dielectric studies on 2-methyl-2-nitro-propoane under high-pressure. *Z. Natur. A. J. Phys. Sci.*, **50**, 502–4. [239]

Butler, L. G., Cory, D. G., Dooley, K. M., Miller, J. B. and Garroway, A. N. (1992). NMR imaging of anisotropic solid-state chemical-reactions using multiple-pulse line-narrowing techniques and H-1 T1-weighting. *J. Am. Chem. Soc.*, **114**, 125–35. [238]

Buttar, D., Charlton, M. H., Docherty, R. and Starbuck, J. (1998). Theoretical investigations of conformational aspects of polymorphism. Part 1: *o*-acetamidobenzamide. *J. Chem. Soc., Perkin Trans. 2*, 763–72. [154, 170, 180]

Byrn, S. R. (1982). *Solid-state chemistry of drugs*, pp. 7–10. Academic Press, New York, U. S. A. [5, 240]

Byrn, S. R., Curtin, D. Y. and Paul, I. C. (1972). The X-ray crystal structures of the yellow and white forms of dimethyl 3,6-dichloro-2,5-dihydroxyterephthalate and a study of the conversion of the yellow form to the white form in the solid state. *J. Am. Chem. Soc.*, **94**, 890–8. [178, 215]

Byrn, S.R., Pfeiffer, R., Ganey, M., Hoiberg, C. and Poochikian, G. (1995). Pharmaceutical solids: a strategic approach to regulatory considerations. *Pharm. Res.*, **12**, 945–54. [253, 256]

Byrn, S. R., Pfeiffer, R. R. and Stowell, J. G. (1999). *Solid state chemistry of drugs*, 2nd edn, SSCI, Inc. West Lafayette, IN. [5, 133, 134f, 240, 241, 242, 243, 244, 249, 251, 254, 256]

Byrn, S. R., Xu, W., Newman, A. W. (2001). Chemical reactivity in solid-state pharmaceuticals: formulation implications. *Adv. Drug Deliv. Revs.*, **48**, 115–36. [243, 255]

Byrn, S. R., Sutton, P. A., Tobias, B., Frye, J. and Main, P. (1988). The crystal structure solid state NMR spectra and oxygen reactivity of five crystal forms of prednisolone tert-butylacetate. *J. Am. Chem. Soc.*, **110**, 1609–14. [135]

Byrne, J. F. and Kurz, P. F. (1967). Metal free phthalocyanine in the new x-form. Patent US 3 357 989 (1967); DE 1 619 654 (1967); GB 1 169 901. [266]

Cabana, B. E., Willhite, L. E. and Bierwagen, M. E. (1969). Pharmacokinetic evaluation of the oral adsorption of different ampicillin preparations in beagle dogs. *Antimicrob. Agents Chemother.*, **9**, 35–41. [246]

Cady, H. H. (1974). The PETN-DiPEHN-TriPEON system. *LA-4486-MS Los Alamos Scientific Laboratory* U. S. Atomic Energy Commission Contract W-7405-ENG. 36. [284, 295, 296]

Cady, H. H. and Smith, L. C. (1962). Studies on the polymorphs of HMX. *Los Alamos Scientific Laboratory* LAMS-2652, Chemistry (TID-4500 17th ed.) based on Contract W-7405-ENG 36 with the U. S. Atomic Energy Commission. [278, 279t, 280]

Cady, H. H. and Larson, A. C. (1975). Pentaerythritrol tetranitrate II: Its crystal structure and transformation to PETNI, an algorith for refinement of crystal structures with poor data. *Acta Crystallogr. B*, **31**, 1864–9. [284]

Cady, H. H., Larson, A. C. and Cromer, D. T. (1963). The crystal of α-HMX and a refinement of the structure of β-HMX. *Acta Crystallogr.*, **16**, 617–23. [279t]

Cailleau, H., Luty, T., Le Cointe, M. and Lemée-Cailleau, M. -L. (1997). Cooperative mechanism at the neutral-to-ionic transition *Biuletyn Instytutu Chemii Fizycnej I Teoretycznej Politechniki Wroclawskiej*, **5**, 19–34. [196, 197f, 197]

Caira, M. R. (1998). Crystalline polymorphism of organic compounds. *Topics in Curr. Chem.*, **198**, 163–208. [240]

Caira, M. R., Peinaar, E. W. and Lotter, A. P. (1996). Polymorphism and pseudopolymorphism of the antibacterials nitrofurantoin. *Mol. Cryst. Liq. Cryst., Sci. Technol. A*, **278**, 241–64. [5]

Calleja, F. J. B., Arche, A. G., Ezquerra, T. A., Cruz, C. S., Batallan, F., Frick, B. and Cabarcos, E. L. (1993) Structure and properties of ferroelectric copolymers of poly (vinylidene flouride). *Adv. Polym. Sci.*, **108**, 1–48. [28]

Cameroni, R., Coppi, G., Gamberini, G. and Forni, F. (1976). Dissolution and enzymic hydrolysis of chloramphenicol palmitic and stearic esters. *Farmaco Ed. Prat.*, **31**, 615–24. [247]

Cammenga, H. K. and Hemminger, W. F. (1990). *Labo*, **21**, 7. [148]

Capiomont, A., Bordeaux, D. and Lajzérowicz-Bonneteau, J. (1972). Crystal structure of the tetragonal form of the nitroxide 2,2,6,6-tetramethylpiperidine 1-oxyl. *Comptes Rendu Academie Sciences, Paris, Series C.*, **275**, 317–20. [202]

Capiomont, A., Lajzerowicz, J., Legrand, J.-F. and Zeyen, C. (1981). Structure of the ferroelectric–ferroplastic phase of tanane (neutron diffraction). Evaluation of the lattice energy. *Acta Crystallogr. B*, **37**, 1557–60. [202]

Cardew, P. T. and Davey, R. (1982). *Tailoring of Crystal Growth*. Institute of Chemical Engineers, North Western Branch, Symposium Papers, Number 2 (ISBN 090663623X). pp. 1.1–1.8. [44]

Cardew, P. T. and Davey, R. J. (1985). The kinetics of solvent-mediated phase transformations. *Proc. R. Soc. of London, A*, **398**, 415–28. [75, 268]

Cardew, P. T., Davey, R. J. and Ruddic, A. J. (1984). Kinetics of polymorphic solid-state transformations. *J. Chem. Soc., Faraday Trans.*, **80**, 659–68. [75]

Cardozo, R. L. (1991) Enthalpies of combustion, formation, vaporization and sublimation of organics. *AIChE J.*, **37**, 290–8. [168]

Carles, M., Eloy, D., Pujol, L. and Bodot, H. (1987). Photochromic and thermochromic salicylideneamines: Isomerization in the crystal, infrared identities and conformational influence. *J. Mol. Struct.*, **156**, 43–58. [218]

Carlin, R. L. (1989). *Magnetochemistry*. Springer, Berlin. [203]

Carlson, K. D., Wang, H. H., Beno, M. A., Kini, A. M. and Williams, J. M. (1990). Ubiquitous superconductivity near 4K in salts of the BEDT-TTF/I system: is there a common source? *Mol. Cryst. Liq. Cryst.*, **181**, 91–104. [79]

Carlton, R. A., Difeo, T. J., Powner, T. H., Santos, I. and Thomson, M. D. (1996). Preparation and characterization of polymorphs for an LTD4 antagonist, RG 12525. *J. Pharm. Sci.*, **85**, 461–7. [108]

Carper, W. R., Davis, L. P. and Extine, M. W. (1982). Molecular structure of 2,4,6-trinitrotoluene. *Journal of Physical Chemistry*, **86**, 459–62. [288, 289t]

Carstensen, J. T. (1977). *Pharmaceutics of solids and solid dosage forms.* John Wiley & Sons: New York. [243]

Cash, D. J. (1981). Exclusion chromatography of anionic dyes. Anomalous elusion peaks due to reversible aggregation. *J. Chromatogr.*, **209**, 405–12. [81]

Cassoux, P., Valade, L., Kobayashi, H., Kobayashi, A., Clark, R. A. and Underhill, A. E. (1991). Molecular metals and superconductors derived from metal complexes of 1,3-dithiol-2-thione-1,5-dithiotale(dmit). *Coord. Chem. Rev.*, **110**, 115–60. [80]

Ceolin, R., Agafonov, V., Gonthier-Vassal, A., Swarc, H., Cense, J. M. and Ladure, P. (1995). Solid-state studies on crystalline and glassy flutamide—Thermodynamic evidence for dimorphism. *J. Therm. Anal.*, **45**, 1277–84. [254]

Ceolin, R., Agafonov, V., Louer, D., Dzyabchenko, V. A., Toscani, S. and Cense, J. M. (1996). Phenomenology of polymorphism. 3. p,t diagram and stability of piracetam polymorphs. *J. Solid State Chem.*, **122**, 186–94. [58, 74]

Cerius2 Molecular Modeling Package, MSI Inc. San diego CA, U. S. A. [84, 181]

Chaka, A. M., Zaniewski, R., Youngs, W., Tessier, C. and Klopman, G. (1996). Predicting the crystal structure of organic molecular materials. *Acta Crystallogr. B*, **52**, 165–83. [185]

Chamot, E. M. and Mason, C. W. (1973). *Handbook of chemical microscopy*. Vol. 1, 3rd edn. Principles and Use of Microscopes and Accessories. Physical Methods for the Study of Chemical Problems, Wiley-Interscience, New York. [22, 46, 48, 95]

Chan, F. C., Anwar, J., Cernik, R., Barnes, R. and Wilson, R. M. (1999). *Ab initio* structure determination of sulfathiazole polymorph V from synchroton x-ray powder diffraction data. *J. Appl. Crystallogr.*, **32**, 436–41. [111]

Chandra, A. K., Lim, E. C. and Ferguson, J. (1958). Absorption and fluorescence spectra of crystalline pyrene. *J. Chem. Phys.*, **28**, 765–8. [231]

Chang, L. C., Caira, M. R. and Guillory, J. K. (1995). Solid-state characterization of dehydroepiandrosterone. *J. Pharm. Sci.*, **84**, 1169–79. [249]

Chang, S. J., Hahn, Th. and Klee, W. E. (1984). Nomenclature and generation of three-periodic nets, the vector method. *Acta Crystallogr. A*, **40**, 42–50. [55]

Chapman, S. J. and Whitaker, A. (1971).X-ray powder diffraction data for some azo pigments. *J. Soc. of Dyers Colourists*, **87**, 120–1. [273]

Chemburkar, S. R., Bauer, J., Deming, K., Spiwek, H., Patel, K., Morris, J., Henry, R., Spanton, S., Dziki, W., Porter, W., Quick, J., Bauer, P., Donaubauer, J., Narayanan, B. A., Soldani, M., Riley, D. and McFarland, K. (2000). Dealing with the impact of rintoavir polymorphs on the late stages of bulk drug process development. *Org. Process Res. Dev.*, **4**, 413–7. [90, 148, 249, 252]

Chen, X. F., Liu, S. H., Duan, C. Y., Xu, Y. H., You, X. Z., Ma, J. and Min, N. B. (1998). Synthesis, crystal structure and triboluminescence of 1,4-dimethylpyridinium tetrakis (2-thenoyltrifluoroacetonato)europate. *Polyhedron*, **17**, 1883–9. [222]

Chen, X., Zhuang, P. and Ren, S. (1990). Study on crystal transformation of C. I. Pigment red 57:1. *Huadong Huagong Xueyuan Xuebao*, **16**, 434-9. [*CA* **115**: 10900a] [261t]

Cheng, S. Z. D., Li, C. Y., Calhoun, B. H., Zhu, L. and Zhou, W. W. (2000). Thermal analysis: the next two decades. *Thermochim. Acta*, **355**, 59–68.

Chiang, L.-C., Caira, M. R. and Guillory, J. K. (1995). Solid state characterization of dehydroepiandrosterone. *J. Pharm. Sci.*, **84**, 1169–79. [148]

Chiarelli, R. and Rassaat, A. (1991). Magnetic properties of some biradicals of D_{2d} symmetry. In *Magnetic molecular materials* (ed. Gatteschi, *et al.*), pp. 191–202. Kluwer Academic Publishers, Dordrecht, The Netherlands. [198]

Chick, M. C. and Thorpe, B. W. (1970). Polymorphism in 2,4,5-trinitrotoluene Report 382. Department of Supply, Australian Defence Scientific Service, Defence Standards Laboratories, Maribyrnong, Victoria. [291]

Chick, M. C. and Thorpe, B. W. (1971). Polymorphism in 2,4,5-trinitrotoluene. *Aust. J. Chem.*, **24**, 191–5. [291, 295]

Chickos, J. (1987). Heats of sublimation. In *Molecular structure and energetics.* Vol. 2. *Physical measurements* (ed. J. F. Liebman and A. Greenberg), pp. 67–150. VCH Publishers, New York, U. S. A. [38, 153, 168, 291]

Chickos, J. S., Braton, C. M., Hesse, D. G. and Liebman, J. F. (1991). Estimating entropies and enthalpies of organic compounds. *J. Org. Chem.*, **56**, 927–38. [32, 38]

Chikaraishi, Y., Otsuka, M. and Matsuda, Y. (1995). Dissolution behavior of piretanide polymorphs at various temperatures and pHs. *Chem. Pharm. Bull.*, **43**, 1966–9. [244]

Chin, D. N. (1999). Improving the efficiency of predicting hydrogen-bonded organic molecules. *Trans. Am. Cryst. Assoc.*, **33(ACA Transactions)**, 33–43. [185]

Chin, D. N., Palmore, G. T. R. and Whitesides, G. M. (1999*a*). Predicting crystalline packing arrangements of molecules that form hydrogen-bonded tapes. *J. Am. Chem. Soc.*, **121**, 2115–22. [185]

Chin, D. N., Zerkowski, J. A., MacDonald, J. C. and Whitesides, G. M. (1999*b*). Strategies for the design and assembly of hydrogen-bonded aggregates in the solid state. In *Molecular assemblies in the solid state* (ed. J. K. Whitesell), pp. 185–253. Wiley, Chichester. [185]

Choi, C. S. and Boutin, H. P. (1970). Study of the crystal structure of β-cyclotetramethylenetetranitramine by neutron diffraction. *Acta Crystallogr. B*, **26**, 1235–40. [279t]

Choi, C. S. and Bulusu, S. (1974). The crystal structure of dinitropentamethylenete-tramine. *Acta Crystallogr. B*, **30**, 1576–80. [286]

Choi, C. S. and Prince, E. (1972). The crystal structure of cyclotrimethylene-trinitramine. *Acta Crystallogr. B*, **28**, 2857–62. [283]

Choi, C. S. and Prince, E. (1979). 1,2,3-Triaminoguanidinium nitrate by neutron diffraction. *Acta Crystallogr. B*, **35**, 761–3. [284]

Cholerton, T. J., Hunt, J. H., Klinkert, G. and Martin-Smith, M. (1984). Spectro-scopic studies on ranitidine: its structure and the influence of temperature and pH. *J. Chem. Soc., Perkin Trans. 2*, 1761–6. [127f, 148, 252]

Chunwachirasiri, W., West, R. and Winokur, M. J. (2000) Polymorphism, struc-ture, and chromism in poly(di-n-octylsilane) and poly(di-n-decylsilane). *Macro-molecules*, **33**, 9720–31. [28, 214]

Churchill, W. S. (1950). *The Second World War*. Vol. III, p. 814. Houghton Mifflin Co. Boston.

Ciba, Ltd. (1965) Dutch Patent Application 6,405,130. *Chem. Abstr.*, **63**, 15103. [262]

Ciechanowicz, M., Skapski, A. C. and Troughton, P. G. H. (1976). The crystal structure of the orthorhombic form of hydridodicarbonylbis(triphenylphosphine)-iridium(I): successful location of the hydride hydrogen atom from x-ray data. *Acta Crystallogr. B*, **32**, 1673–80. [44]

Ciechanowicz-Rutkowska, M. (1977). An independent investigation of the crystal structure of 1,8-dinitronaphthalene (orthorhombic form) at 22 and 97 °C. *J. Solid State Chem.*, **22**, 185–92. [295]

Ciurczak, E. W. (1987). Uses of near-infrared spectroscopy in pharmaceutical analysis. *J. Appl. Spectrosc.*, **23**, 147–63. [130]

Clark, G. R., Waters, J. M. and Waters, T. N. (1975). Crystal and molecular struc-ture of brown form of bis(N-methyl-2-hydroxy-1-naphthaldiminato)copper(II). *J. Inorg. Nucl. Chem.*, **37**, 2455–8. [163]

Clark, G. R., Waters, J. M., Waters, T. N. and Williams, G. J. (1977). Polymor-phism in Schiff base complexes of copper(II): the crystal and molecular structure of a second brown form of bis(N-methyl-2-hydroxy-1-naphthaldiminato)copper(II). *J. Inorg. Nucl. Chem.*, **39**, 1971–5. [163]

Clas, S. D., Dalton, C. R. and Hancock, B. C. (1999). Differential scanning calorime-try: application in drug development. *Pharm. Sci. Technol. Today*, **2**, 311–20. [250]

Cleverly, B. and Williams, P. P. (1959). Polymorphism in substituted barbituric acid. *Tetrahedron* **7**, 277–88. [255]

Clydesdale, G., Roberts, K. J. and Docherty, R. (1994*a*). Computational studies of the morphology of molecular crystals through solid-state intermolecular force calculations using the atom-atom method. In *Colloid and surface engineering: controlled particle, droplet and bubble formation* (ed. D. J. Wedlock), pp. 95–135. Butterworth Heineman, London. [93]

Clydesdale, G., Roberts, K. J. and Docherty, R. (1994*b*). Modeling the morphology of molecular crystals in the presence of disruptive tailor-made additives. *J. Cryst. Growth*, **135**, 331–40. [93]

Clydesdale, G., Roberts, K. J. and Docherty, R. (1996). HABIT95—a program for predicting the morphology of molecular crystals as a function of the growth environment. *J. Cryst. Growth*, **166**, 78–83. [93, 181]

Clydesdale, G., Roberts, K. J. and Walker, E. M. (1997). The crystal habit of molecular materials: a structural perspective. In *Theoretical aspects and computer modeling of the molecular solid state* (ed. A. Gavezzotti), pp. 203–32. John Wiley & Sons, Chichester. [47, 93]

Cobbledick, R. E. and Small, R. W. H. (1974). Crystal structure of the δ-form of 1,3,5,7-tetranitro-1,3,5,7-tetraazacyclooctane (δ-HMX). *Acta Crystallogr. B*, **30**, 1918–22. [279t, 279]

Cobbledick, R. E. and Small, R. W. H. (1975). Crystal structure of the complex formed between 1,3,5,7-tetranitro-1,3,5,7-tetraazacyclooctane (HMX) and N,N-dimethylformamide (DMF). *Acta Crystallogr. B*, **31**, 2805–8.

Cohen, M. D. and Schmidt, G. M. J. (1964*a*). Topochemistry. Part I. A survey. *J. Chem. Soc.*, 1966–2000. [22, 188, 218]

Cohen, M. D., Hirschberg, Y. and Schmidt, G. M. J. (1964*b*). Topochemistry. Part VII. The photoactivity of anils of salicylaldehydes in rigid solutions. *J. Chem. Soc.*, 2051–9. [218]

Cohen, M. D., Hirschberg. Y. and Schmidt, G. M. J. (1964*c*). Topochemistry. Part VIII. The effect of solvent, temperature and light on the structure of anils by hydroxynaphthaldehydes. *J. Chem. Soc.*, 2060–7. [218]

Cohen, M. D., Schmidt, G. M. J. and Flavian, S. (1964*d*). Topochemistry. Part VI. Experiments on photochromy and thermochromy of crystalline anils of salicylaldehydes. *J. Chem. Soc.*, 2041–51. [234]

Cohen, M. D., Schmidt, G. M. J. and Sonntag, F. I. (1964*e*). Topochemistry. Part II. The photochemistry of *trans*-cinnamic acids. *J. Chem. Soc.*, 2000–13. [22]

Cohen, R., Ludmer, Z. and Yakhot, V. (1975). Structural influence on the excimer emission from a dimorphic crystalline stilbene. *Chem. Phys. Lett.*, **34**, 271–4. [232, 233f]

Colapietro, M., Domenicano, A., Marciante, C. and Portalone, G. (1984). Angular ring distortions in benzene derivatives. *Acta Crystallogr. A*, **40**, C98–9. [115, 154]

Colapietro, M., Domenicano, A., Portalone, G., Schulz, G. and Hargittai, I. (1984*a*). Molecular structure and ring distortion of *p*-dicyanobenzene in the gas phase and in the crystal. *J. Mol. Struct.*, **112**, 141–57. [154]

Colapietro, M., Domenicano, A., Portalone, G., Torrini, I., Hargittai, I. and Schultz, G. (1984*b*). Molecular structure and ring distortions of p-diisocyanobenzene in the gaseous phase and in the crystal. *J. Mol. Struct.*, **125**, 19–32. [154]

Coleman, L. B., Cohen, M. J., Sandman, D. J., Yamagishi, F. G., Garito, A. F. and Heeger, A. J. (1973). Superconducting fluctuations and the Peierls instability in an organic solid. *Solid State Commun.*, **12**, 1125–32. [189]

Colthup, N., Daly, L. H. and Wiberly, S. E. (1990). *Introduction to infrared and Raman spectroscopy*. Academic Press, London. [131]

Connick, W., May, F. G. J. and Thorpe, B. W. (1969). Polymorphism in 2,4,6-trinitrotoluene. *Austral. J. Chem.*, **22**, 2685–8. [291, 292f]

Cooke, P. M. (1998). Chemical microscopy. *Anal. Chem.*, **70**, 385R–423R. [95]

Coombes, D. S., Nagi, G. K. and Price, S. L. (1997). On the lack of hydrogen bonds in the crystal structure of alloxan. *Chem. Phys. Lett.*, **265**, 532–7. [184]

Cooper, P. W. and Kurowski, S. R. (1996). *Introduction to the Technology of Explosives* VCH, New York. [275, 288]

Cooper, W., Edmonds, J. W., Wudl, F. and Coppens, P. (1974). The 2-2'-bi-1,3-dithiole. *Cryst. Struct. Commun.*, **3**, 23–6. [52f]

Coppens, P., Cooper, W. F., Kenny, N. C., Edmonds, J. W., Nagel, A. and Wudl, F. (1971). Crystal and molecular structure of the aromatic sulfur compound 2,2'-bi-1,3-dithiole. d-Orbital participation in bonding. *J. Chem. Soc. D*, 889–90. [52f]

Coppens, P., Maly, K. and Petricek, V. (1990). Composite crystals: what are they and why are they so common in the organic solid state? *Mol. Cryst. Liq. Cryst.*, **181**, 81–90. [79]

Coppens, P., Fomitchev, D. V., Carducci, M. D. and Culp, K. (1998). Crystallography of molecular excited states. Transition-metal nitrosyl complexes and the study of transient species. *J. Chem. Soc., Dalton Trans.*, 865–72. [30, 234]

Cordes, A. W., Haddon, R. C., Hicks, R. G., Oakley, R. Y. and Palstra, T. T. M. (1992a). Preparation and solid-state structures of (cyanophenyl)dithia- and (cyanophenyl)diselenadiazolyl. *Inorg. Chem.*, **31**, 1802–8. [78]

Cordes, A. W., Haddon, R. C., Hicks, R. G., Oakley, R. T., Palstra, T. T. M., Schneemeyer, L. F. and Waszczak, J. V. (1992b). Polymorphism of 1,3-phenyl bis(diselenadiazolyl). Solid-state structure and electronic properties of β-1,3-[(Se$_2$N$_2$C)C$_6$H$_4$(CN$_2$Se$_2$)]. *J. Am. Chem. Soc.*, **114**, 1729–32. [78]

Cornelissen, J. P., Haasnoot, J. G., Reedijk, J., Faulmann, C., Legros, J.-P., Cassoux, P. and Negrey, P. J. (1992). Crystal structures and electrochemical properties of two phases of tetrabutylammonium bis(1,3-dithiole-2-thione-4,5-diselenolato)-nickelate (III). *Inorg. Chim. Acta*, **202**, 131–9. [191]

Cornelissen, J. P., Pomarede, B., Spek, A. L., Reefman, D., Haasnoot, J. G. and Reedijk, J. (1993). Two phases of [Me$_4$N][Ni(dmise)$_2$]$_2$: synthesis, crystal structures, electrical conductivities and intermolecular overlap calculations of α and β-tetramethylammonium bis[bis(2-selenoxo-1,3-dithiole-4,5-dithiolato)nickelate], the first conductors based on the M(C$_3$S$_4$Se) system. *Inorg. Chem.*, **32**, 3720–6. [191]

Cornell, W. D., Cieplak, P., Bayly, C. I., Gould, I. R., Merz, K. M. Jr., Ferguson, D. M., Spellmeyer, D. C., Fox, T., Caldwell, J. W. and Kollman, P. A. (1995). A second generation force field for the simulation of proteins, nucleic acids and organic molecules. *J. Am. Chem. Soc.*, **117**, 5179–97. [166]

Corradini, P. (1973). X-ray studies of conformation: observation of different geometries of the same molecule. *Chem. Ind. (Milan)*, **55**, 122–9. [2, 155, 155f, 157]

Cox, S. R. and Williams, D. E. (1981). Representation of the molecular electrostatic potential by a net atomic charge model. *J. Comput. Chem.*, **2**, 304–23. [154]

Craig, D. Q. M., Royall, P. G., Kett, V. L. and Hopton, M. L. (1999). The relevance of the amorphous state to pharmaceutical dosage forms: glassy drugs and freezedried systems. *Int. J. Pharm.*, **179**, 179–207. [253]

Cramer, F., Sprinzl, R., Furgac, N., Freist, W., Saenger, W., Manor, P., Sprinzll, L. and Sternback, H. (1974). Crystallization of yeast phenylalanine transfer RNA: polymorphism and studies of sulfur-substituted mercury binding derivatives. *Biochim. Biophys. Acta*, **349**, 351–65. [19]

Craven, B. M., Vizzini, E. A. and Rodriguez, M. M. (1969). The crystal structures of two polymorphs of 5,5'-diethyl barbituric acid. *Acta Crystallogr. B*, **25**, 1978–92. [184]

Cromer, D. T., Lee, K.-Y. and Ryan, R. R. (1988). Structures of two polymorphs of 1,1'-dinitro-3,3'-azo-1,2,4-triazole. *Acta Crystallogr. C*, **44**, 1673–4. [294]

Crookes, D. L. (1985). Aminoalkyl furan derivative. U. S. Patent #4,521,431. [144, 256]

Crosby, J., Pittam, J. D., Wright, N. C. A., Ohashi, M. and Mrukami, K. (1999). Impact of crystallisation behavior on the development of the azole antifungal ZD0870. Presented at 1st International Symposium on 'Aspects of Polymorphism & Crystallisation – Chemical Development Issues', Hinckley, Leicestershire, UK, Scientific Update. [256]

Crottaz, O., Kubel, F., Schmid, H. (1997) Jumping crystals of the spinels of $NiCr_2O_4$ and $CuCr_2O_4$. *J. Mater. Chem.*, **7**, 143–6. [223]

Crystallics, B. V. (Amsterdam) (2000). High throughput polymorph screening. Presented at 2nd International Symposium on Polymorphism and Crystallisation—Chemical Development Issues, Chester, UK. (Organized by Scientific Update, UK). [92]

Csöregh, I., Brehmer, T., Bombicz, P. and Weber, E. (2001). Halogen. . .halogen versus OH. . .O supramolecular interactions in the crystal structures of a series of halogen and methyl substituted *cis*-9,10-diphenyl-9,10-dihydroanthracene-9,10-diols. *Cryst. Eng.* **4**, 343–57. [272]

Curry, C. J., Rendle, D. F. and Rogers, A. (1982). Pigment analysis in the forensic examination of paints. I. Pigment analysis by x-ray powder diffraction. *J. Forensic Sci.*, **22**, 173–7. [260t, 261t]

Curtin, D. Y. and Engelmann, J. H. (1972). Intramolecular oxygen-nitrogen benzoyl migration of 6-aroyloxyphenanthridines. *J. Org. Chem.*, **37**, 3439–43. [7]

Curtin, D. Y. and Paul, I. C. (1981). Chemical consequences of the polar axis in organic solid-state chemistry. *Chem. Rev.*, **81**, 525–41. [207]

Curtin, D. Y., Paul, I. C., Duesler, E. N., Lewis, T. W., Mann, B. J. and Shiau, W.-I. (1979). Studies of thermal reactions in the solid state. *Mol. Cryst. Liq. Cryst.*, **50**, 25–42. [237]

Dähne, L. and Biller, E. (1998a). Excitonic interactions in dye arrays. *J. Inf. Recording*, **24**, 171–7. [231]

Dähne, L. and Biller, E. (1998b). Color variation in highly oriented dye layers by polymorphism of dye aggregates. *Adv. Mater.*, **10**, 241–5. [231]

Dähne, S. and Kulpe, S. (1977). *Structural principles of unsaturated organic compounds* Abhandlungen der Akademie der Wissenschaften der DDR. Akademie-Verlag, Berlin. [258]

Dainippon (1982). Preparation of imidazolone series orange color pigment—by heating aqueous suspension of α-type diazonitrochloroaniline coupled with acetoacetylaminobenzoimidazolone. Patent JP 57-141 457; *Chem. Abstr.*, **98**, 55548p. [260t]

Dainippon (1986). Japanese patent 272 697-A. [261t]

Dainippon (1988). Japanese patent 358 810-A. [261t]

Dainippon (1994). δ-type and anthrone blue pigment—is a mixture containing titanium dioxide and is useful as car paint having good weatherability, crystal stability and dispersibility. Patent WO 9605255-A1. [262t]

Dainippon (1996a). Japanese patent 09194752-A. [261t]

Dainippon (1996b). Japanese patent 09227791-A. [261t]

Dainippon (1996c). Japanese patent 092411524-A. [261t]

Dainippon (1996d). Japanese patent 09268529-A. [261t]

Dainippon (1997a). New crystalline monoazo lake color useful for printing inks—having specified diffraction peaks at various angles in x-ray diffraction pattern. Japanese patent 09194752-A.

Dainippon (1997b). New crystal type red monoazo lake dye—for use as a barium lake dye substituent and as printing inks, coating etc. Japanese patent 09227791-A.

Daniels, F., Williams, J. W., Bender, P., Alberty, R. A., Cornwell, C. D. and Harriman, J. E. (1970). *Experimental physical chemistry*, 7th edn. p. 53. McGraw-Hill, New York. [168]

Davey, R. J. (1993). General discussion. *Faraday Discuss.*, **95**, 160–2. (See also Cardew and Davey, 1982.) [44]

Davey, R. J., Maginn, S. J., Andrews, S. J., Buckley, A. M., Cottier, D., Dempsey, P., Rout, J. E., Stanley, D. R. and Taylor, A. (1993). Stabilization of a metastable crystalline phase by twinning. *Nature*, **366**, 248–50. [92]

Davey, R. J., Maginn, S. J., Andrews, S. J., Black, S. N., Buckley, A. M., Cottler, D., Dempsey, P., Plowman, R., Rout, J. E., Stanley, D. R. and Taylor, A. (1994). Morphology and polymorphism in molecular crystals: terephthalic acid. *J. Chem. Soc., Faraday Trans.*, **90**, 1003–9. [47, 52f, 63, 113, 114f, 116f]

Davey, R. J., Blagden, N., Potts, G. D. and Docherty, R. (1997). Polymorphism is molecular crystals: Stabilization of a metastable form by conformational mimicry. *J. Am. Chem. Soc.*, **119**, 1767–72. [253]

Davey, R. J., Blagden, N., Righini, S., Allison, H., Quayle, M. J. and Fuller, S. (2000). Crystal polymorphism as a probe for molecular assembly during nucleation from solutions: the case of 2,6-dihydroxybenzoic acid. *Cryst. Growth Des.*, **1**, 59–65. [72]

David, R. and Giron, D. (1994). Crystallization. *Handbook Powder Technol.*, **9**, 193–241. [5]

David, W. I. F., Shankland, K. and Shankland, N. (1998). Routine determination of molecular crystal structures from powder diffraction data. *J. Chem. Soc., Chem. Commun.*, 931–2. [111]

Davies, E. S. and Hartshorne, N. H. (1934). Identification of some aromatic notro-compounds by optical crystallographic methods. *J. Chem. Soc.*, 1830–6. [212]

d'Avignon, D. A. and Brown, T. L. (1981). ^{14}N and ^{2}H nuclear quadrupole resonance spectra of anthranilic acid. *J. Phys. Chem.*, **85**, 4073–9. [59]

Day, J. and McPherson, A. (1991). Characterization of two crystal forms of cytochrone c from *Valida membranaefaciens*. *Acta Crystallogr. B*, **47**, 1020–2. [86, 86f]

de Diesbach, H. and von der Weid, E. (1927). Quelques sels complexes des o-dinitrils avec cuivre et la pyridine. *Helv. Chim. Acta*, **10**, 886–8. [265]

de Ilarduya, M. C. T., Martin, C., Goni, M. M. and Martinez-Oharriz, M. C. (1997). Polymorphism of sulindac: Isolation and characterization of a new polymorph and three new solvates. *J. Pharm. Sci.*, **86**, 248–51. [5]

de Jong, A. W. K.(1922*a*). Über die Einwirkung des Lichtes auf die zimtsaüren und über die Kontitution der Truxillsaüren. *Berichte*, **55**, 463–74. [234]

de Jong, A. W. K. (1922*b*). Über die konstitution der Truxill- und Truxinsaüren und uber die Einwirkung des Sonnenlichtes auf die zimtsaüren und zimtsaüre salze. *Berichte*, **56**, 818–32. [234]

de Matas, M., Edwards, H. G. M., Lawson, E. E., Shields, L. and York, P. (1998). FT-Raman spectroscopic investigation of a pseudopolymorphic transition in caffeine hydrate. *J. Mol. Struct.*, **440**, 97–104. [5,132]

de Wet, F. N., Gerber, J. J., Lotter, A. P., van der Watt, J. G. and Dekker, T. G. (1998). A study of the changes during heating of paracetamol. *Drug Dev. Ind. Pharm.*, **24**, 447–53. [249]

DeClercq, J. P., Germain, G. and van Meerssche, M. (1972). 21-β-Acetoxy-17α-hydroxy-4-pregnene-3,11,20-trione, $C_{23}H_{30}O_6$. *Cryst. Struct. Commun.*, **1**, 59–62. [161]

Decurtins, S., Shoemaker, C. B. and Wockman, H. H. (1983). Structure of a new modification of bromobis(N, N-diethyldithiocarbamato)iron(III), $Fe[S_2CN(C_2H_5)_2]_2Br$. *Acta Crystallogr. C*, **39**, 1218–21. [198]

Deeley, C. M., Spragg, R. A. and Threlfall, T. L. (1991). A comparison of Fourier transform infrared and near-infrared Fourier transform Raman spectroscopy for quantitative measurements: an application in polymorphism. *Spectrochim. Acta A*, **47**, 1217–23. [132]

Deene, W. A. and Small, R. W. H. (1971). Refinement of the structure of rhombohedral acetamide. *Acta Crystallogr. B*, **27**, 1094–8. [184]

Deffet, L. (1942). *Repertoire des Composés Organique Polymorphes*. Editions Desoer, Liége. [11]

DeJong, E. J. (1979). Nucleation: a review. In *Industrial crystallization 78* (7th Symposium, Warsaw) (ed. E. J. DeJong and S. J.), pp. 3–17. North-Holland, Amsterdam, The Netherlands. [68]

Desiraju, G. R. (1983).Intermolecular proton transfers in the solid state. Conversion of the hydroxyazo into the quinone hydrazone tautomer of 2-amino-3-hydroxy-6-phenylazopyridine. X-ray crystal structures of the two forms. *J. Chem. Soc., Perkin Trans. 2*, **1983**, 1025–30. [84]

Desiraju, G. R. (1989). Crystal engineering—the design of organic solids. *Material science monographs*, Vol. 54, Elsevier, Amsterdam, The Netherlands. [154, 199]

Desiraju, G. R. and Sarma, J. A. R. P. (1986). The chloro-methyl exchange rule and its violations in the packing of organic molecular solids. *Proc. Ind. Acad. Science*, **96**, 599–605. [272]

Desiraju, G. R. and Steiner, T. (1999). *The weak hydrogen bond*. Oxford University Press, Oxford. [154]

Desiraju, G. R., Paul, I. C. and Curtin, D. Y. (1977). Conversion in the solid state of the yellow to the red form of 2-(4′-methoxyphenyl)-1,4-benzoquinone. X-ray crystal structures and anisotropy of the rearrangement. *J. Am. Chem. Soc.*, **99**, 1594–601. [215]

Deuschel, W., Honigmann, B., Jettmar, W. and Schroeder, H. (1964). Quinacridone pigments. British patent 923,069. *Chem. Abstr.*, **61**, 8447. [261t, 262]

Devilliers, M. M., van der Watt, J. G. and Lotter, A. P. (1991). The interconversion of the polymorphic forms of chloramphenicol palmitate (CAP) as a function of environmental temperature. *Drug Dev. Ind. Pharm.*, **17**, 1295–303. [248]

Dick, J. J. (1995) Shock-wave behavior in explosive monocrystals. *J. Phys. IV*, **5**, 103–6. [275]

Dickinson, C. and Holden, J. R. (1977). Crystal structures of hexanitrodiphenol amine and its potassium salt. *American Crystallographic Association Series 2*, 5, 55 (Abstracts, American Crystallographic Association Summer Meeting, August, 1977, East Lansing, Michigan, Abstract PA7). [295]

Di Martino, P., Guyot-Hermann, A. M., Conflant, P., Drach, M. and Guyot, J. C. (1996). A new pure paracetamol for direct compression: The orthorhombic form. *Int. J. Pharmaceutics* **128**, 1–8. [256]

Di Martino, P., Scoppa, M., Joivís, E., Palmieri, G. F., Andres, C., Pourcelot, Y. and Martelli, S. (2001). The spray drying of acetazolamide as a method to modify crystal properties and to improve compression behavior. *Int. J. Pharmaceutics*, **213**, 209–21. [256]

Ding, J., Herbst, R., Praefcke, K., Kohne, B. and Saenger, W. (1991). A crystal that hops in phase transition, the structure of trans, trans, anti, trans, trans-perhydropyrene. *Acta. Crystallogr. B*, **47**, 739–42. [223]

Dittmer, D. C. (1997). 'No-solvent' organic synthesis. *Chem. Ind.*, October 6, 779–84. [238]

Dobler, M. (1984). 18-crown-6: only a simple molecule? *Chimia*, **38**, 415–21. [154]

Dollimore, D. and Phang, P. (2000). Thermal analysis. *Anal. Chem.*, **72**, 27R–36R. [108]

Domenicano, A., Schultz, G., Hargittai, I., Colapietro, M., Portalone, G., George, P. and Bock, C. W. (1990). Molecular structure in nitrobenzene in the planar and orthogonal conformations. A concerted study by electron diffraction, x-ray crystallography and molecular orbital calculations. *Struct. Chem.*, **1**, 107–22. [115t]

Domenicano, A. and Hargittai, I. (1993). Gas/crystal structural differences in aromatic molecules. *Acta Chimica Hungarica—Models in Chemistry*, **130**, 347–62. [154]

Donohue, J. (1974). *The structure of the elements*. Wiley, New York. [18]

Donohue, J. (1985). Revised space group frequencies for organic compounds. *Acta Crystallogr. A*, **41**, 203–4. [183]

Dorset, D. L. (1996). Electron crystallography. *Acta Crystallogr. B*, **52**, 753–69. [112]

Dorset, D., McCourt, M. P., Gao, L. and Voigt-martin, I. G. (1998). Electron crystallography of small molecules: criteria for data collection and strategies for structure solution. *J. Appl. Crystallogr.*, **31**, 544–53. [112]

Draguet-Brughmans, M., Bouche, R., Flandre, J. P. and van den Bulcke, A. (1979). Polymorphism and bioavailability of pentobarbital. *Pharm. Acta Helv.*, **54**, 140–5. [244, 244f]

Dromzee, Y., Chiarelli, R., Gambarelli, S. and Rassat, A. (1996). Dupeyredioxyl (1,3,5,7-tetramethyl-2,6-diazaadamantane-N-N'-dioxyl). *Acta Crystallogr. C*, **52**, 474–7. [198, 199f]

Drucker, C. (1925). *Hand- und Hiflsbuch zur Ausführung physikochemischer Messungen Akademische Verlagsgesellschaft*. pp. 228–33. *m. b. H.*, Leipzig. [147]

Duax, W. L. and Norton, D. A. (1975). *Atlas of steroid structure*, Vol. 1, Plenum Press, New York. [160]

Duax, W. L., Fronkowiak, M. D., Griffin, J. F. and Rohrer, D. C. (1982*a*). A comparison between crystallographic data and molecular mechanics calculations on the side chain and backbone conformations of steroids. In *Intermolecular dynamics* (ed. J. Jortner and B. Pullman), p. 502. D. Reidel, Dordrecht. [160]

Duax, W. L., Griffin, J. F. and Weeks, C. M. (1982*b*). *Atlas of steroid structure*, Vol. 2, Plenum Press, New York. [160]

Duax, W. L., Numazawa, M., Osawa, Y., Strong, P. D. and Weeks, C. M. (1981). Conformational analysis of steroids: polymorphic forms of 17-β-Acetoxy-6β-bromo-4-androsten-3-one. *J. Org. Chem.*, **46**, 2650–5. [161]

Duebel, R., Schui, F. and Wester, N. (1984). Crystal β-form of pigment red 53 : 1— with more yellow tone, for coloring paints, plastics and printing inks. Hoechst Patent EP 97913. [260t]

Dunitz, J. D. (1979). *X-ray analysis and the structure of organic molecules*. Cornell University Press, Ithaca, New York. [46, 112, 151, 154]

Dunitz, J. D. (1991). Phase transitions in molecular-crystals from a chemical standpoint. *Pure Appl. Chem.*, **63**, 177–85. [4, 29, 30, 223, 237]

Dunitz, J. D. (1995). Phase changes and chemical reactions in molecular crystals. *Acta Crystallogr. B*, **51**, 619–31. [3, 14]

Dunitz, J. D. (1996). Thoughts on crystals as supermolecules. In *The crystal as a supramolecular entity. Perspectives in supramolecular chemistry*, Vol. 2 (ed. G. R. Desiraju), pp. 1–30. Wiley, Chichester. [4, 166, 183]

Dunitz, J. D. and Bernstein, J. (1995). Disappearing polymorphs. *Acc. Chem. Res.*, **28**, 193–200. [70, 77, 90, 92]

Dunitz, J. D. and Gavezzotti, A. (1999). Attractions and repulsions in molecular crystals: what can be learned from the crystal structures of condensed ring aromatic hydrocarbons? *Acc. Chem. Res.*, **32**, 677–84. [153, 169]

Dunitz, J. D., Filippini, G. and Gavezzotti, A. (2000). Molecular shape and crystal packing: a study of $C_{12}H_{12}$ isomers, real and imaginary. *Helvet. Chim. Acta*, **83**, 2317–35. [184]

Dzyabchenco, A. V. (1984). Theoretical structures of crystalline benzene. II. Verification of atom-atom potentials. *Zh. Strukt. Khim.*, **25**, 57–62. [185]

Dzyabchenco, A. V. (1987). Theoretical structures of benzene crystals. VI. Global searches in bisystem structural classes. *Zh. Strukt. Khim.*, **28**, 59–65. [185]

Dzyabchenco, A. V. and Bazilevskii, M. V. (1985). Theoretical structures of crystalline benzene. II. Calculation of transition state. *Zh. Strukt. Khim.*, **26**, 78–84. [185]

Dzyabchenco, A. V., Pivinia, T. S. and Arnautova, E. A. (1996). Prediction of structure and density of organic nitramines. *J. Mol. Struct.*, **378**, 67–82. [185]

Eastes, J. W. (1956). Preparation of phthalocyanine pigments. American Cyanamid Patent US 2 770 629. [266]

Ebert, A. A., Jr. and Gottlieb, H. B. (1952). Infrared spectra of organic compounds exhibiting polymorphism. *J. Am. Chem. Soc.*, **74**, 2806–10. [125, 266]

Edwards, G. (1950). Vapor pressure of 2,4,6-Trinitrotoluene. *Trans. Faraday Soc.*, **46**, 423–7. [291]

Eistert, B., Weygand, F. and Csendes, E. (1952). Polymorphism of the chalcones. *Chem. Berichte*, **85**, 164–8. [255]

Ellern, A., Bernstein, J., Becker, J. Y., Zamir, S., Shahal and Cohen, S. A. (1994). New polymorphic modification of tetrathiafulvalene. Crystal structure, lattice energy and intermolecular interactions. *Chem. Mater.*, **6**, 1378–85. [52f, 271]

Ellison, R. D. and Holmberg, R. W. (1960). Cell dimensions and space group of 1,1-diphenyl-2-picrylhydrazine. *Acta Crystallogr.*, **13**, 446–7. [203]

Encyclopedia Brittanica (1798). [20]

Engelberg, A. B. (1999). Special patent provisions for pharmaceuticals: Have they outlived their usefulness? A political legislative and legal history of U. S. Law and observations for the future. *IDEA: J. Law Technol.*, **39**, 389. [298, 300]

Enokida, T. and Hirohashi, R. (1991). A new crystal of copper phthalocyanine synthesized with 1,8-diaza-bicyclo-(5,4,0)undecane-7. *Mol. Cryst. Liq. Cryst.*, **195**, 265–79. [207, 266, 267]

Enokida, T. and Hirohashi, R. (1992). Electrophotographic dual-layered photoreceptors incorporating copper phthalocyanines. *J. Imag. Sci. Technol.*, **36**, 135–41. [207]

Ephraim, F. (1923). Über die löslichkeit von kobaltiaken. *Berichte Deutsches Chemische Gesellescaft*, **56**, 1530–42. [239]

Erk, P. (1998) *Proc. 2nd Int. Symp. Phthalocyanines*, Edinburgh. [262t]

Erk, P. (1999). Crystal design from molecular to application properties. In *From molecules and crystals to materials* (ed. D. Braga, F. Grepioni and A. G. Orpen), Vol. C538 of *NATO ASI Series*, pp. 143–61. Kluwer Academic Publishers, Dordrecht. [185, 258]

Erk, P. (2000). Private communication. [259, 264]

Erk, P. (2001*a*). Crystal design of organic pigments—a prototype discipline of materials science. *Curr. Opin. Solid State Mater. Sci.*, **5**, 155–160. [259]

Erk, P. (2001*b*). In preparation [264, 270]

Erlenmeyer, E., Jr., Brakow, C. and Herz, O. (1907). Isomeric cinnamic acids. *Berichte*, **40**, 653–63. [234]

Eshkova, Z. I., Moiseeva, S. S., Konysheva, L. I., Bir, E. Sh. and Demina, L. V. (1976). X-ray study of binary systems on the basis of linear quinacridone. *FATIPEC Cong.*, **13**, 270–4. [261t]

Espenbetov, A. A., Antipin, M. Yu., Struchkov, Yu. T., Philippov, V. A., Trirel'son, V. G., Ozerov, R. P. and Svetlov, B. S. (1984). Structure of 1,3-propanethiol trinitrate (β-modification), $C_3H_5N_3O_9$. *Acta Crystallogr. C*, **40**, 2096–8. [284]

Etter, M. C. (1985). Aggregate structures of carboxylic acids and amides. *Israel J. Chem.*, **25**, 312–9. [55]

Etter, M. C. (1990). Encoding and decoding hydrogen-bond patterns of organic compounds. *Acc. Chem. Res.*, **23**, 120–6. [54, 55, 61]

Etter, M. C. (1991). Hydrogen bonds as design elements in organic chemistry. *J. Phys. Chem.*, **95**, 4601–10. [54, 55, 61, 66, 69]

Etter, M. C. and Siedle, A. R. (1983). Solid state rearrangement of (phenylazo-phenyl)palladium hexafluoroacetylacetonate. *J. Am. Chem. Soc.*, **105**, 641–3. [219]

Etter, M. C., Kress, R. B., Bernstein, J. and Cash, D. J. (1984). Solid-state chemistry and structures of a new class of mixed dyes. Cyanine-oxonol. *J. Am. Chem. Soc.*, **106**, 6921–7. [81, 93]

Etter, M. C., MacDonald, J. C. and Bernstein, J. (1990*a*). Graph-set analysis of hydrogen-bond patterns in organic crystals. *Acta Crystallgr. B*, **46**, 256–62. [55, 56, 77]

Etter, M. C., Urbanczyk-Lipkowska, Z., Zia-Ebrahimi, M. and Panunto, T. W. (1990*b*). Hydrogen bond-directed cocrystallization and molecular recognition properties of diarylureas. *J. Am. Chem. Soc.*, **112**, 8415–26. [55, 56, 77]

Etter, M. C., Huang, K. S., Frankenbach, G. M. and Adsmond, D. (1991). Control of symmetry and asymmetry in hydrogen-bonded nitroaniline materials. In *Materials for nonlinear optics, chemical perspectives* (ed. S. R. Marder, J. E. Sohn and G. D. Stuckty), ACS Symposium Series, **455**, 446–55. [207]

European Patent Office (1997). Decision of 27 may 1997, Case Number T 0475/92—3.3.4, Application Number 83301951.6, Publication Number 0091787, IPC C07K 3/12. [307]

Evans, R. M. (1974). *The perception of color.* Wiley, New York. [257]

Even, J. and Bertault, M. (2000). Transformations in molecular crystals and chemical reactions: a physico-chemical approach. *Condens. Matter News*, **8**, 9–21. [223, 237]

Ewald, P. P. (1962). The beginnings. In *Fifty years of X-ray diffraction* (ed. P. P. Ewald) pp. 36. N. V. A. Oosthoek's Uitgeversmaatschappij, Utrecht. [10]

Fábián, L. and Kálmán, A. (1999). Volumetric measure of isostructurality. *Acta Crystallogr. B*, **55**, 1099–1108. [271]

Faigman, D. L. (1999). *Legal alchemy. The use and misuse of science and the law.* W. H. Freeman and Co., New York. [298]

Fairrie, G. (1925). Sugar. Fairrie and Company Ltd., Liverpool. [79]

Falini, G., Albeck, S., Weiner, S. and Addadi, L. (1996). Control of aragonite or calcite polymorphism by mollusk shell macromolecules. *Science*, **271**, 67–9. [19]

Fanconi, B. M., Gerhold, G. A. and Simpson, W. T. (1969). Influence of exciton phonon interaction on metallic reflection from molecular crystals. *Mol. Cryst. Liq. Cryst.*, **6**, 41–81. [231]

Farbwerke vorm Meister Lucius & Bruening (1902). Verfahren zur Herstellung eines roten, besonders zur Bereitung von Farblacken geeigneten Monoazofarbstoffes aus o-chlor-*m*-toluidin-*p*-sulfosaeure und *beta*-naphthol. Deutsches Reichspatent DRP 145908. [260t]

Farmer, V. C. (1957). Effects of grinding during the preparation of alkali-halide disks on the infrared spectra of hydroxylic compounds. *Spectrochim. Acta*, **8**, 374–89. [128]

Farnum D. G., Mehta, G., Moore, G. I. and Siegal, F. P. (1974). Attempted Reformatskii of benzonitrile 1,4-dioxo-3,6-diphenylpyrrolo[3,4-C]pyrrole, a lactam analog of pentalene. *Tetrahedron Lett.*, **29**, 2549–52. [271]

Fawcett, T. G. (1987). Great than the sum of its parts: a new instrument. *Chemtech*, 564–9. [148]

Fawcett, T. G., Martin, E. J., Crowder, C. E., Kincaid, P. J., Strandjord, A. J., Blazy, J. A., Newman, R. A. and Armentrout, D. N. (1986). Analyses of multi-phase pharmaceuticals using simultaneous differential scanning calorimetry and x-ray diffraction. *Adv. in X-ray Anal.*, **29**, 323–32. [148]

Fawcett, T. G., Harris, W. C., Newman, R. A., Whiting, L. F. and Knoll, F. J. (1989). Combined thermal analyzer and X-ray diffractiometer. U. S. patent #4,821,303. [148]

Federal Register (2000). United States Government Printing Office, Superintendent of Documents, P.O. Box 371954, Pittsburgh, PA 15250–7954. (http://www.gpo.gov) [256]

Ferraris, J., Cowan, D. O., Walatka, V. and Perlstein, J. H. (1973). Electron transfer in a new, highly conducting donor-acceptor compound. *J. Am. Chem. Soc.*, **95**, 948–9. [189]

Ferraro, J. R. and Nakamoto, K. (1994). *Introductory Raman spectroscopy.* Academic Press, London. [131]

Ferro, D. R., Bruckner, S., Meille, S. V. and Ragazzi, M. (1992). Energy calculations for isotactic polypropylene—a comparison between models of the alpha-crystalline and gamma-crystalline structures. *Macromolecules*, **25**, 5231–5. [182]

FIAT (1948). Field Information Agency Technical, 'German Dyestuffs and Intermediates,' *Final Report* 1313, Vol. III. Dyestuff Research, Technical Industrial

Intelligence Division, U. S. Dept. Commerce, Washington, DC, pp. 434–63. [262t, 266, 270, 271]

Fiebich, K. and Mutz, M. (1999). Evaluation of calorimetric and gravimetric methods to quantify the amorphous content of desferal. *J. Therm. Anal. Calorim.*, **57**, 75–85. [254]

Filippini, G., Gavezzotti, A. and Novoa, J. J. (1999). Modeling the crystal structure of the 2-hydronitronylnitroxide radical (HNN): observed and computer-generated polymorphs. *Acta Crystallogr. B*, **55**, 543–55. [183]

Findlay, A. F. (1951). *The phase rule and its applications* (9th edn) (ed. A. N. Campbell and N. O. Smith), pp. 7–19. Dover Publications, New York, U. S. A. [4, 9, 23, 24f, 30, 37]

Findlay, W. P. and Bugay, D. E. (1998). Utilization of Fourier Transform Raman spectroscopy for the study of pharmaceutical crystal forms. *J. Pharm. and Biomed. Anal.*, **16**, 921–30. [132]

Firsich, D. W. (1984). Energetic material separations and specific polymorph preparation via thermal gradient sublimation. *J. Hazard. Mater.*, **9**, 133–7. [294]

Fischer, P., Zolliker, P., Meier, B. H., Ernst, R. R., Hewat, A. W., Jorgensen, J. D. and Rotella, F. J. (1986). Structure and dynamics of terephthalic acid from 2 to 300K. I. High resolution neutron diffraction evidence for a temperature dependent order-disorder transition—a comparison of reactor and pulsed neutron source powder techniques. *J. Solid State Chem.*, **61**, 109–25. [115t]

FIZ. (2001). Fachinformationszentram Karlsruhe, Geselleschaft für wissenschaft lichtechische Information mbH, Herman-vontelmholtz – Platz 1, D-76344 Eggenstein-Leopoldschafen, Germany. (http://www.fiz-karlsruhe.de). [18]

Fletton, R. A., Lancaster, R. W., Harris, R. K., Kenwright, A. M., Packer, K. J., Waters, D. N. and Yeadon, A. (1986). A comparative spectroscopic investigation of two polymorphs of 4′-methyl-2′-nitroacetanilide using solid-state infrared and high-resolution solid-state nuclear magnetic resonance spectroscopy. *J. Chem. Soc. Perkin Trans. 2*, 1705–9. [224]

Flores, L. and Jones, F. (1972). Physicothermal stabilities of dye solids. II. Phase transitions and melting behavior of some disperse and vat dyes. *J. Soc. Dyers Colour.*, **88**, 101–6. [260t]

Foltz, M. F. (1994). Thermal stability of ε-hexanitrohexaazaisowurtzitane in an estane formulation. *Propell. Explos. Pyrot.*, **19**, 63–9. [282]

Foltz, M. F., Coon, C. L., Garcia, F. and Nichols III, A. L. (1994a). The thermal stability of the polymorphs of hexanitrohexaazaisowurtzitane. 1. *Propell., Explos. Pyrot.*, **19**, 19–25. [282]

Foltz, M. F., Coon, C. L., Garcia, F. and Nichols III, A. L. (1994b). The thermal stability of the polymorphs of hexanitrohexaazaisowurtzitane. 1. *Propell., Explos. Pyrot.*, **19**, 133–44. [282]

Fomitchev, D. V., Bagley, K. A. and Coppens, P, (2000). The first crystallographic evidence for side-on coordination of N_2 to a single metal center in a photoinduced metastable state. *J. Am. Chem. Soc.*, **122**, 532-3. [30]

Forster, A., Gordon, K., Schmierer, D., Soper, N., Wu, V. and Rades, T. (1998). 12. Characterization of two polymorphic forms of Ranitidine-HCl. *Internet J. Vib. Spectra*, **2**, 1–12. (http://www.ijvs.com/volume2/edition2/section2.html) [127f]

Förster, T. and Kasper, K. (1954). Ein Konzentrationsumschlag der Fluoreszenz. *Z. Phys. Chem. Neue Folge*, **1**, 275–7. [231]

Foster, K. R. and Huber, P. W. (1999). *Judging science. scientific knowledge and the federal courts*. MIT Press, Cambridge, MA. [298]

Foster, R. (1969). *Organic charge transfer molecules*. Academic Press, London. [153]

Foxman, B. M., Goldberg, P. L. and Mazurek, H. (1981). Confrontational polymorphism of $Ni(NCS)_2[P(CH)_2CH_2CN)_3]_2$. Crystallographic study of three polymorphs. *J. Inorg. Chem.*, **20**, 4368–75. [163]

Francis, C. V. and Tiers, G. V. D. (1992). Straight-chain carbamyl compounds for 2nd harmonic-generation. *Chem. Mater.*, **4**, 353–8. [213]

Francis, F. and Piper, S. H. (1939). The higher *n*-aliphatic acids and their methyl and ethyl esters. *J. Am. Chem. Soc.*, **61**, 577–81. [98]

Frankenheim, M. L. (1835). *Die Lehre von der Cohäsion*. August Schulz, Breslau.

Frankenheim, M. L. (1839). Ueber die Isomerie.*Praktische Chem.*, **16**, 1–14. [20]

Fraxedas, J., Caro, J., Santiso, J., Figueras, A., Gorostiza, P. and Sanz, F. (1999*a*). Molecular organic thin films of *p*-nitrophenyl nitronyl nitroxide: surface morphology and polymorphism. *Phys. Status Solidi. B: Basic Res.*, **215**, 859–63. [147, 201]

Fraxedas, J., Caro, J., Santiso, J., Figueras, A., Gorostiza, P. and Sanz, F. (1999*b*). Polymorphic transformation observed on molecular organic thin films: *p*-nitrophenyl nitronyl nitroxide radical. *Europhys. Lett.*, **48**, 461–7. [147, 201]

Free, M. L. and Miller, J. D. (1994). Effect of sample and incident beam areas on quantitative spectroscopy. *Appl. Spectros.*, **48**, 891–3. [128]

Freer, S. T. and Kraut, J. (1965). Crystal structures of D,L-homocysteine thiolactone hydrochloride: two polymorphic forms and a hybrid. *Acta Crystallgr.*, **19**, 992–1002. [30, 79]

Freyer, A. J., Lowe-Ma, C. K., Nissan, R. A. and Wilson, W. S. (1992) Synthesis and explosive properties of dinitropicrylbenzimidazoles and the 'Trigger Linkage' in dinitropicrylbenzotriazoles. *Austral. J. Chem.*, **45**, 525–39. [294, 295]

Friedländer, P. (1879). IX. Krystallographische Untersuchung einiger organischen Verbindungen. *Z. Kristallogr.*, **3**, 168–79. [288, 291]

Froelich, H. (1989). Beta-crystal modification of an azo pigment. Hoechst EP 320774 A2. [261t]

Frommer, J. (1992). Scanning Tunneling Microscopy and atomic force microscopy in organic chemistry. *Angw. Chem., Int. Ed. Engl.*, **31**, 1298–328. [147]

Frydman, L., Oliviery, A. C., Diaz, L. E., Frydman, B., Schmidt, A. and Vega, S. (1990). A ^{13}C solid-state NMR study of the structure and dynamics of the polymorphs of sulfanilamide. *Mol. Phys.*, **70**, 563–79. [140]

Fryer, J. R. (1997). Pigments: myth shape and structure. *Surf. Coat. Int.*, 421–6. [79, 259]

Fryer, J. R., McKay, R. B., Mather, R. R. and Sing, K. S. (1981). The technological importance of the crystallographic and surface properties of copper phthalocyanine pigments. *J. Chem. Technol. Biotechnol.*, **31**, 371–87. [265, 268]

Fu, J.-H., Rose, J., Tam, M. F. and Wang, B.-C. (1994). New crystal forms of a μ-class glutathione 5-transferase from rat liver. *Acta Crystallogr. D*, **50**, 219–22. [85, 86f]

Fuji, O., Takano, M, Sakatami, T and Iwamoto, E. (1980). Yellow pigment with good heat and light resistance – is bistetrachloroisoindoline-1-one-3-ylidine phenylene-1,4-diamine cdp. Toyo Soda Mfg. Co. Ltd. Patent JP 55-12106; *Chem. Abstr.*, **93**, 27823c. [261t]

Furedi-Milhofer, H., Garti, N. and Kamyshny, A. (1999) Crystallization from microemulsions—a novel method for the preparation of new crystal forms of aspartame. *J. Cryst. Growth*, **199**, 1365–70. [305]

Fyfe, C. A. (1983). Solid state NMR for Chemists. Guelph, Ontario, CFC. [133]

Gallagher, H. G. and Sherwood, J. N. (1996). Polymorphism, twinning and morphology of crystals of 2,4,6-trinitrotoluene grown from solution. *J. Chem. Soc., Faraday Trans.*, **92**, 2107–16. [288, 289, 289t, 291, 292f, 293f]

Gallagher, H. G., Roberts, K. J., Sherwood, J. N. and Smith, L. A. (1997). A theoretical examination of the molecular packing, intermolecular bonding and crystal morphology of 2,4,6-trinitrotoluene in relation to polymorphic structural stability. *J. Mater. Chem.*, **7**, 229–35. [291, 293f]

Gallier, J., Toudic, B., Delugeard, Y., Cailleau, H., Gourdji, M., Peneau, A. and Guibe, L. (1993). Chlorine-nuclear-quadrupole-resonance study of the neutral-to-ionic transition tetrathiafulvalence-chloranil. *Phys. Rev. C*, **47**, 11688–95. [197]

Gao, D. and Williams D. E. (1999). Molecular packing groups and *ab initio* crystal structure prediction. *Acta Crystallogr. A*, **55**, 621–7. [185]

Gao, P. (1996). Determination of the composition of delaviridine mesylate polymorph and pseudopolymorph mixtures using C-13 CP/MAS NMR. *Pharm. Res.*, **13**, 1095–104. [5]

Garside, J. (1985). Industrial crystallization from solution. *Chem. Eng. Sci.*, **40**, 3–26. [68]

Garside, J. and Davey, R. J. (1980). Secondary contact nucleation: kinetics, growth and scale-up. *Chem. Eng. Commun.*, **4**, 393–424. [68]

Garti, N. and Sato, K. (1988). *Crystallization and polymorphism of fats and fatty acids*. Marcel Dekker, Inc., New York, U. S. A. [27]

Garti, N. and Zour, H. (1997). The effect of surfactants on the crystallization and polymorphic transformation of glutamic acid. *J. Cryst. Growth*, **172**, 486–98. [27]

Gaultier, J., Hauw, C. and Bouas H. (1976). Crystal and molecular structure of a new tetracyclic hydrocarbon: 'lepidopterene'. *Acta Crystallogr. B* **32**, 1220–3. [164]

Gavezzotti, A. (1983). The calculation of molecular volumes and the use of volume analysis in the investigation of structured media and of solid-state organic reactivity. *J. Am. Chem. Soc.* **105**, 5220–5. [178]

Gavezzotti, A. (1985). Molecular free surface: a novel method of calculation and its uses in conformational studies and in organic crystal chemistry. *J. Am. Chem. Soc.*, **107**, 962–7. [178]

Gavezzotti, A. (1991). Generation of possible crystal structures from molecular structure for low-polarity organic compounds. *J. Am. Chem. Soc.*, **113**, 4622–9. [154, 183]

Gavezzotti, A. (1994*a*). Are crystal structures predictable? *Acc. Chem. Res.*, **27**, 309–14. [151, 183, 185]

Gavezzotti, A. (1994*b*). *PROMET3. A program for the generation of possible crystal structures from the molecular structure of organic compounds*. University of Milano, available upon request. [183, 186]

Gavezzotti, A. (1996). Polymorphism of 7-dimethylaminocyclopenta[*c*]coumarine: Packing analysis and generation of trial crystal structures. *Acta Crystallogr. B* **52**, 201–8. [183]

Gavezzotti, A. (1997). Computer simulations of organic solids and their liquid-state precursors. *Faraday Discussions*, **106** (Solid-state chemistry: new opportunities from computer simulations), 63–77. [183]

Gavezzotti, A. and Desiraju, G. R. (1988). A synthetic analysis of packing energies and other packing parameters for fused-ring aromatic hydrocarbons. *Acta Crystallogr. B*, **44**, 427–34. [85]

Gavezzotti, A. and Flippini, G. (1994). Geometry of the intermolecular X-H$\cdots$Y (X, Y = N, O) hydrogen bond and the calibration of empirical hydrogen-bond potentials. *J. Phys. Chem.*, **98**, 4831–7.

Gavezzotti, A. and Flippini, G. (1995). Polymorphic forms of organic crystals at room conditions: thermodynamic and structural implications. *J. Am. Chem. Soc.*, **117**, 12299–305. [16, 32, 49, 166, 168, 183]

Gavezzotti, A., Flippini, G., Kroon, J., van Eijck, B. P. and Klewinghaus, P. (1997). The crystal polymorphism of tetrolic acid ($CH_3C\equiv CCOOH$): a molecular dynamics study of precursors in solution, and in crystal structure generation. *Chem. Eur. J.*, **3**, 893–9. [183]

Gdanitz, R. J. (1992). Prediction of molecular crystal structures by Monte Carlo simulated annealing without reference to diffraction data. *Chem. Phys. Lett.*, **190**, 391–6. [185]

Gdanitz, R. J. (1997). *Ab initio* prediction of possible molecular crystal structures. In *Theoretical aspects and computer modeling of the molecular solid state* (ed. A. Gavezzotti), pp. 185–99. Wiley, Chichester. [182]

Gdanitz, R. J. (1998). *Ab initio* prediction of molecular crystal structure. *Curr. Opin. Solid State Mater. Sci.*, **3**, 414–18. [182, 185]

Geladi, P. and Kowalski B. R. (1986). Partial least-squares regression: a tutorial. *Anal. Chim. Acta*, **185**, 1–17. [130]

Gemperline, P. J. and Boyer, N. T. (1995). Classification of near-infrared spectra using wavelength distances: comparison to the Mahalonobis distance and residual variance methods. *Anal. Chem.*, **67**, 160–6. [130]

Genck, W. J. (2000). The effects of mixing on scale-up—how crystallization and precipitation react. *Chem. Process.*, **63**, 47. [256]

George, R. S., Cady, H. H., Rogers, R. N. and Rohwer, R. K. (1965). Solvates of octahydro-1,3,5,7-tetranitro-1,3,5,7-tetrazocaine (HMX). Relatively stable monosolvates. *Ind. Eng. Chem. Prod. Res. Dev.,* **4**, 209–14. [270]

Gerber, J. J., Caira, M. R. and Lotter, A. P. (1993). Structures of the conformational polymorphs of the cholesterol-lowering drug Probucol. *J. Crystallogr. Spectrosc. Res.*, **23**, 863–9. [160]

Geuther, A. (1883). VII. Ueber die Constitution der Doppel verbindungen von Salzen der sulfon saüren mit neutralen schwefelsäureräthern, und über die constitution der sulfate, sowie über den Grund ihrer Dimorphie. *Ann. Chem.*, **218**, 288–302. [22]

Ghinescu, I., Dragonic, V.-A., Saidac, S. and Crustescu, E. (1984). Interprinderea de Coloranti 'Colorom' Patent RO 83 912. [260t]

Giacovazzo, C. (ed.) (1992). *Fundamentals of crystallography*. International Union of Crystallography, Oxford. [112]

Gibbs, J. W. (1876). On the equilibrium of heterogeneous substances. *Trans. Connecticut Acad. Arts Sci.*, **3**, 108–248. [29]

Gibbs, J. W. (1878). On the equilibrium of heterogeneous substances. *Trans. Connecticut Acad. Arts Sci.*, **3**, 343–524. [29]

Gibbs, T. R. and Popolato, A. (1980). *Los Alamos scientific laboratory explosive property data*. University of California Press, Berkeley. [279t]

Gibson, K. D. and Scheraga, H. A. (1995). Crystal packing without symmetry constraints. 2. Possible crystal packings of benzene obtained by energy minimization from multiple starts. *J. Phys. Chem.*, **99**, 3765–73. [185]

Gieren, A. and Hoppe, W. (1971). X-ray crystal structure analysis of bisphthalocyanatouranium(IV). *J. Chem. Soc., Chem. Commun.*, 413–4. [266]

Gigg, J., Gigg, R., Payne, S. and Conant, R. (1987). The allyl group for protection in carbohydrate chemistry. Part 21. ($\pm$)-1,2:5,6- and ($\pm$)-1,2:3,4-di-O-isopropylidene-myo-inositol. Unusual behavior of crystals of ($\pm$)-3,4-di-O-acetyl-1,2,5,6-tetra-O- benzyl-myo-inositol. *J. Chem. Soc., Perkin Trans. I*, 2411–4. [223]

Gilardi, R. D. and Butcher, R. J. (1998*a*). A new class of flexible energetic salts. 3. The crystal structure s of the 3,3-dinitroazetidinium dinitramide and 1-isopropyl-3,3-dinitroazetidinium dinitramide salts. *J. Chem. Crystallogr.*, **28**, 163–9. [287]

Gilardi, R. D. and Butcher, R. J. (1998*b*). A new class of flexibile energetic salts, part 5: The structures of two hexammonium polymorphs and the ethan-1,2-diamonium salts of dinitramide. *J. Chem. Crystallogr.*, **28**, 673–81. [286]

Gilardi, R., Flippen-Anderson, J., George, C. and Butcher, R. J. (1997). A new class of flexible energetic salts: The crystal structures of the ammonium, lithium, potassium, and cesium salts of dinitramide. *J. Am. Chem. Soc.*, **119**, 9411–6. [287]

Gilliland, G. L., Tung, M., Blakeslee, D. M. and Ladner, J. (1994). The biological macromolecule crystallization database, version 3.0: new features, data and NASA archive for protein crystal growth data. *Acta Crystallogr. D*, 408–13. [19]

Girlando, A., Marzola, F., Pecile, C. and Torrance, J. B. (1983). Vibrational spectroscopy of mixed stack organic semiconductors: neutral and ionic phases of tetrathiafulvalene-chloranil (TTF-CA) charge transfer complex. *J. Chem. Phys.*, **79**, 1075–85. [196]

Giron, D. (1981). Polymorphism. *Labo-Pharma - problems et techniques*, **307**, 151–60. [240]

Giron, D. (1988). Impacts of solid-state reactions on medicaments. *Mol. Cryst. Liq. Cryst.*, **161**, 77–100. [252]

Giron, D. (1990). Thermal-analysis in pharmaceutical routine analysis. *Acta Pharm. Jugoslav.*, **40**, 95–157. [250]

Giron, D. (1995). Thermal analysis and calorimetric methods in the characterisation of polymorphs and solvates. *Thermochim. Acta*, **248**, 1–59. [105, 106, 108f, 109f, 109, 242, 250, 256]

Giron, D. (1997). Thermal analysis of drugs and drug products. In *Encyclopedia of pharmaceutical technology*, Vol. 15: *Thermal analysis of drugs and drug products to unit processes in pharmacy: fundamentals* (ed. J. Swarbrick and J. C. Boylan), pp. 1–79. Marcel Dekker, New York. [242, 243, 250, 253]

Giron, D. (1998). Contribution of thermal methods and related techniques to the rational development of pharmaceuticals—Part I. *Pharma. Sci. Technol. Today*, **1**, 191–9. [250]

Giron, D. (1999a). Thermal analysis, microcalorimetry and combined techniques for the study of pharmaceuticals. *J. Therm. Anal. Calorim.*, **56**, 1285–304. [250, 256]

Giron, D. and Goldbronn, C. (1997). Use of DSC and TG for identification and quantification of the dosage form. *J. Therm. Anal.*, **48**, 473–83. [250]

Giron, D., Edel, B. and Piechon, P. (1990). X-ray quantitative determination of polymorphism in pharmaceuticals. *Mol. Cryst. Liq. Cryst.*, **187**, 557–67. [252]

Giron, D., Draghi, M., Goldbronn, C., Pfeffer, S. and Peichon, P. (1997). Study of the polymorphic behavior of some local anesthetic drugs. *J. Therm. Anal.*, **49**, 913–27. [251]

Giron, D., Piechon, P., Goldbronn, S. and Pfeffer, S. (1999). Thermal analysis, microcalorimetry and combined techniques techniques for the study of the polymorphic behaviour of a purine derivative. *J. Therm. Anal. Calorim.*, **57**, 61–73. [250]

Giron-Forest, D. (1984). Anwedung der thermischen analyse in der pharmazie. *Pharmazeutische Industry*, **46**, 851–9. [250]

Glasstone, S. (1940). *Text-book of Physical Chemistry*, pp. 465–70. MacMillan and Co., London. [30]

Glaxo Inc. v. Boehringer-Ingelheim Corp., U.S. District Court (Docket No. 3 : 95CV01342) May 14, 1996. [301]

Glaxo Inc. v. Geneva, Pharmaceuticals, *et al.* 1994, 1996; Nos. 94–1921, 94–4589 and 96–3489, D. NJ). Glaxo, Inc. *et al.*, v. Torpharm Inc. *et al.*, NO. 95 C 4686 (ND IL Eastern Div., May 18, 1997; No. Civ. AMD 96-455, Nov. 4, 1998). [301]

Glaxo, Inc. *et al.*, v. Torpharm Inc. *et al.* (1998), 153 F.3d 1366. [301]

Glusker, J. P., Lewis, M. and Rossi, M. (1994). *Crystal structure analysis for chemists and biologists*. VCH Publishers, New York. [46, 112, 147, 151]

Goetz, F. and Brill, T. B. (1979). Laser Raman spectra of α-, β-, χ-, and δ-octahydro-1,3,5,7-tetranitro-1,3,5,7-tetrazocine and their temperature dependence. *J. Phys. Chem.*, **83**, 340–6. [280]

Golovina, N. I., Titkov, A. N., Raevskii, A. V. and Atovmyan, L. O. (1994). Kinetics and mechanism of phase transitions in the crystals of 2,4,6-trinitrotoluene and benzotrifuroxane. *J. Solid State Chem.*, **113**, 229–38. [289t, 290f, 291]

Goto, H., Fujinawa, T., Asahi, H., Ogata, H., Miyajima, S. and Maruyama, Y. (1996). Crystal structures and physical properties of 1,6-diaminopyrene-*p*-chloranil (DAP-CHL) charge transfer complex. Two polymorphs and their unusual electrical properties. *Bull. Chem. Soc. Jpn.*, **69**, 85–93. [82, 195]

Grabar, D. G. and McCrone, W. C. (1950). Antabuse (tetraethyl thiuram disulfide). *Anal. Chem.*, **22**, 620–1. [14]

Grabar, D. G., Rauch, F. C. and Fanelli, A. J. (1969) Observation of a solid-solid polymorphic transformation in 2,4,6-trinitrotoluene. *J. Phys. Chem.*, **73**, 3514–16. [288]

Graeber, E. J. and Morosin, B. (1974) The crystal structures of 2,2',4,4',6,6'-hexanitroazobenzene (HNAB) Forms I and II. *Acta Crystallogr. B* **30**, 310–17. [158, 294]

Graham. J. A., Grim, N. M. and Fately, W. G. (1985). Fourier Transform infrared photoacoustic spectroscopy of condensed-phase samples. In *Fourier transform infrared spectroscopy* Vol. 4 (ed. J. R. Ferraro and L. J. Basil), pp. 345–92. Academic Press, New York. [129]

Grainger, C. T. and McConnell, J. F. (1969). Crystal structure of 1-*p*-nitrobenzeneazo-2-naphthol (parared) from overlapped twin-crystal data. *Acta Crystallogr. B*, **25**, 1962–70. [260t]

Grant, D. J. W. and Brittain, H. G. (1995). Solubility of Pharmaceutical Solids. In *Physical characterization of Pharmaceutical solids* (ed. H. G. Brittain), pp. 321–86. Marcel Dekker, New York. [244]

Grant, D. J. W. and Higuchi, T. (1990). Solubility Behavior of Organic Compounds. In *Techniques of Chemistry*, Vol. 21 (ed. W. H. Jr. Saunders). John Wiley and Sons, New York. [111, 244]

Graser, F. and Hädicke, E. (1980). Crystal structure and perylene-3,4:9,10-bis(dicarboxymid) pigments. *Justus Liebigs Annalen der Chemie*, 1994–2011. [263]

Graser, F. and Hädicke, E. (1984). Crystal structure of perylene-3,4:9,10-bis(dicarboxymid) pigments. 2. *Justus Liebigs Annalen der Chemie*, 483–94. [263]

Green, B. S. and Knossow, M. (1981). Lamellar twinning explains the nearly racemic composition of chiral, single crystals of hexahelicene. *Science*, **214**, 795–7. [30]

Grell, J., Berstein, J. and Tinhofer, G. (2002). Investigation of hydrogen bond patterns: The graph set approach. Part I. General ideas and their application to crystal structures with exactly one molecule in the asymmetric unit. *Crystallogr. Rev.*, in press. [58]

Grell, J., Bernstein, J. and Tinhofer, B. (1999). Graph-set analysis of hydrogen-bond patterns: some mathematical concepts. *Acta Crystallogr. B*, **55**, 1030–43. [55, 57, 58]

Griesser, U. (2000). Private communication to the author. [115, 125, 126f]

Griesser, U. J. and Burger, A. (1993). The polymorphic drug substances of the European Pharmacopoeia. Part 8. Thermal analytical and FTIR-microscopic investigations of etofylline crystal forms. *J. Pharm. Sci.*, **61**, 133–43. [128]

Griesser, U. J. and Burger, A. (1999). Statisitical aspects of the occurrence of crystal forms among organic drug substances. *Abstracts, XVIII Congress and General Assembly of the International Union of Crystallography*, Glasgow, Scotland. [241, 242]

Griesser, U. J., Burger, A. and Mereiter, K. (1997). The polymorphic drug substances of the European pharmacopoeia. 9. Physicochemcial properties and crystal structure of acetazolamide crystal forms. *J. Pharma. Sci.*, **86**, 352–8. [55, 58]

Griesser, U. J., Szelagiewicz, M., Hofmeier, U. C., Pitt, C. and Cianferani, S. (1999). Vapor pressure and heat of sublimation of crystal polymorphs. *J. Therm. Anal. Calorim.*, **57**, 45–60. [35]

Griesser, U. J., Auer, M. E. and Burger, A. (2000). Micro-thermal analysis, FTIR- and Raman-microscopy of (R,S)-proxyphyliine crystal forms. *Microchem. J.*, **65**, 283–92. [250, 252]

Griffiths, C. H. and Monahan, A. R. (1976). Polymorphism and spectroscopic characterization of an azo-pigment. *Mol. Cryst. Liq. Cryst.*, **33**, 175–87. [78, 260t]

Groth, P. H. R. (1906a). *An Introduction to chemical crystallography* (trans. H. Marshall), pp. 28–31. Gurney & Jackson, London. [23, 75]

Groth, P. H. R. (1906b). *Chemische kristallographie. Erster Teil. Elemente— anorganische verbindungen ohne salzcharakter—einfache und komplexe halogenide, cyanide und azide der metalle, nebst den zugehörgen alkylverbindungen* [Chemical crystallography. First part. Elements—nonionic inorganic compounds—simple and complex metallic halides, cyanides and azides, together with their accompanying alkyl derivatives]. W. Engelemann, Leipzig. [10, 18, 75]

Groth, P. H. R. (1908). *Chemische kristallographie. Zweiter teil.Die anorganischen oxo- und sulfosalze* [Chem. crystallography. Second part. Inroganic oxy- and sulfo salts]. W. Engelemann, Leipzig. [10, 18]

Groth, P. H. R. (1910). *Chemische kristallographie. Dritter teil. Aliphatische und Hydroaromatische Kohlenstoffverbindungen* [Chemical crystallography. Third part. Aliphatic and hydroaromatic carbon compounds]. W. Engelemann, Leipzig. [10, 25]

Groth, P. H. R. (1917). *Chemische kristallographie. Vierter teil. Aromatische kohlenstoffverbindungen mit einem benzolringe* [Chemical crystallography. Fourth part. Aromatic hydrocarbons with only one benzene ring]. W. Engelemann, Leipzig. [10, 25, 291, 293f]

Groth, P. H. R. (1919). *Chemische kristallographie. Fünfter Teil. Aromatische kohlenstoffverbindungen mit meheren benzolringen heterocyclische verbindungen.*

[Chemical crystallography. Fifth part. Aromatic hydrocarbons with multiple benzene rings. Heterocyclic compounds] pp. 104–5. Engelmann, Leipzig. [10, 13f, 25, 212]

Grubb, P. W. (1986). *Patents in chemistry and biotechnology*. Clarendon Press, Oxford. [297]

Grunenberg, A., Henck, J.-O. and Siesler, H. W. (1996). Theoretical derivation and practical applications of energy/temperature diagrams as an instrument in preformulation studies of polymorphic drug substances. *Int. J. Pharm.*, **129**, 147–58. [33, 34f, 35f, 38, 106, 132]

Grunenberg, A., Keil, B. and Henck, J.-O. (1995). Polymorphism in binary mixtures, as exemplified by nimodipine. *Int. J. Pharm.*, **118**, 11–21. [135f, 136t]

Gruno, M., Wulff, H. and Pflegel, P. (1993). Polymorphism of benzocaine. *Pharmazie*, **48**, 834–7. [5]

Gu, C. H. and Grant, D. J. W. (2001). Estimating the relative stability and hydrates from heats of solution and solubility data. *J. Pharm. Sci.*, **90**, 1277–87. [111]

Gu, W. (1993). Factor analysis of phase transitions and conformational changes in pentaerythritol tetrastearate. *Anal. Chem.*, **65**, 823–7. [125]

Gu, X. J. and Jiang, W. (1995). Characterization of polymorphic forms of fluconazole using Fourier transform Raman spectroscopy. *J. Pharm. Sci.*, **84**, 1438–41. [132]

Gudmunsdottir, A. D., Lewis, T. J., Randall, L. H., Scheffer, J. R, Rettig, S. J., Trotter, J. and Wu, C. H. (1996). Geometric requirements of hydrogen abstractabity and 1,4-biradical reactivity in the Norrish/Yang type II reaction: Studies based on the solid state photochemistry and X-ray crystallography of medium-sized ring and macrocyclic diketones. *J. Am. Chem. Soc.*, **118**, 6167–84. [237]

Guillory, J. K. and Erb, D. M. (1985). Using solution calorimetry to quantitate binary mixtures of three crystalline forms of sulfamethoxazole. *Pharm. Manufacturing*, **2**, 30–3. [109]

Guillory, J. K. (1999). Generation of polymorphs, hydrates, solvates, and amorphous solids. In *Physical characterization of pharmaceutical solids* (ed. H. G. Brittain), pp. 183–226. Marcel Dekker, New York. [254]

Guinot, S. and Leveiller, F. (1999). The use of MTDSC to assess the amorphous phase content of a micronised drug substance. *Int. J. Pharm.*, **192**, 63–75. [255]

Guo, Y. S., Byrn, S. R. and Zografi, G. (2000). Physical characteristics and chemical degradation of amorphous quinapril hydrochloride. *J. Pharm. Sci.*, **89**, 128–43. [254]

Guzman, J. and Largo-Cabrerizo, J. (1978). Polymorphism in the 2-(4-morpholinothio)benzothiazole. *J. Heterocycl. Chem.*, **15**, 1531–3. [158]

Haaland, D. M. and Thomas, E. V. (1998). Partial least squares methods for spectral analysis. 1. Relation to other quantitative calibration methods and the extraction of qualitative information. *Anal. Chem.*, **60**, 1193–202. [130]

HABIT95, QCPE program 670. Quantum Chem. Program Exchange (QCPE), Creative Arts Building 181, Indiana University, Bloomington, Indiana 47405.

Hädicke, E. and Graser, F. (1986a). Structures of eleven perylene-3,4:9,10-bis(dicarboxymid) pigments. *Acta Crystallogr. C*, **42**, 189–95. [260t, 263, 264]

Hädicke, E. and Graser, F. (1986b). Structures of three perylene-3,4:9,10-bis(dicarboxymid) pigments. *Acta Crystallogr. C*, **42** 195–8. [263]

Hadjoudis, E., Kariv, E. and Schmidt, G. M. J. (1972). Solid-gas interactions. II. Solid-state cis-trans isomerization of alkoxycinnamic acids in iodine vapor. *J. Chem. Soc., Perkin Trans. II*, 1056–60. [238]

Hagemann, J. W. and Rothfus, J. A. (1993). Transitions of saturated monoacid triglycerides—modeling conformational change at glycerol during alpha ->beta' -> beta transformation. *J. Am. Oil Chem. Soc.*, **70**, 211–7. [28]

Hager, O., Foces-Foces, C., Llamas-Saiz, A. L. and Weber, E. (1998). Temperature-dependent phase transitions in two crystalline host-guest complexes derived from mandelic acid. *Acta Crystallogr. B*, **54**, 82–93. [237]

Hagler, A. T. and Bernstein, J. (1978). Conformational Polymorphism II. Crystal energetics by computational substitution. Further evidence for the sensitivity of the method. *J. Am. Chem. Soc.*, **100**, 6349–53. [181]

Hagler, A. T. and Lifson, S. (1974). Energy functions for peptides and proteins .II. Amide hydrogen bond and calculation of amide crystal properties. *J. Am. Chem. Soc.*, **96**, 5327–35. [153, 167]

Hagler, A. T., Huler, E. and Lifson, S. (1974). Energy functions for peptides and proteins. I. Deviation of a consistent force field including the hydrogen bond from amide crystals. *J. Am. Chem. Soc.*, **96**, 5319–27. [153]

Hahn, T. and Klapper, H. (1992). Point groups and crystal classes, Chapter 10, Vol. A: *Space group symmetry, International Tables for Crystallography* (3rd ed.) (ed. T. Hahn) pp. 752–892. International Union of Crystallography, Kluwer Academic Publishers, Dordrecht, The Netherlands. [46]

Hähnle, R. and Optiz, K. (1976). Hoechst AG Patent DE 2 524 187. [260t]

Haleblian, J. and McCrone, W. C. (1969). Pharmaceutical applications of polymorphism. *J. Pharm. Sci.*, **58**, 911–29. [4, 5, 26, 240]

Haleblian, J. K. (1975). Characterization of habits and crystalline modification of solids and their pharmaceutical applications. *J. Pharm. Sci.*, **64**, 1269–88. [240]

Hall, P. (1971). Thermal decomposition and phase transitions in solid nitramides. *Trans. Faraday Soc.*, **67**, 556–62. [283, 286]

Hall, R. C., Paul, I. C. and Curtin, D. Y. (1988). Structures and interconversion of polymorphs of 2,3-dichloroquinazirin. Use of second harmonic generation to follow the change of a centrosymmetric to a polar structure. *J. Am. Chem. Soc.*, **110**, 2848–54. [213]

Hall, S. R., Kolinsky, P. V., Jones, R., Allen, S., Gordon, P., Bothwell, B., Bloor, D., Norman, P. A., Hursthouse, M., Karaulov, A., Baldwin, J., Goodyear, M. and Bishop, D. (1986). Polymorphism and nonlinear optical activity in organic crystals. *J. Cryst. Growth*, **79**, 745–51. [207, 212]

Haller, T. M., Rheingold, A. L. and Brill, T. B. (1983). The structure of the complex between octahydro-1,3,5,7-tetranitro-1,3,5,7-tetrazocine (HMX) and N, N-dimethylformamide (DMF), $C_4H_8N_8O_8.C_3H_7NO$. A second polymorph. *Acta Crystallogr. C*, **39**, 1559–63. [279]

Haman, C. and Wagner, H. (1971). Textures of evaporated copper phthalocyanine films. *Kristall und Technik*, **6**, 307–20. [268]

Hamilton, W. C. and Ibers, J. A. (1968) *Hydrogen bonding in solids*, pp. 19–21. W. A. Benjamin, New York, U. S. A. [55]

Hamzaoui, F., Baert, F. and Wojcik, G. (1996). Electron-density study of *m*-nitrophenol in the orthorhombic structure. *Acta Crystallogr. B* **52**, 159–64. [81, 212]

Han, J., Gupte, S. and Suryanarayanan, R. (1998). Application of pressure differential scanning calorimetry in the study of pharmaceutical hydrates. II. Ampicillin trihydrate. *Int. J. Pharm.*, **170**, 63–72. [148, 251]

Hancock, B. C. and Parks, M. (2000). What is the true solubility advantage for amorphous pharmaceuticals? *Pharm. Res.*, **17**, 397–404. [253]

Hancock, B. C. and Zografi, G. (1997). Characteristics and significance of the amorphous state in pharmaceutical systems. *J. Pharm. Sci.*, **86**, 1–12. [253]

Hantzsch, A. (1907*a*). Concerning chromoisomers. *Angew. Chem.*, **20**, 1889–92. [214]

Hantzsch, A. (1907*b*). Yellow, red, green, violet and colorless salts from dinitro compounds. *Ber. Dtsch. Chem. Ges.*, **40**, 1533–55. [214]

Hantzsch, A. (1908). Concerning chromo isomers. *Zeitschrift und Zentralblatt für Technische Chem.*, **20**, 1889. [214]

Hao, Z. and Iqbal, A. (1997). Some aspects of organic pigments. *Chem. Soc. Rev. [London]*, **26**, 203–13. [257, 258, 261t, 274]

Hao, Z., Schloeder, I. and Iqbal, A. (1999*a*). 1,4-Diketo-3,6-bis-4-chloro-phenyl-pyrrolopyrrole in β-modification—useful as pigment with more yellow-red nuance than α-modification is. Prepd. by acid hydrolysis of soluble carbamate. Derivation and preparation by cooling. Ciba-Geigy Ltd. EP 690058 A1. [271]

Hao, Z., Iqbal, A. and Herren, F. (1999*b*). 1,4-Diketo-3,6-bis-4-chloro-phenyl-pyrrolopyrrole in β-modification—useful as pigment with more yellow-red nuance than α-modification. Prepd. by acid hydrolysis of soluble carbamate. Derivation and preparation by cooling. Ciba-Geigy Ltd. EP 690057 B1. [271]

Hao, Z., Iqbal, A. and Herren, F. (1999*c*). 1,4-Diketo-3,6-bis-4-chloro-phenyl-pyrrolopyrrole in β and γ modification, prepd. by acid hydrolysis of soluble carbamate. Derivation and preparation by cooling. Ciba-Geigy Ltd. EP 690059 B1. [271]

Harano, Y. and Oota, K. (1978). Measurement of crystallization of potassium bromate from its quiescent aqueous solution by differential scanning calorimeter. Homogeneous nucleation rate. *J. Chem. Eng. J.*, **11**, 159–61. [70]

Hardy, G. E., Zink, J. I., Kaska, W. C. and Baldwin, J. C. (1978). Structure and triboluminescence of polymorphs of $(PH_3P)_2C$. *J. Am. Chem. Soc.*, **100**, 8001–2.

Hardy, G. E., Kaska, W. C., Chandra, B. P. and Zink, J. I. (1981). Triboluminescence-structure relationships in polymorphs of hexaphenylcarbodiphosphorane and anthranilic acid, molecular crystals, and salts. *J. Am. Chem. Soc.*, **103**, 1074–9. [59, 221]

Hargittai, I. (1992). Gas-phase electron diffraction. In *Accurate molecular structures—their determination and importance* (ed. A. Domenicano and I. Hargittai), Oxford University Press, Oxford, pp. 95–125.

Hargittai, I. and Levy, J. B. (1999). Accessible geometric changes. *Struct. Chem.*, **10**, 387–9. [152, 152t]

Harper, J. K. and Grant, D. M. (2000). Solid state C-13 chemical shift tensors in terpenes. 3. Structural characterization of polymorphous verbenol. *J. Am. Chem. Soc.*, **122**, 3708–14.

Harris, K. D. M. and Thomas, J. M. (1991). Probing polymorphism and reactivity in the organic solid-state using ^{13}C NMR spectroscopy: studies of *p*-formyl-trans-cinnamic acid. *J. Solid State Chem.*, **93**, 197–205. [140]

Harris, R. K. (1985). Quantitative aspects of high-resolution solid-state nuclear magnetic resonance spectroscopy. *The Analyst*, **110**, 649–55. [142]

Harris, R. K. (1993). State-of-the-art for solids. *Chem. Br.*, 601–4. [133]

Harris, R. K., Yeung, R. R., Lamont, R. B., Lancester, R. W., Lynn, S. M. and Staniforth, S. E. (1997). 'Polymorphism' in a novel antiviral agent: Lamivudine. *J. Chem. Soc., Perkin Trans. 2*, 2653–9. [128t]

Hartauer, K. J., Miller, E. and Guillory, J. K. (1992). Diffuse reflectance infrared Fourier transform spectroscopy for the quantitative analysis of mixtures of polymorphs. *Int. J. Pharmacol.*, **85**, 163–74. [129, 130]

Hartley, H. (1902). *Polymorphism. An historical account.* pp. 15. Holywell Press, Oxford. [19, 21, 22]

Hartman, P. M. and Hartman, P. (eds.) (1973). *Crystal growth—an introduction.* North Holland, Amsterdam. [47]

Hartshorne, N. H. and Stuart, A. (1960). *Crystals and the polarizing microscope.* Arnold, London. [95]

Hartshorne, N. H. and Stuart, A. (1964). *Practical optical crystallography*, pp. 1–46. American Elsevier, New York, N. Y. [48, 95]

Hasegawa, M. (1986). Topochemical photopolymerization of diolefin crystals. *Pure Appl. Chem.*, **58**, 1179–88. [236]

Haselbach, E. and Heilbronner, E. (1968). Electronic structure and physical chemical properties of azo compounds. Part XIV. The conformation of benzalaniline. *Helv. Chim. Acta*, **51**, 16–34. [226]

Hatada, M., Jancarik, J., Graves, B. and Kim, S. -H. (1985). Crystal structure of aspartame, a peptide sweetener. *J. Am. Chem. Soc.*, **107**, 4279–82. [305, 307]

Hayashi, Y. and Sakaguchi, I. (1981). Dioxazine violet pigment with stable β-crystalline form—prepared by heating metastable dioxazine violet pigment with aromatic cdp. which is sparingly soluble in water. Sunmitomo Chem. Co. Ltd Patent DE 3 031 444 A1. [260t]

Hayashi, Y. and Sakaguchi, I. (1982). Stable dioxazine violet pigment production from metastable form—by heating aqueous suspension with aliphatic or alicyclic ketone or acetate. Sunmitomo Chem. Co. Ltd Patent DE 3 211 607 A1. [260t]

Hayward, I. P., Batchelder, D. N. and Pitt, G. D. (1994). Applications of Raman spectroscopy and imaging to industrial quality control. *The Analyst*, **22**, M22–M28. [133]

Hegedus, B. and Görög, S. (1985). The polymorphism in cimetidine. *J. Pharm. Biomed. Anal.*, **3**, 303–13. [73, 255]

Hemminger, W. and Höhne, G. (1984). Calorimetry: fundamentals and practice. Verlag Chemie, Weinheim. [111]

Henck, J.-O. and Kuhnert-Brandstätter, M. (1999). Demonstration of the terms enantiotropy and monotropy in polymorphism research exemplified by flurbiprofen. *J. Pharm. Sci.*, **88**, 103–8. [33]

Henck, J.-O., Bernstein, J., Ellern, A. and Boese, R. (2001). Disappearing and reappearing polymorphs. The case of benzocaine : picric acid. *J. Am. Chem. Soc.*, **123**, 1834–41. [252]

Henck, J.-O., Finner, E. and Burger, A. (2000). Polymorphism of tedisamil dihydrochloride. *J. Pharm. Sci.*, **89**, 1151–9. [249]

Henck, J.-O., Griesser, U. J. and Burger, A. (1997). Polymorphism of drug substances. An economic challenge? *Die Pharmazeutische Industrie*, **59**, 165–9. [5, 240]

Henck, J.-O., Bernstein, J., Ellern, A. and Boese, R (2001). Disappearing and reappearing polymorphs. The case of benzocaine : picric acid. *J. Am. Chem. Soc.*, **123**, 1834–41. [92, 103f]

Hendrickson, B. A., Preston, M. S. and York, P. (1995). Processing effects on crystallinity of cephalexin—characterization by vacuum microbalance. *Int. J. Pharm.*, **118**, 1–10. [5]

Henson, B. F., Asay, B. W., Sander, R. K., Son, S. F., Robinson, J. M. and Dickson, P. M. (1999). Dynamic measurement of the HMX beta-delta phase transition by second harmonic generation. *Phys. Rev. Lett.*, **82**, 1213–16. [280]

Herbst, W. and Hunger, K. (1997). *Industrial organic pigments*, 2nd edn. VCH, Weinheim. [257, 258, 259, 262t, 263, 265, 266, 267f, 268, 269, 270, 271]

Herbstein, F. H. (1971). Crystalline π-molecular compounds. Chemistry, spectroscopy and crystallography. In *Perspectives in structural chemistry*, Vol. 4. (ed. J. D. Dunitz and J. A. Ibers), pp. 166–395. John Wiley & Sons, New York. [189, 195]

Herbstein, F. H. (2000). How precise are measurements of unit-cell dimensions from single crystals? *Acta Crytallogr. B*, **56**, 547–57. [147]

Herbstein, F. H. (2001) Varieties of polymorphism. In *Advances in structure analysis*. eds R. Kucel and J. Hasek, Czech and Slovak Crystallographic Association, Prague, pp. 114–54. (http://www.xray.ce/ecm/book/) [8, 29, 113, 115t, 49]

Herrera, M. L. and Rocha, F. J. M. (1996). Effects of sucrose ester on the kinetics of polymorphic transition in hydrogenated sunflower oil. *J. Am. Oil Chem. Soc.*, **73**, 321–6. [28]

Hertel, E. (1931). Kristallstruktur. *Z. Electrochem.*, **37**, 536–8. [234]

Hertel, E. and Römer, G.H. (1930). Der strukturelle Aufbau organischer Molekülverbindungen mit zwei- und eindimensionalen Abwechselungsprinzip. *Z.*

Phys. Chem. Abteilung B: Chemie der Elementarprozesse, Aufbau der Materie, **11**, 77–89. [288]

Herz, A. (1974). Dye-dye interactions of cyanines in solution and at silver bromide surfaces. *Photogr. Sci. Eng.*, **18**, 323–35. [229]

Hibbert, H. (1914). Nitroglycerin. *Z. Sprengstoffw.*, **9**, 305–7. [284]

Higginson, P. (2000) Automated crystal screening technology. Presented at 2nd International Symposium on Polymorphism and Crystallisation—Chemical Development Issues, Chester, UK. (Organized by Scientific Update, UK) [92]

Hill, S. A., Jones, K. H., Seager, H. and Taskis, C. B. (1975). Dissolution and bioavailability of the anhydrate and trihydrate forms of ampicillin. *J. Pharm. Pharmacol.*, **27**, 594–8. [246]

Hirshfeld, F. L. and Mirsky, K. (1979). The electrostatic term in lattice-energy calculations: acetelene, carbon dioxide and cyanogen. *Acta Crytallogr. A*, **35**, 366–70. [166]

Hitchens, A. L. and Garfield, F. M. (1941). Basic lead styphnate and a process of making it. Western Cartridge U.S. Patent 2 265 230. [294]

Hoard, M. S. and Elakovich, S. D. (1996). Grinding-induced polymorphism in the aporphine alkaloid magnoflorine. *Phytochemistry*, **43**, 1129–33. [128]

Hoechst (1982). European Patent 97 913. [260t]

Hoechst (1988). European Patent 320 774. [261t]

Hofmann, D. M. W. and Lengauer, T. (1997). A discrete algorithm for crystal structure prediction of organic molecules. *Acta Crystallogr. A*, **35**, 225–35. [185]

Hofmann, D. M. W. and Lengauer, T. (1999). Prediction of crystal structures of organic molecules. *J. Mol. Struct. (Theochem)*, **474**, 13–23. [185]

Höhener, A. and Smith, R. E. (1987). New crystalline form of sulphonated diazo dye—for dyeing and printing wool, polyamide, etc., more easily filtered than amorphous form. Ciba-Geigy AG Patent EP 222 697. [260t]

Holden, J. R., Du, Z. and Ammon, H. L. (1993). Prediction of possible crystal structures for C-, H-, N-, O-, and F-containing organic compounds. *J. Comput. Chem.*, **14**, 422–37. [183]

Holston Defense Corporation (Eastman Kodak, Kingsport, TN) (1962). Phys. and chemical properties of RDX and HMX. Control No. 20-P-26 Series B. [278, 279t, 280]

Hong, W., Barton, R. J., Robertson, B. E., Weil, J. A. and Brown, K. C. (1991). Crystal and molecular structures of 2 polymorphs of 2,2-di(*p*-nitrophenyl)-1-picrylhydrazine dichloromethane. *Can. J. Chem.*, **68**, 1306–14. [203]

Honigmann, B. (1964). Modification and physical-particle shapes of organic pigments. *Farbe Lack*, **70**, 789–91. [268]

Honigmann, B. (1966). Crystal properties of organic pigments. *J. Paint Technol.*, **38**, 77–84. [258]

Honigman, B., Lenné, H. U. and Schrödel, R. (1965). Relations between the structures of the modifications of the platinum and copper phthalocyanines and some chloride derivatives. *Z. Kristallogr.*, **122**, 185–205. [268]

Horn, D. and Honigman, B. (1974). Polymorphism of copper phthalicyanine. *XII. Fatipec Kongress*, Garmisch-Parkinkirchen, Germany, Mai, 1974, pp. 181–9. [266, 267, 268, 269f]

Hostettler, M., Birkedal, H. and Schwarzenbach, D. (2001). Polymorphs and structures of mercuric iodide. *Chimia*, **55**, 541–545. [20]

Huang, K.-S., Britton, D., Etter, M. C. and Byrn, S. R. (1995). Polymorphic characterization and structural comparisons of the non-linear optically-active and inactive forms of two polymorphs of 1,3-bis(m-nitrophenyl)urea. *J. Mater. Chem.*, **5**, 379–83. [77]

Huang, K.-S., Britton, D., Etter, M. C. and Byrn, S.R. (1996). Synthesis, polymorphic characterization and structural comparisons of the non-linear optically active and inactive forms of polymorphs of 3-(nitroanilino)cycloalk-2-en-1-ones. *J. Mater. Chem.*, **6**, 123–9. [212]

Huber, P. W. (1991). *Galileo's revenge: junk science in the courtroom.* Basic Books Division of Harper Collins Publishers, New York. [66, 297]

Hughes, O. S., Hursthouse, M. B., Threlfall, T. and Tavener, S. (1999). A new polymorph of sulfathiazole. *Acta Crystallogr. C*, **55**, 1831–3. [81]

Hulliger, J. (1994). Chemistry and crystal growth. *Angew. Chem. Int. Ed. Engl.*, **33**, 143–62. [256]

Hultgren, R. (1936). An x-ray study of symmetrical trinitrotoluene and cyclotrimethylnitramine. *J. Chem. Phys.*, **4**, 84. [288]

Hunger, K. (1999). The effect of crystal structure on color application properties of organic pigments. *Rev. Prog. Coloration and Related Topics*, **29**, 71–84. [258, 259, 268]

ICDD (2001). *Power diffraction file.* International Center for Diffraction Data, Campus Boulevard, Newton Square, Pa. (http://www.icdd.com). Available software for powder diffraction may be found through this web site: http://www.icdd.com/resources/websites.htm. [17, 18, 119]

Imaeda, K., Enoki, T., Mori, T., Inokuchi, H., Sasaki, M., Nakasuji, K. and Murata, I. (1989). Electronic properties of new organic conductors based on 2,7-bis(methylthio)-1,6-dithiapyrene (MTDTRY) with TCNQ and *p*-benzoquinone derivatives. *Bull. Chem. Soc. Jpn*, **62**, 372–9. [191]

Imahori, S. and Hirako, S. (1976). β-crystalline phase perylene pigment—prepn by condensing with xylidine derived in organic solvent. Mitsubishi Chemical Industries Co. Ltd. Patent JP 51-7025; *Chem. Abstr.*, **84**, 166251*s*. [261t]

Inabe, T., Goto, H., Fujinawa, T., Ahashi, H., Ogata, H., Miyajima, S. and Maruyama, Y. (1996). Unusual electrical properties of 1,6-diaminopyrene charge-transfer complex crystals. *Mol. Cryst. Liq. Cryst.*, **284**, 283–90. [82, 193]

International Commission on Harmonization. (1996). Q2B validation of analytical procedures: methodology. [122]

International Tables for X-ray crystallography. Vol. A. Space-group symmetry. (1987). Hahn, T. (ed). 2nd edition, revised. D. Reidel: Dordrecht. [115t, 115]

Ismailsky, W. A. (1913). Thesis, Technische Hochschule Dresden, 1913. [257]

Ito, S., Nishimura, M., Kobayashi, Y., Itai, S. and Yamamoto, K. (1997). Characterization of polymorphs and hydrates of GK-128, a serotonin(3) receptor antagonist. *Int. J. Pharm.*, **151**, 133–43. [5]

Ito, T. (1950). *X-ray studies on polymorphism*, pp. 111–21. Tokyo: Maruzen Co. Ltd. [288]

IUCr (2001) Available software for powder diffraction may be found through this web site: http://www.iucr.ac.uk/sincris-top/logiciel/. [119]

Jacewicz, V. W. and Nayler, J. H. C. (1979). Can metastable crystal forms 'disappear'? *J. Appl. Cryst.*, **12**, 396–7. [91]

Jacob, G., Toupet, L., Ricard, L. and Cagnon, G. (1999), Private communication to CSD. [282]

Jacobsen, C. S. and Torrance, J. B. (1983). Behavior of charge-transfer absorption upon passing through the neutral-ionic phase transition. *J. Chem. Phys.*, **78**, 112–5. [195, 196]

Jaffe, E. E. (1992). Quinacridone and some of its derivatives. *JOCCA-Surf. Coat. Int.*, **75**, 24–31. [259]

Jaffe, E. E. (1996). *Encyclopedia of chemical technology*, 4th edn, Vol. 19. Wiley, New York. [258]

Jakeway, S. C., de Mello, A. J. and Russell, E. L. (2000). Miniaturized total analysis systems for biological analysis. *Fresenius J. Anal. Chem.*, **366**, 525–39. [147]

Jameson, G. B., Oswald, H. R. and Beer, H. R. (1984). Structural phase transition in dihalo(N, N'-di-*tert*-butyldiazabutadiene)nickel) complexes. Structures of bis[dibromo(N, N'-di-*tert*-butyldiazabutadiene)nickel] and dibromo(N, N'-di-*tert*-butyldiazabutadiene)nickel. *J. Am. Chem. Soc.*, **106**, 1669–75. [237]

Janczak, J. and Kubiak, R. (1992). Crystal and molecular structures of metal-free phthalocyanines, 1,2-dicyanobenzene tetramers. II. α Form. *J. Alloys Compd*, **190**, 121–4. [270]

Jang, M.-S., Nakamura, T., Takashige, M. and Kojima, S. (1980). Crystal growth and polymorphism of 2,2,6,6-tetramethyl-piperidino oxy. *Jpn J. Appl. Phys.*, **19**, 1413–14. [202]

Jaslovsky, G. S., Egyed, O., Holly, S. and Hegedus, B. (1995). Investigation of the morphological composition of cimetidine by FT-Raman spectroscopy. *Appl. Spectrosc.*, **49**, 1142–5. [130]

Jeffrey, G. A. and Saenger, W. (1991). *Hydrogen bonding in biological structures*. Springer-Verlag, Berlin. [60, 61, 154]

Jelinek, Z. K., Maly, J. and Pizl, J. (1964). Veränderungen der krystallinischen delta-modifikation con indanthren-blau RS unter dem electronenmikroskop. *Proceedings, Third European Regional Conference on Electron Microscopy*, Czechoslovak Academy of Sci., Prague, pp. 357–8. [271]

Jelley, E. E. (1936). Spectral absorption and fluorescence of dyes in the molecular state. *Nature*, **138**, 1009–10. [230]

Jenkins, R. and Snyder, R. L. (1996). Introduction to X-ray powder diffractometry. In *Chemical Analysis: A Series of monographs on analytical chemistry and its applications*, Vol. 138 (ed. J. D. Winefordner), pp. 324–35. Wiley-Interscience, New York. [17, 18, 115, 117, 147]

Jensen, W. B. (1998). Logic, history and the chemistry textbook. *J. Chem. Educ.*, **75**, 817–28. [4]

Jensen, W. B. (2001). Generalizing the phase rule. *J. Chem. Educ.*, **78**, 1567–9. [30]

Joachim, J., Opota, D. O., Joachim, G., Reynier, J. P., Monges, P. and Maury, L. (1995). Effect of solvate formation on lyoavailability of lorazepam during wet granulation. *STP Pharma Sci.*, **5**, 486–8. [5]

Johansson, D., Bergenstaahl, B. and Lundgren, E. (1995). Wetting off at crystals by triglyceride oil and water. 1. The effect of additives. *J. Am. Oil Chem. Soc.*, **72**, 921–31. [28]

Jovanovic, O., Karlovic, D. J. and Jakovljevic, J. (1995). Chocolate precrystallization—a review. *Acta Alimentaria*, **24**, 225–39. [28]

Julian, Y. and McCrone, W. C. (1971). Accurate use of hot stages. *Microscope*, **19**, 225–34. [95]

Kachi, S., Terada, M. and Hashimoto, H. (1998). Effects of amorphous and polymorphs of PF1022A, a new antinematode drug, an Angiostrongylus costaricensis in mice. *Japanese J. Pharmacol.*, **77**, 235–45. [245]

Kahn, O., Garcia, Y., Létard, J. F. and Mathoniere, C. (1999). Hysteresis and memory effect in supramolecular chemistry. In *Supramolecular engineering of synthetic metallic materials*, Vol. C518 of *NATO ASI Series* (ed. J. Veciana, C. Rovira and D. B. Amabilino), pp. 127–44. Kluwer Academic Publishers, Dordrecht. [197, 198]

Kahr, B. and McBride, M. (1992). Optically anomalous crystals. *Angew. Chem., Int. Ed. Engl.*, **31**, 1–26. This reference contains a historical account of the phenomenon of optical anomalies, a field which went dormant for nearly half a century for reasons similar to those involving activity in the field of polymorphism. [10, 214]

Kalinkova, G. N. and Hristov, S. (1996). Infrared spectroscopic and thermal-analysis of different modifications of calcium valproate. *Vib. Spectrosc.*, **11**, 1443–9. [5]

Kalinkova, G. N. and Stoeva, S. (1996). Polymorphism of azlocillin sodium. *Int. J. Pharmacol.*, **135**, 111–4. [245]

Kalipharma, Inc. v. Bristol-Meyers Company, No. 88 CIV. 4640; 707 F. Supp. 741 [301, 303]

Kálmán, A., Parkanyi, L. and Argay, Gy. (1993*a*). On the isostructuralism of organic molecules in terms of Kitaigorodskii's early perceptions. *Acta Chim.-Hungar. Models Chem.*, **130**, 279–98. [271]

Kálmán, A., Parkanyi, L. and Argay, Gy. (1993*b*) Classification of the isostructurality of organic molecules in the crystalline state. *Acta Crystallogr. B*, **49**, 1039–49. [271]

Kaneko, F., Sakashita, H., Kobayashi, M., Kitagawa, Y. Matsuura, Y. and Suzuki, M. (1994*a*). Double-layered polytypic structure of the B form of octadecanoic acid, $C_{18}H_{36}O_2$. *Acta Crystallogr. C*, **50**, 245–7. [129]

Kaneko, F., Sakashita, H., Kobayashi, M., Kitagawa, Y., Matsuura, Y. and Suzuki, M. (1994*b*). Double-layered polytypic structure of the E form of octadecanoic acid, $C_{18}H_{36}O_2$. *Acta Crystallogr. C*, **50**, 247–50. [129]

Kaneko, K., Shirai, O., Miyamoto, H., Kobayashi, M. and Suzuki, M. (1994c). Oblique infrared transmission spectroscopic study on the E → C and B → C phase transitions of stearic acid: effects on polytypic structure. *J. Phys. Chem.*, **98**, 2185–91. [129]

Kanters, J. A., de Koster, A., van Geerstein, M. and van Dijck, L. A. (1985). Structures of modification I of cortisone acetate, 21-acetoxy-17α-hydroxy-4-pregnene-3,11,20-trione, $C_{23}H_{30}O_6$. *Acta Crystallogr. C*, **41**, 760–3. [160]

Karfunkel, H. R. and Gdanitz, R. J. (1992). *Ab initio* prediction of possible crystal structures for general organic molecules. *J. Comput. Chem.*, **13**, 1171–83. [185]

Karfunkel, H. R. and Leusen, F. J. J. (1992). Practical aspects of predicting possible crystal structures on the basis of molecular information only. *Speedup*, **6**, 43–50. [185]

Karfunkel, H. R., Leusen, F. J. J. and Gdanitz, R. J. (1994). The *ab initio* prediction of yet unknown molecular crystal structures by solving the crystal packing problem. *J. Comput.-Aided Mol. Des.*, **1**, 177–85. [185]

Karfunkel, H., Wilts, H., Hao, Z. M., Iqbal, A., Mizuguchi, J. and Wu, Z. J. (1999). Local similarity in organic crystals and the non-uniqueness of X-ray powder patterns. *Acta Crystallogr. B*, **55**, 1075–89. [115]

Karpowicz, R. J. and Brill, T. B. (1982). The β-δ transformation of HMX: its thermal analysis and relationship to propellants. *Am. Inst. Aeronaut. Astronaut. J.*, **20**, 1586–91. [280]

Karpowicz, R. J., Sergio, S. T. and Brill, T. B. (1983). β-Polymorph of hexahydro-1,3,5-trinitro-*s*-triazine. A Fourier Transform infrared spectroscopy study of an energetic material. *Ind. Eng. Chem. Prod. Res. Dev.*, **22**, 363–5. [283]

Karpowicz, R. J. and Brill, T. B. (1984). Comparison of the molecular structure of hexahydro-1,3,5-trinitro-*s*-triazine in the vapor, solution and solid phases. *J. Phys. Chem.*, **88**, 348–52. [283]

Katan, C. and Koenig, C. (1999). Charge-transfer vaiation caused by symmetry breaking in a mixed-stack organic compound: TTF-2, 5Cl₂BQ. *J. Phys. Condens. Matter*, **11**, 4163–77. [195]

Kato, R., Kobayashi, H., Kobayashi, A., Moriyama, S., Nishio, Y., Kajita, K. and Sasaki, W. (1987). A new ambient-pressure superconductor, κ-(BEDT-TTF)₂I₃. *Chem. Lett.*, 507–10. [79, 192]

Katrusiak, A. (1990). High pressure X-ray diffraction study on the structure and phase-transition of 1,3-cyclohexanedione crystals. *Acta Crystallogr. B*, **46**, 246–56. [239]

Katrusiak, A. (1991). High pressure X-ray diffraction studies on organic crystals. *Cryst. Res. Tech.*, **26**, 523–31. [239]

Katrusiak, A. (1995). High pressure X-ray diffraction of pentaerythritol. *Acta Crystallogr. B*, **51**, 873–9. [239]

Katrusiak, A. (2000). Conformational transformation coupled with the order-disorder phase transition in 2-methyl-1,3-cyclohexanedione crystals. *Acta Crystallogr. B*, **56**, 872–81. [237]

Katrusiak, A. and Szafranski, M. (1996). Structural phase transitions in guanidimium nitrate. *J. Mol. Struct.*, **378**, 205–23. [220]

Kaupp, G., Schmeyers, J., Haak, M., Marquardt, T. and Hermann, A. (1996). AFM in organic solid state reactions. *Mol. Cryst. Liq. Cryst. Sci. Technol. Sect. A*, **276**, 315–37. [238]

Kawamura, T. (1987). New diazo compound. Crystal polymorph used for yellow diazo pigment—has good transparency and tinting strength when used for printing inks. Toyo Ink Patent JP 62153353. [261t]

Kazmaier, P. M. and Hoffmann, R. (1994) A theoretical study of crystallochromy. Quantum interference effects in the spectra of perylene pigments. *J. Am. Chem. Soc.*, **116**, 9684–91. [264]

Kedem, K. and Bernstein J. (2001). Unpublished. [156]

Kellens, M., Meeussen, W. and Reynaers, H. (1992). Study of the polymorphism and the crystallization kinetics of tripalmitin: a microscopic approach. *J. Am. Oil Chem. Soc.*, **69**, 906–11. [28]

Keller, A. and Cheng, S. Z. D. (1998). The role of metastability in polymer phase transitions. *Polymer*, **39**, 4461–87. [28]

Kelley, P. F., Man, S.-M., Slawin, A. M. Z. and Waring, K. W. (1999). Preparation of a range of copper complexes of diphenylsulfimide: x-ray crystal structures of $[Cu(Ph_2SNH)_4]Cl_2$ and $[Cu_4(\mu_4\text{-}O)(\mu\text{-}Cl)_6(Ph_2SNH)_4]$. *Polyhedron*, **18**, 3173–9. [164]

Kelly, J. J. and Giambalvo, V. A. (1966). 2,9-dimethylquinacridone in a 'yellow' crystalline form. American Cyanamid Co. Patent US 3 264 300. [261t]

Kelly, P. F., Slawin, A. M. Z., Waring, K. W. and Wilson, S. (2001). Further investigation into the isomerism of Cu(II) complexes of diphenylsulfimide: the preparation and x-ray crystal structure of $\{[Cu(Ph_2SNH)_4][CuBr_2(Ph_2SNH)_2]Br_2\}$, the first example of a $\{[CuL_4][CuX_2L_2]X_2\}$ structure. *Inorg. Chim. Acta*, **312**, 201–4. [76]

Kelly, P. F., Slawin, A. M. Z. and Waring, K. W. (1997). Preparation and crystal structures of two forms of trans-$[CuCl_2\{N(H)SPh_2\}_2]$, an unusual example of square planar/pseudo-tetrahedral isomerism in a neutral copper(II)complex. *J. Chem. Soc., Dalton Trans.*, 2853–4. [76, 164]

Kennard, O. (1993). From data to knowledge—use of the Cambridge Structural Database for studying molecular interactions. *Supermol. Chem.*, **1**, 277–95. [16]

Khalafallah, N., Khalil, S. A. and Moustafa, M. A. (1974). Bioavailability of determination of two crystal forms of sulfameter in humans from urinary excretion data. *J. Pharm. Sci.*, **63**, 861–4. [246]

Khalil, S. A., Moustafa, M. A., Ebian, A. R. and Motawi, M. M (1972). GI absorption of two crystal forms of sulfameter in man. *J. Pharm. Sci.*, **61**, 1615–17. [246]

Khoshkhoo, S. and Anwar, J. (1993). Crystallization of polymorphs: the effect of solvent. *J. Phys. D: Appl Phys.*, **26**, B90–B93. [92]

Kiaoka, H. and Ohya, K. (1995). Pseudopolymorphism and phase stability of 7-piperidino-1,2,3,5-tetrahydroimidazo-[2,1-B]quinazolin-2-one (DN-9693). *J. Therm. Anal.*, **44**, 1047–56. [5]

Kiers, C. Th., de Boer, J. L., Olthof, R. and Spek, A. L. (1976). The crystal structure of a 2,2-diphenyl-1-picrylhydrazyl (DPPH) modification. *Acta Crystallogr. B*, **32**, 2297–305. [203]

Kikuchi, K., Honda, Y., Ishikawa, Y., Saito, K. Ikemoto, I., Murata, K., Hiroyuki, A., Ishiguro, T. and Kobayashi, K. (1988). Polymorphism and electrical conductivity of the organic superconductor $(DMET)_2AuBr_2$. *Solid State Commun.*, **66**, 405–8. [191]

Kim, H. S., Jeffrey, G. A. and Rosenstein, R. D. (1968). The crystal structures of the κ form of D-mannitol. *Acta Crystallogr. B*, **24**, 1449–55. [73]

Kim, J.-H., Park, Y.-C., Yim, Y.-J. and Han, J.-S. (1998). Crystallization behavior of hexanaitrohexaazaisowurtzitane at 298 K and quantitative analysis of mixtures by FTIR. *J. Chem. Eng. Jpn.*, **31**, 478–81. [282]

Kim, S. H., Quigley, G. J., Suddath, F. L., McPherson, A., Sneden, D., Kin, J. J., Weinzirl, J. and Rich, A. (1973). X-ray crystallographic studies of polymorphic forms of yeast phenylalanine transfer RNA. *J. Mol. Biol.*, **75**, 421–8. [19]

Kimura, K., Hirayama, F. and Uekama, K. (1999). Characterization of tolbutamide polymorphs (Burger's forms II and IV) and polymorphic transition behavior. *J. Pharm. Sci.*, **88**, 385–91. [244]

Kimura, K., Hirayama, F., Arima, H. and Uekama, K. (2000). Effects of age-ing on crystallization, dissolution, and absorption characteristics of amorphous tolbutamide-2-hydroxypropyl-beta-cyclodextrin complex. *Chem. Pharm. Bull.*, **48**, 646–50. [254]

King, M. V., Bello, J., Pgnatano, E. H. and Harder, D. (1962). Crystalline forms of bovine pancreatic ribonuclease. Some new modifications. *Acta Crystallogr.*, **15**, 144–7. [19]

King, M. V., Magdoff, B. S., Adelman, M. B. and Harker, D. (1956). Crystalline forms of bovine pancreatic ribonuclease: Techniques of preparation, unit cells and space groups. *Acta Crystallogr.*, **9**, 460–5. [19]

Kinoshita, M. (1994). Ferromagnetism of organic radical crystals. *Jpn. J. Appl. Phys.*, **33**, 5718–33. [197, 201]

Kishimoto, S. and Naruse, M. (1987). The 'bundling' phenomenon in aspartame crystallisation. *Chem. Ind.*, 16 Feb., 127–8. [305]

Kishimoto, S. and Naruse, M. (1988). A process development for the bundling crystallization of aspartame. *J. Chem. Technol. Biotechnol.*, **43**, 71–82. [305]

Kishimoto, S., Nagashima, N., Naruse, M. and Toyokura, K. (1989). The 'bundle-like' crystals in aspartame crystallization. *Process Technol. Proc.*, **6**, 511–14. [305, 307]

Kistenmacher, T. J., Emge, T. J., Bloch, A. N. and Cowan, D. O. (1982). Structure of the red, semiconductor form of 4, 4′,5, 5′-tetramethyl-$\Delta^{2,2'}$-bis-1, 3-diselenole and 7,7,8,8-tetracyano-p-quinodimethane. *Acta Crystallogr. B*, **38**, 1193–9. [190]

Kitaigorodskii, A. I. (1961). *Organic chemical crystallography*, Consultants Bureau, New York. [153, 154, 178, 182, 183, 284]

Kitaigorodskii, A. I. (1970). General view on molecular packing. *Advances in Structure Research by Diffraction Methods*, **3**, 173–247. [154]

Kitaigorodskii, A. I. (1973a). *Molecular crystals and molecules*, pp. 2–3, 163–7, 184–90. Academic Press, New York. [32, 166]

Kitaigorodskii, A. I. (1973*b*). General view on molecular packing. In *Advances in structure research by diffraction methods*, Vol. 3 (eds. R. Brill and R. Mason), pp. 173–247. Pergamon Press, Oxford. [32, 166]

Kitamura, M. (1989). Polymorphism in the crystallization of L-glutamic acid. *J. Cryst. Growth*, **96**, 541–6. [70, 74]

Kitamura, S., Chang, L. C. and Guillory, J. K. (1994). Polymorphism of mefloquine hydrochloride. *Int. J. Pharm.*, **101**, 127–44. [5, 243]

Kitaoka, H., Wada, C., Moroi, R. and Hakusui, H. (1995). Effect of dehydration on the formation of levofloxacin pseudopolymorphs. *Chem. Pharm. Bull.*, **43**, 649–53. [5]

Klaproth, M. H. (1798). *Bergmannische J.* I, 294–9. [cited by Partington, J. R. (1964). *A History of Chemistry*, Vol. 4, pp. 203. MacMillan & Co., London] [20]

Klebe, G., Graser, F., Hädicke, E. and Berndt, J. (1989). Crystallochromy as a solid state effect: correlation of molecular conformation, crystal packing and colour in perylene-3,4:9,10-bis(carboximide) pigments. *Acta Crystallogr. B*, **45**, 69–77. [263, 264]

Klein, J., Lehmann, C. W., Schmidt, H. W. and Maier, W. F. (1998). Combinatorial material libraries on the microgram scale with an example of hydrothermal synthesis. *Angew. Chem., Int. Ed. Engl.*, **37**, 3369–72. [148]

Klug, H. P. and Alexander, L. E. (1974). *X-ray diffraction procedures for polycrystalline and amorphous materials.* John Wiley & Sons, New York. [117, 253, 254]

Knudsen, B. I. (1966). Copper phthalocyanine. Infrared absorption spectra of polymorphic modifications. *Acta Chem. Scand.*, **20**, 1344–50.

Kobayashi, A., Kato, R., Kobayashi, H., Moriyama, S., Nishio, Y., Kajita, K. and Sasaki, W. (1987). Crystal and electronic structure of a new molecular superconductor—(BEDT-TTF)$_2$I$_3$. *Chem. Lett.*, 459–62. [192]

Kobayashi, M., Matsumoto, Y., Ishida, A., Ute, K. and Hatada, K. (1994). Polymorphic structures and molecular vibrations of linear oligomers of polyoxymethylene studied by polarized infrared and Raman spectra measured on single crystals. *Spectrochim. Acta A*, **50**, 1605–17. [128]

Kobayashi, N. and Ando, H. (1988*a*). Red monoazo lake pigment Dainippon Inc. Patent JP63-225 666; *Chem. Abstr.*, **110**, 116666*r*. [261t]

Kobayashi, N. and Ando, H. (1988*b*). Monoazo lake for printing inks etc.—prepared by diazotizing 2-amino-5-methyl benzenesulphonic acid, coupling with 2-hydroxy-3-naphtholic acid and adding barium chloride solution. Dainippon Inc. Patent JP63-225 667; *Chem. Abstr.*, **110**, 77504*q*. [261t]

Kobayashi, N. and Ando, H. (1988*c*). New monoazo lake dye —comprises barium salt of 1-(4-methyl-2-sulphonylphenyl azo)2-hydroxy-3-naphthoic acid. Dainippon Inc. Patent JP63-225 668; *Chem. Abstr.*, **110**, 116665*q*. [261t]

Kobayashi, Y., Ito, S., Itai, S. and Yamamoto, K. (2000). Physicochemcial properties and bioavailability of carbamazepine polymorphs and dihydrate. *Int. J. Pharm.*, **193**, 137–46. [248]

Koch, U., Kuhnt, R., Lück, M. and Modrow, H.-W. (1987*a*). C. I. disperse yellow 42 α- or β-modification production—by adding salt to hot aqueous alkaline solution and acidifying at specified temperatures. Martin-Luther Universität, Halle-Wittenberg, Patent DD 251 359. [260t]

Koch, U., Modrow, H.-W. and Wallascheck, G. (1987*b*). New metastable α-modification of C. I. disperse yellow 23 production—by conversion to color stable β-modification by dissolving crude product in aqueous sodium hydroxide, cooling filtered solution etc. Martin-Luther Universität, Halle-Wittenberg, Patent DD 251 359. [260t]

Kofler, A. (1941). Thermal analysis with a hot-stage microscope. *Z. Phys. Chem. Abteilung A*, **187**, 201–10. [102]

Kofler, A. and Kofler, A. (1948). The estimation of the melting points of unstable modifications of organic materials. *Monatshefte Chem.*, **78**, 13–22. [98]

Kofler, L. and Kofler, A. (1954). Thermo-mikro-methoden zur Kennzeichnung organischer Stoffe und stoffgemiche. (Thermomicromethods for the study of organic compounds and their mixtures). Innsbruck, Wagner. A 1980 translation of this book by Walter C. McCrone is available from McCrone Associates, Inc. [8, 15, 95, 98, 240]

Kofler, L. and Winkler, H. (1950*a*). A new method for the analysis of mixtures of pharmaceuticals. *Archivos Farmaceuticos*, **283**, 176–83. [104]

Kofler, L. and Winkler, H. (1950*b*). A rapid method for constructing melting diagrams. II. *Monatshefte Chem.*, **81**, 746–50. [104]

Kohno, Y., Ueda, K. and Imamura, A. (1996). Molecular dynamics simulations of initial decomposition process on the unique N−N bond in nitramines in the crystalline state. *J. Phys. Chem.*, **100**, 4701–12. [279t, 281]

Koizumi, S. and Matsunaga, Y. (1972). Polymorphism and physical properties of the 1,6-diaminopyrene-*p*-chloroanil and related molecular complexes. *Bull. Chem. Soc. Jpn.*, **45**, 423–8. [82]

Kolinsky, P. V. (1992). New materials and their characterization for photonic device applications. *Opt. Eng.*, **31**, 1676–84. [207]

Komai, A., Shirane, N., Ito, Y. and Terui, S. (1977). Copper phthalocyanine in ρ crystal form—which is redder than α type and used for textile printing, resin coloring and in printing ink. Nippon Shokubai Kagaku Kogyo, Ltd. Patent DE 2 659 211. [266]

Komorski, R. A. (ed.) (1986). High resolution NMR spectroscopy of synthetic polymers in bulk. In *Methods in stereochemical analysis series*, Vol. 7. (series ed. A. P. Marchand). VCH, Deerfield Beach, Fla. [133]

Koshelev, V. I., Shelyapin, O. P., Shtanov, N. P., Moroz, V. A., Kovalenk, S. A. and Paramoaova, L. N. (1987). 4, 4′-dibenzamido-1, 1′-dianthraquinonyl and its polymorphic modifications. *J. Appl. Chem. USSR* **60**, 559–62; English translation of *Zh. Prikladnoi Khim.*, (Leningrad), **60**, 596–9. [261t]

Kozin, (1964) *Zhurnal Prikladnoi Khimii*, **5**, 324. [295]

Kraut, J. and Jensen L. H. (1963). Refinement of the crystal structure of adenosine-5′-phosphate. *Acta Crystallogr.*, **16**, 79–88. [159]

Krc, J. Jr. (1955). Crystallographic properties of primary explosives. Quarterly Progress Report No. 2, Armour Research Foundation, Chicago, Illinois. [279t]

Kress, R. B. and Etter, M. C. (1989). Personal communication to the author.

Krishman, K. and Ferraro, J. R. (1982). *Fourier transform infrared spectroscopy*, Vol. 4. Academic Press, New York. [125]

Krishnan, M., Namasivayan, V., Lin, R. S., Pal, R. and Burns, M. A. (2001). Microfabricated reaction and separation systems. *Curr. Opin. Biotechnol.*, **12**, 92–98. [147]

Kritl, A., Srcic, S., Vrecer, F., Sustar, B. and Vojnovic (1996). Polymorphism and pseudopolymorphism—influencing the dissolution properties of the guanine derivative acyclovir. *Int. J. Pharm.*, **139**, 231–5. [5]

Kronick, P. L. and Labes, M. M. (1961). Organic semiconductors. V. Comparison of measurements on single-crystal and compressed microcrystalline molecular complexes. *J. Chem. Phys.*, **35**, 2016–9. [82, 193]

Kronick, P. L., Scott, H. and Labes, M. M. (1964). Composition of some conducting complexes of 1,6-diaminopyrene. *J. Chem. Phys.*, **40**, 890–4. [82, 193]

Kruse, H. and Sommer, K. (1976). Color stable modification of monoazo dye of benzene series—made by heating in aqueous or organic suspension. Hoechst AG Patent DE 2 520 577. [260t]

Kubiak, R. and Janczak, J. (1992) Crystal and molecular structures of metal-free phthalocyanines, 1,2-dicyanobenzene tetramers. I. β Form. *J. Alloys Compd.*, **190**, 117–20. [270]

Kuhnert-Brandstätter, M. (1962). The Kofler method in chemical microscopy. *Microchem. J. Symp. Ser.*, **2**, 221–31. [249]

Kuhnert-Brandstätter, M. (1965). The status and future of chemical microscopy. *Pure Appl. Chem.*, **10**, 133–44. [241]

Kuhnert-Brandstätter, M. (1971). Thermomicroscopy in the analysis of pharmaceuticals. In *International series of monographs in analytical chemistry*, Vol. 45 (ed. R. Belcher and M. Freiser). Pergamon, Oxford. [8, 11, 26, 95, 97f, 97, 98, 99f, 101f, 102, 104, 154, 160, 241, 249]

Kuhnert-Brandstätter, M. (1973). Polymorphie von arzneistoffen und ihre bedeutung in der pharmazeutischen technologie. *Informationsdienst A.P.V.*, **19**, 73–90. [243]

Kuhnert-Brandstätter, M. (1975) Polymorphism in pharmaceuticals. *Pharmazie in Unserer Zeit*, **4**, 131–7. [240]

Kuhnert-Brandstätter, M. (1982). Thermomicroscopy of organic compounds. In *Comprehensive analytical chemistry*, Vol. 16 (ed. G. Svehla), pp. 329–513, Elsevier, Amsterdam. [95, 100f, 101f, 102, 104]

Kuhnert-Brandstätter, M. (1996). Thermoanalytical methods and their pharmaceutical applications. *Pharmazie*, **51**, 443–57. [105, 250]

Kuhnert-Brandstätter, M. and Junger, E. (1976). I. R. Spektroskopische Untersuchungen polymorphen Kristallmodifikationen von Alkoholen und Phenolen. *Spectrochim. Acta Part A*, **23**, 1453–61. [125]

Kuhnert-Brandstätter, M. and Martinek, A. (1965). Statistics on polymorphism of drugs. *Mikrochim. Acta. [Wien]*, **5–6**, 909–19. [241]

Kuhnert-Brandstätter, M. and Moser, I. (1979). On the polymorphism of dapsone and ethambutol dihydrochloride. *Mikrochim. Acta I*, 125–36. [255]

Kuhnert-Brandstätter, M. and Riedmann, M. (1989) Thermal analytical and infrared spectroscopic investigations on polymorphic organic compounds-III. *Mikrochim. Acta [Wien]*, *1*, 373–485. [128, 260t]

Kuhnert-Brandstätter, M. and Sollinger, H. W. (1989). Thermal analytical and infrared spectroscopic investigations on polymorphic organic compounds. V. *Mikrochim. Acta*, **3**, 125–36 and references therein. [32]

Kuleshova, L. N. and Zorky, P. M. (1980). Graphical enumeration of hydrogen-bonded structures. *Acta Crystallogr. Sect. B*, **36**, 2113–15. [55]

Kumar, V. S. S., Kuduva, S. S. and Desiraju, G. R. (1999). Pseudopolymorphs of 3,5-dintrosalicylic acid. *J. Chem. Soc. Perkin Trans.*, 1069–73. [255]

Labana, S. S. and Labana, L. L. (1967). Quinacridones. *Chem. Rev.*, **67**, 1–18. [262]

Lahav, M. and Leiserowitz, L. (1993). Tailor-made auxiliaries for the control of nucleation growth and dissolution of two- and three-dimensional crystals. *J. Phys. D: Appl. Phys.*, **26**, B22–B31. [47, 93, 252]

Lahav, M., Green, B. S. and Rabinovich, D. (1979). Asymmetric synthesis via reactions in chiral crystals. *Acc. Chem. Res.*, **12**, 191–7. [235]

Laine, E., Pirttimaki, J. and Rajala, R. (1995). Thermal studies on polymorphic structures of ibopamin. *Thermochim. Acta*, **248**, 205–16. [256]

Lancaster, R. (1999). Lamuvidine—development with a sting in the tail! Presented at the 1st International Symposium on Aspects of Polymorphism & Crystallisation—Chemical Development Issues, Hinckley, Leicestershire, UK, June, 1999. Organized by Scientific Update, Wyvern Cottage, High Street, Mayfield, East Sussex, TN20 6AE, UK, http://www.scientificupdate.co.uk [128f, 137f]

Landers, A. G., Brill, T. B. and Marino, R. A. (1981). Electronic effects and molecular motion in β-octahydro-1,3,5,7-teranitro-1,3,5,7-tetrazocine based on ^{14}N nuclear quadrupole resonance spectroscopy. *J. Phys. Chem.*, **85**, 2618–23. [281]

Landers, A. G., Apple, T. M., Dybowski, C. and Brill, T. B. (1985). ^{1}H nuclear magnetic resonance of α-hexahydro-1,3,5-trinitro-*s*-triazine (RDX) and the α-, β-, γ-, and δ-polymorphs of octahydro-1,3,5,7-tetranitro-1,3,5,7-tetrazocine (HMX). *Magn. Reson. Chem.*, **23**, 158–60. [280]

Landes, B. G., Malanga, M. T. and Thill, B. P. (1990). Polymorphism in syndiotactic polystyrene. *Adv. X-ray Anal.*, **33**, 433–44. [148]

Langkilde, F. W., Sjoeboem, J., Tekenbergs-Hjelte, L. and Mark, J. (1997). Quantitative FT-Raman analysis of two crystal forms of a pharmaceutical compound. *J. Pharm. Biomed. Anal.*, **15**, 687–96. [132]

Latimer, W. M. and Rodebush, W. H. (1920). Polarity and ionization from the standpoint of the Lewis theory of valence. *J. Chem. Soc.*, **42**, 1419–33. [53]

Law, K. Y. (1993). Organic photoconductive materials—recent trends and developments. *Chem. Rev.*, **93**, 449–86. [204, 205, 262t, 268]

Law, K. Y., Tarnawsky, J. I. W. and Popovic, Z. D. (1994). Azo pigments and their intermediates—a study of the structure-sensitivity relationship of photogenerating bisazo pigments in bilayer xerographic devices. *J. Imag. Sci. Technol.*, **38**, 118–24. [204]

Le Cointe, M., Lemee-Cailleau, M. H., Cailleau, H., Toudic, B., Toupet, L., Heger, G., Moussa, F., Schweiss, P., Kraft, K. H. and Karl, N. (1995a). Symmetry breaking and structural changes at the neutral-to-ionic transition tetrathiafulvalene-p-chloranil. *Phys. Rev., Sect B*, **51**, 3374–86. [196, 197]

Le Cointe, M., Gallier, J., Cailleau, H., Gourdji, M., Peneau, A. and Guibe, L. (1995b). ^{35}Cl NQR study on TTF-CA crystals: symmetry lowering and hysteresis at the neutral-to-ionic transition. *Solid State Commun.*, **94**, 455–9. [197]

Le Cointe, M., Lemee-Cailleau, M. H., Cailleau, H. and Toudic, B. (1996). Structural aspects of the neutral-to-ionic transition in mixed stack charge-transfer complexes. *J. Mol. Struct.*, **374**, 147–53. [197]

Lee, K. and Gilardi, R. (1993). NTO polymorphs. *Mat. Res. Soc. Symp. Proc.*, **296**, 237–42. [285]

Legrand, J. F., Lajzerowicz, J., Lajzerowicz-Bonneteau, J. and Capiomont, A. (1982). Ferroelastic and ferroelectric phase transition in a molecular crystal: tanane. 3.-From *ab initio* computation of the intermolecular forces to statistical mechanics of the transition. *J. Phys.*, **43**, 1117–25. [202]

Legros, J.-P. and Valade, L. (1988). The highly conducting metal complex TTF[Pd(dmit)$_2$]$_2$. *Solid State Commun.*, **68**, 599–604. [80, 191]

Legros, J.-P. and Kvick, A. (1980). Deformation electron density of α-glycine at 120 K. *Acta Crystallogr. B*, **36**, 3052–9. [56]

Lehmann, M. S., Koetzle, T. F. and Hamilton, W. (1972). Precision neutron diffraction structure determination of protein and nucleic acid components. VIII. Crystal and molecular structure of the β-form of the amino acid, L-glutamic acid. *J. Cryst. Mol. Struct.*, **2**, 225–33. [50]

Lehmann, O. (1877a). Über die dimorphie des hydrochinons und des paranitrophenols. *Z. Kristallogr.*, **1**, 43–8. [22, 95]

Lehmann, O. (1877b). Ueber physikalische isomerie. *Z. Kristallogr.*, **1**, 97–131. [22, 95, 96f]

Lehmann, O. (1885). XX. Mikrokrystallographische Untersuchungen. *Z. Kristallogr.*, **10**, 321. [234]

Lehmann, O. (1888). Molekularphysik, Engelmann, Leipzig. [95, 102]

Lehmann, O. (1891). *Die Krystallanalyse oder die chemische Analyse durch Beobachtung der Krystallbildung mit Hülfe des Mikroskps*. Wilhelm Engelmann, Leipzig. [22, 23, 95]

Lehmann, O. (1910). Das Kristallisationsmikroskop, Vieweg, Braunschweig. [95]

Leonidov, N. B. (2000). History of the development of the concept of polymorphism of chemical materials (brief outline). *Mendeleev Chem. J.*, **41**, 7–22. [*Zhurn. Ross. Khim. Ob-va im. D. I. Mendeleeva*, **41**, 10–21] [19]

LePage, Y. and Donnay, G. (1976). Refinement of the crystal structure of low-quartz. *Acta Crystallogr.*, **32**, 2456–9. [115]

Leung, S. S., Padden, B. E., Munson, E. J. and Grant, D. J. W. (1998a). Solid-state characterization of two polymorphs of aspartame hemihydrate. *J. Pharm. Sci.*, **87**, 501–7. [305]

Leung, S. S., Padden, B. E., Munson, E. J. and Grant, D. J. W. (1998*b*). Hydration and dehydration behavior of aspartame hemihydrate. *J. Pharm. Sci.*, **87**, 508–13. [305]

Leusen, F. J. J. (1996). *Ab initio* prediction of polymorphs. *J. Cryst. Growth*, **166**, 900–3. [185, 263]

Leusen, F. J. J., Pinches, M. R. S., Austin, N. E., Maginn, S. J., Lovell, R., Karfunkel, H. R. and Paulus, F. F. (1996), to be published (as cited by Leusen, 1996). [263]

Levene, P. A. (1935). *J. Biol Chem.*, **108**, 419. [70]

Levy, G. C., Lichter, R. L. and Nelson, G. L. (1980). *Carbon-13 nuclear magnetic resonance spectroscopy.* John Wiley & Sons, New York. [133]

Lewis, J. P., Sewell, T. D., Evans, R. B. and Voth, G. A. (2000). Electronic structure calculation of the structures and energies of the three polymorphic forms of crystalline HMX. *J. Phys. Chem., Sect. B*, **104**, 1009–13. [281]

Lewis, N. (2000). Shedding some light on crystallization issues. Lecture transcript from the first international symposium on aspects of polymorphism and crystallization—chemical development issues. *Org. Process Res. Dev.*, **4**, 407–12. [237]

Leznoff, C. C. and Lever, A. P. B. (eds.) (1996). *Phthalocyanines—properties and applications*, Vol. 4, VCH, New York. [Vols 1–3 published 1989, 1992 and 1993 respectively]. [265]

Leznoff, D. B., Rancurel, C., Sutter, J.-P., Rettig, S. J., Pink, M., Paulsen, C. and Kahn, O. (1999). Ferromagnetic interactions and polymorphism in radical-substituted gold phosphine complexes. *J. Chem. Soc., Dalton Trans.*, 3593–9. [202]

Li, Y. H., Han, J., Zhang, G. G. Z., Grant, D. W. J. and Suryanarayanan, R. (2000). *In situ* dehydration of carbamazepine dihydrate: A novel technique to prepare amorphous anhydrous carbamazepine. *Pharm. Dev. Technol.*, **5**, 257–66. [254]

Lieberman, H. (1935). Über die Bildung von Chinakridonen aus p-Di-arylmeno-terephtalsäuren. *Annalen*, **518**, 245–59. [259]

Lieberman, H., Kirchhoff, H., Gliksman, W., Loewy, L., Gruhn, A., Hammerich, T. H., Anitschkoff, N. and Schulze, B. (1935). Transformation products of succinyl-succinic esters. VI. Formation of quinacridones from p-diarylaminoterephthalic acids. *Annalen*, **518**, 245–59. [259]

Lieberman, H. F., Davey, R. J. and Newsham, D. M. T. (2000). Br···Br and Br···H interactions in action: Polymorphism, hopping, and twinning in 1,2,4,5-tetrabromobenzene. *Chem. Mater.*, **12**, 490–4. [219, 223]

Liebman, M. N. (1982). Correlation of structure and function in biologically active small molecules and macromolecules by distance matrix partitioning. Griffin, J. F. and Duax, W. L. (eds). *Mol. Struct. Biol. Act. Proc. Meet.*, Elsevier: New York, pp. 193–212. [156]

Lima-de-Faria, J. (ed.) (1990). *Historical atlas of crystallography*, pp. 68–9. Kluwer Academic Publishers, Dordrecht, The Netherlands. This reference contains a well-documented historical account of the development of the microscope for the study of crystals. [19, 20, 25]

Lin, S. Y. (1992). Isolation and solid-state characteristics of a new crystal form of indomethacin. *J. Pharm. Sci.*, **81**, 572–6. [148]

Lincke, G. (2000). A review of thirty years of research on quinacridones. X-ray crystallography and crystal engineering. *Dyes and Pigments*, **44**, 101–22. [259]

Lincke, G. and Finzel, H.-U. (1996). Studies on the structure of alpha-quinacridone. *Cryst. Res. Technol.*, **31**, 441–52. [263]

Lindenbaum, S. and McGraw, S. E. (1985), *Pharm. Manufacturing*, Jan., 27–30. [109, 111t]

Lindenbaum, S., Rattie, E. S., Zyber, G. E., Miller, M. E. and Ravin, L. J. (1985). Polymorphism of Auranofin. *Int. J. Pharmacol.*, **26**, 123–32. [111]

Lingafelter, E. C., Simmons, G. L., Morosin, B., Scheringer, C. and Freiburg, C. (1961). The crystal structure of the α-form of bis-(N-methylsalicylaldiminato)-copper. *Acta Crystallogr.*, **14**, 1222–5. [163]

Lin-Vien, D., Colthup, N. B., Fatley, W. G. and Grasselli. (1991). *Infrared and Raman characteristic frequencies of organic molecules*. Academic Press, New York. [132]

Litvinov, I. A., Struchkov, Yu. T., Arbuzov, B. A., Makarova, N. A. and Mukmenev, E. T. (1985) Crystal and molecular structure and quantum-chemical calculation for the stable β-modification of 1,2,3-propanetriol trinitrate $C_3H_5(ONO_2)_3$. Translated from *Isvestia Akad Nauk SSSR, Ser. Khim.* 1558–63. [284]

Löbbert, G. (2000). Phthalocyanines, from *Ullmann's Encyclopedia of Industrial Chemistry*, 6th edn. Wiley-VCH Verlag, GmbH, Weinheim; web access: http://www.interscience.wiley.com/ullmanns. [265, 266, 268]

Lockhart, T. P. and Manders, W. F. (1986). Solid State ^{13}C NMR probe for organotin (IV) structural polymorphism. *Inorg. Chem.*, **25**, 583–5. [141]

Loisel, C., Keller, G, Lecq, G., Bourgaux and Ollivon, M. (1998). Phase transitions and polymorphism of cocoa butter. *J. Am. Oil Chem. Soc.*, **75**, 425–39. [148, 237]

Lommerse, J. P. M. and Motherwell, W. D. S. (1999). In preparation. [186]

Lommerse, J. P. M., Motherwell, W. D. S., Ammon, H. L., Dunitz, J. D., Gavezzotti, A., Hofmann, D. W. M., Leusen, F. J. J., Mooij, W. T. M., Price, S. L., Schweizer, B., Schmidt, M. U., van Eijck, B. P., Verwer, P. and Williams, D. E. (2000). A test of crystal structure prediction of small organic molecules. *Acta Crystallogr. B*, **56**, 697–714. [148, 167, 182, 186, 264]

Lord, R. C. and Yu, N. Y. (1970*a*). Laser-excited Raman Spectroscopy of biomolecules. I. Native lysozyme and its constituent amino acids. *J. Mol. Biol.*, **50**, 509–24. [224]

Lord, R. C. and Yu, N. Y. (1970*b*). Laser-excited Raman Spectroscopy of biomolecules. II. Native ribonuclease and α-chymotrypsin. *J. Mol. Biol.*, **5**, 203–13. [224]

Lotz, B. (2000). What can polymer crystal structure tell about polymer crystallization processes? *Eur. Phys. J. E*, **3**, 185–94. [28]

Loutfy, R. O. (1981). Bulk optical properties of phthalocyanine pigment particles. *Can. J. Chem.*, **59**, 549–54. [270]

Loutfy, R. O., Hor, A.-M., Hsiao, C. K., Baranyi, G. and Kazmaier, P. (1988). Organic photoconductive materials. *Pure Appl. Chem.*, **60**, 1047–54. [270]

Lovinger, A. J., Forrest, S. R., Kaplan, M. L., Schmidt, P. H. and Venkatesan, T. (1984). Structural and morphological investigation of the development of electrical conductivity in ion irradiated thin films of an organic material. *J. Appl. Phys.*, **55**, 476–82. [263]

Lowe-Ma, C. (2000). Private Communication to the author. [287, 295]

Lowe-Ma, C. K., Nissan, R. A. and Wilson, W. S. (1989). The synthesis and properties of picryldinitrobenzimidales and the 'Trigger Linakge' in picryldinitrobenzotriazoles. *NWC Technical Publication 7008* Naval Weapons Center, China Lake, Ca. U. S. A. [294]

Lowe-Ma, C. K., Fischer, J. W. and Willer, R. L. (1990). Structures of four related 4,5,6,7-tetrahydro-1,2,5-oxadiazolo[3,4-b]pyrazines. *Acta Crystallogr. C*, **46**, 1853–9. [294]

Ludlam-Brown, I. and York, P. (1993). The crystalline modification of succinic acid by variations in crystallization conditions. *J. Phys. D: Appl. Phys.*, **26**, B60–B65. [92]

MacDonald, J. C. (1993). Hydrogen-bonded aggregates: imidazole as a hydrogen-bond director with applications toward the design of solid-state materials. *Ph.D. Thesis*, University of Minnesota, Minneapolis, Minnesota. [58]

Macicek, J. and Yordanov, A. (1992). BLAF—a robust program for tracking out admittable Bravais lattices from the experimental unit-cell data. *J. Appl. Crystallogr.*, **25**, 73–80. [115]

MacLean, E. J., Tremayne, M., Kariuki, B. M., Harris, K. D. M., Iqbal, A. F. M. and Hao, Z. M. (2000). Structural understanding of a polymorphic system by structure solution and refinement from powder X-ray diffraction data: the alpha and beta phases of the latent pigment DPP-Boc. *J. Chem. Soc. Perkin Trans. 2*, 1513–19. [271]

Maeda, K. (1991). Photochromism in organized media. *J. Synth. Chem. Jpn.*, **49**, 554–65. [216]

Main, P., Cobbledick, R. E. and Small, R. W. H. (1985). Structure of the fourth form of 1,3,5,7-tetranitro-1,3,5,7-tetraazacyclooctane (δ-HMX) $2C_4H_8N_8O \cdot O \cdot 5H_2O$. *Acta Crystallogr. C*, **41**, 1351–4. [279t]

Mallard, E. (1876). Explication des phenomes optiques anomaux. *Annales Mines*, **10**, 60–196. [21]

Mallard, F. E. (1879). *Traité de cristallographie géometrie et physique.* [Treatise on Geometrical Crystallography and Physics]. Dunod, Paris. [21]

Manger, C. W. and Struve, W. S. (1958). Organic pigments. E. I. du Pont de Nemours and Company U. S. Patent 2,844,581. [262]

Marchetti, A. P., Salzberg, C. D. and Walker, E. I. P. (1976). The optical properties of crystalline 1, 1′-diethyl-2, 2′ cyanine iodide. *J. Chem. Phys.*, **64**, 4693–7. [205]

Marsh, R. E. (1984). Concerning a second polymorph of the HMX-DMF complex. *Acta Crystallogr. Sect. C*, **40**, 1632–3. [279]

Martin, D. W. and Waters, T. N. (1973). Conformational influences in copper coordination compounds. IV. Crystal structure of a fourth crystalline isomer of bis(2-hydroxy-N-methyl-1-naphthyl methyleneiminato)copper(II). *J. Chem. Soc. Dalton Trans.*, 2440–3. [163]

Maruyama, S., Ooshima, H. and Kato, J. (1999). Crystal structures and solvent-mediated transformation of taltireline polymorphs. *Chem. Eng. J.*, **75**, 193–200. [252]

Masamura, M. (2000). Error of atomic charges derived from electrostatic potential. *Struct. Chem.*, **11**, 41–5. [166]

Masciocchi, N., Ardizzoia, G. A., La Monica, G., Moret, M. and Sironi, A. (1997). Polymorphism in coordination chemistry. Selective synthesis and *ab-initio* x-ray powder diffraction characterization of two new crystalline phases of solid [Pd(dmpz)₂(Hdmpz)₂]₂ (Hdmpz = 3,5-dimethylpyrazole). *Inorg. Chem.*, **36**, 449–54. [85]

Mataka, S., Moriyama, H., Sawada, T., Takahashi, K., Sakashita, H. and Tashiro, M. (1996) Conformational polymorphism of mechanochromic 5,6-di(*p*-chlorobenzoyl)-1,3,4,7-tetraphenylbenzo[*c*]thiophene. *Chem. Lett.*, 363–4. [221]

Materazzi, S. (1997). Thermogravimetry infra red spectroscopy (TG-FTIR) coupled analysis. *Appl. Spectrosc. Revs.*, **32**, 385–404. [148, 251]

Materazzi, S. (1998). Mass spectrometry coupled to thermogravimetry (TG-MS) for evolved gas characterization: A review. *Appl. Spectrosc. Revs.*, **33**, 189–218. [251]

Mathieu, J. P. (1973). Vibration spectra and polymorphism of chiral compounds. *J. Raman Spectrosc.*, **1**, 47–51. [224]

Matsumoto, S., Matsuhama, K. and Mizuguchi, J. (1999). β Metal-free phthalocyanine. *Acta Crystallogr. C*, **55**, 131–3. [270]

Matsunaga, Y. (1965). Electrical resistivities and the type of bonding in some quinone complexes. *Nature*, **205**, 72–3. [82]

Matsunaga, Y. (1966). Polymorphic forms of the diaminopyrene-p-chloranil and related complexes. *Nature*, **205**, 183–4. [82]

Matsushima, R., Tatemura, M. and Okamoto, N. (1992). Second harmonic generation from 2-arylideneindan-1,3-diones studied by the powder method. *J. Mater. Chem.*, **2**, 507–10. [213]

Matsushita, M. M., Izuoka, A., Sugawara, T., Kobayashi, T., Wada, N., Takeda, N. and Ishikawa, M. (1997). Hydrogen bonded organic ferromagnet. *J. Am. Chem. Soc.*, **119**, 4369–79. [201]

Matthews, J. H., Paul, I. C. and Curtin, D. Y. (1991). Configurational isomerism in crystalline forms of benzophenone anils. *J. Chem. Soc. Perkin Trans. 2*, 113–18. [84]

Mayersohn, M. and Endrenyi, L. (1973). Relative bioavailability of commercial ampicillin formulations. *Can. Med. Assoc. J.*, **109**, 989–93. [246]

Maynard, J. Y. and Peters, H. M. (1991). *Understanding chemical patents*, 2nd ed. Am. Chem. Soc., Washington, D. C. [297]

Mayo, S. L., Olafson, B. D. and Goddard, W. A. (1990). DREIDING: A generic force field for molecular simulations. *J. Phys. Chem.*, **94**, 8897–909. [181]

McCauley, J. A., Varsolona, R. J. and Levorse, D. A. (1993). The effect of polymorphism and metastability on the characterization and isolation of two pharmaceutical compounds. *J. Phys. Sect. D*, **26**, B85–B89. [74]

McClelland, J. F., Luo, S., Jones, R. W. and Seaverson, L. M. (1992). A tutorial on the state-of-the-art of FTIR photoacoustic spectroscopy. *In Photoacoustic and Photothermal Phenomena III*, (ed. D. Bicanic), pp. 113–24. Springer, Berlin. [129]

McClure, R. J. and Craven, B. M. (1974). X-ray data for four crystalline forms of serum albumin. *J. Mol. Biol.*, **83**, 551–5. [19]

McCrone, W. C. (1949). Crystallographic data 25. 2,4,6-trinitrotoluene (TNT). *Anal. Chem.*, **21**, 1583–4. [291, 293f]

McCrone, W. C. (1950*a*). Crystallographic data 36. Cyclotetramethylene tetranitramine (HMX). *Anal. Chem.*, **22**, 1225–6. [278, 279t, 279, 280]

McCrone, W. C. (1950*b*). Crystallographic data 32. RDX Cyclotriamethylenetrinitramine. *Anal. Chem.*, **22**, 954–5. [283]

McCrone, W. C. (1951). Crystallographic data 47. 1,8-dinitronaphthalene I. *Anal. Chem.*, **23**, 1188–9. [295]

McCrone, W. C. (1952). Crystallographic data 54. 2,4,6-2′,4′,6′-Hexanitrodiphenylamine (HND). *Anal. Chem.*, **24**, 592–3. [295]

McCrone, W. C. (1954). Crystallographic data 89. 2,6-dinitrotoluene. *Anal. Chem.*, **26**, 1997–8. [295]

McCrone, W. C. (1957). *Fusion methods in chemical microscopy*. Interscience Publishers, New York, U. S. A. [7, 8, 15, 26, 95, 97, 100, 104, 148, 173, 280, 283, 288, 295]

McCrone, W. C. (1965). Polymorphism. In *Physics and chemistry of the organic solid state*, Vol. 2 (ed. D. Fox, M. M. Labes and A. Weissberger), pp. 725–67. Wiley Interscience, New York, U. S. A. [1, 2, 3, 4, 5, 6, 7, 8, 9, 26, 36, 36f, 37f, 75, 83, 85, 148, 205, 240, 259]

McCrone, W. C. (1967). *Crystallographic Study of SC-101*. Project 883, Chicago, IL. [291]

McCrone, W. C. (1984). Private Communication. [100]

McCrone, W. C. and Adams, W. C. (1955). Crystallographic data 101. Lead styphnate (normal). *Anal. Chem.*, **27**, 2014–15. [294]

McCrone, W. C., McCrone, L. B. and Delly, J. G. (1978). *Polarized light microscopy*. Ann Arbor Science Publishers, Inc. Ann Arbor, MI. [95]

McGeorge, G., Harris, R. K., Chippendale, A. M. and Bullock, J. F. (1996). Conformational analysis by magic-angle spinning NMR spectroscopy for a series of polymorphs of a disperse azobenzene dyestuff. *J. Chem. Soc. Perkin Trans.*, 1733–8. [158]

McIntosh, D. (1919). Crystallization of supersaturated solutions and supercooled liquids. *Trans. Roy. Soc. Can. Sect. 3*, **13**, 265–72. [67]

McKay, R. B., Iqbal, A. and Medinger, B. (1994). In *Technological applications of dispersions* (ed. R. B. McKay). Marcel Dekker, New York. [258]

McKerrow, A. J., Buncel, E. and Kazmaier, P. M. (1993). Molecular modeling of photoactive pigments in the solid state: Investigations of polymorphism. *Can. J. Chem.*, **71**, 390–8. [264]

McLafferty, F. W. (1990). Analytical chemistry: historic and modern. *Acc. Chem. Res.*, **23**, 63–4. [26, 95]

McNaughton, J. L. and Mortimer, C. T. (1975). Differential scanning calorimetry. (Reprinted from IRS; Phys. Chem. Series 2. Vol. 10). Perkin-Elmer, Norwalk, CT. [104]

McPherson, A. (1982). *Preparation and analysis of protein crystals.* pp. 127–59. Wiley, New York. [9, 19]

McPherson, A. (1989). *Preparation and analysis of protein crystals.* Krieger Publishing Company. [19]

McPherson, A. (1998). *Crystallization of biological macromolecules.* [19]

Meguro, T., Kashiwagi, T. and Satow, Y. (2000). Crystal structure of the low-humidity form of aspartame sweetener. *J. Peptide Res.*, **56**, 97–104. [305]

Merck Index. 12th edn on CD ROM. Version 12:1, 1996. Chapman and Hall, London. [13]

Merritt, V. Y. (1978). Organic Photovolataic materials: squarylium and cyanine-TCNQ dyes. *IBM J. Res. Dev.*, **22**, 353–71. [204]

Mesley, R. J., Clements, R. L., Flaherty, B. and Goodhead, K. (1968). The polymorphism of phenobarbitone. *J. Pharm. Pharmacol.*, **20**, 329–40. [255]

Messerschmidt, R.-G. and Harthcock, M. A. (eds.) (1988). *Infrared microspectroscopy, theory and applications.* Marcel Dekker, New York. [129]

Metz, H. J. and Schmidt, M. U. (1999a). Novel crystal modification of C. I. Pigment Red 53 : 2 (γ-phase). Clariant GMBH Patent DE 19827272 A1, EP 965617 A1. [260t]

Metz, H. J. and Schmidt, M. U. (1999b). Novel crystal modification of C.I. Pigment Red 53 : 2 (δ-phase). Clariant GMBH Patent DE 19827273 A1, EP 965616 A1. [260t]

Meyer, R. (1987). *Explosives.* VCH, Weinheim. [275, 279t, 283, 294, 295]

Mighell, A. D. (1976). The reduced cell: its use in identification of crystalline materials. *J. Appl. Crystallogr.*, **9**, 491. [115]

Mighell, A. D. and Stalick, J. K. (1983). *Crystal data. Determinative tables.* Vol. 6, 3rd edn. U. S. National Institute of Science and Technology, JCPDS—Joint Committee on Powder Diffraction Standards, ICDD—International Centre for Diffraction Data. [14]

Miller, J. S. (1998). Polymorphic molecular materials: the importance of tertiary structures. *Adv. Mater.*, **10**, 1553–7. [197, 201]

Miller, J. S. and Epstein, A. S. (1994). Organic and organometallic molecular magnetic materials—designer magnets. *Angew. Chem., Int. Ed. Engl.*, **33**, 385–415. [197]

Miller, J. S. and Epstein, A. S. (1996). Magnets based on the molecular solid state. *Mol. Cryst. Liq. Cryst. Sci. Technol. Sect. A*, **278**, 145–54. [197]

Miller, J. S., Zhang, J. H., Reiff, W. M., Doxon, D. A., Preston, L. D., Reis, A. H., Gebert, E., Extine, M., Troup, J., Epstein, A. J. and Ward, M. D. (1987). Characterization of the charge-transfer reaction between decamethylferrocene and 7,7,8,8-tetracyano-p-quinodimethane (1 : 1). The ^{57}Fe Mössbauer spectra and structures of the paramagnetic and metamagnetic one-dimensional salts of the molecular and electronic structures of [TNCQ]^n ($n = 0, -1, -2$). *J. Phys. Chem.*, **91**, 4344–60. [199]

Miller, P. J., Block, S. and Piermani, G. J. (1991). Effects of pressure on the thermal-decomposition kinetics, chemical-reactivity and phase behavior of RDX. *Combust. Flame*, **83**, 174–84. [283]

Miller, R. L., Sharp, J. H. and Lardon, M. A. Metal phthalocyanines in the x form, especially for use in electrophotography. Xerox Corporation Patent US 3 862 127; DE-OS 1 944 021 (1970); GB 1 268 422 (1972); FR 2 016 641. [266]

Miller, R. S. (1995). *Decomposition, combustion and detonation chem. of energetic materials*. In *Materials research soc. symposium proceedings*, **418** (ed. T. B. Brill, T. P. Russell, W. C. Tao, and R. B. Wardle), pp. 3. Materials Research Soceity, Pittsburgh. [278]

Mislow, K. (1966). *Introduction to stereochemistry*, pp. 33–6. W. A. Benjamin, New York. [152]

Mitra, A. K., Ghosh, L. K. and Gupta, B. K. (1993). Development of methods for the preparation and evaluation of chloramphenicol palmitate ester and its biopharmaceutically effective metastable polymorph. *Drug Dev. Ind. Pharm.*, **19**, 971–80. [246]

Mitscherlich, E. (1820). Sur la relations qui existe entre la forme cristalline et les proportion chimiques. *Ann. Chim. Phys.*, **14**, 172–90. [271]

Mitscherlich, E. (1822). Sur la relation qui existe entre la forme cristalline et les proportions chimiques, I. Mémóire sure les arseniates et les phosphates. *Ann. Chim. Phys.*, **19**, 350–419. [19]

Mitscherlich, E. (1822–23). Über die Körper, welche in zwei verschiedenen krystallisieren Formen. [Considering the materials which can crystallize in two different crystal forms]. *Abhl. Akad. Berlin*, 43–8. [2, 20]

Mitsubishi Chemical Industries (1976). Japanese Patent JA 51- 7025. Beta-crystalline phase perylene pigment preparation. [263]

Mizuguchi, J. (1997). Structural and optical properties of 5,15-diaza-6,16-dihydroxytetrabenzo[b,e,k,n]perylene. *Dyes and Pigments*, **35**, 347–60. [264]

Mizuguchi, J. (1998a). Electronic characterization of N, N'-bis(2-phenylethyl)perylene-3,4:9,10-bis(dicarboximide) and its application to optical disks. *J. Appl. Phys.*, **84**, 4479–85. [264, 265f]

Mizuguchi, J. (1998b). N, N'-bis(2-phenethyl)perylene-3,4:9,10-bis-(dicarboximide). *Acta Crystallogr. C*, **54**, 1479–81. [260t, 264]

Mizuguchi, J. (2001). Form of Pigment Red 149. *Zeitschrifft für Kristallographie*. In press. [263, 264]

Mizuguchi, J., Rochat, A. C. and Rohs, G. (1994). Electronic spectra of 7,14-dithioketo-5,7,12,14-tetrahydroquinolino-[2,3-b]-acridine in solution and in the solid state. *Ber. Bunsenges. Phys. Chem.*, **98**, 19–28. [231]

Mnyukh, Yu. V., Panfilova, N. A., Petropavlov, N. N. and Ukhvatova, N. S. (1975). Polymorphic transitions in molecular crystals. III Transitions exhibiting unusual behavior. *J. Phys. Chem. Solids*, **36**, 127–44. [115]

Möbus, M., Karl, N. and Kobayashi, T. (1992). Structure of perylenetetracarboxylic dianhydride thin films on alkali halide crystal substances. *J. Cryst. Growth*, **116**, 495–504. [263]

Moers, O., Wijaya, K., Lamge, I, Blaschette, A. and Jones, P. G. (2000). Polysulfonamines, CXXVI—hydrogen bonding in crystalline onium dimesylamines: a robust eight-membered ring synthon in coexistence with a third hydrogen binding motif—cyclodimers, supramolecular linkage isomers, and a quintuply interwoven three-dimensional C-H⋯O network. *Z. Naturforsch. Sect. B—a J. Chem. Sci.*, **55**, 738–52. [55]

Momoi, Y., Yamane, M., Yamaguchi, I. and Matsushita, H. (1976). Yellow pigments. Japanese Patent 510 88516 (Application 75-13362). [261t]

Montgomery, L. K., Geiser, U., Wang, H. H., Beno, M. A., Schultz, A. J., Kini, A. M., Carlson, K. D., Williams, J. M., Whitowrth, J. R., Gates, B. D., Cariss, C. S., Pipan, C. M., Donega, K. M., Wenz, C., Kwok, W. K. and Crabtree, G. W. (1988). How well do we understand the synthesis of (ET)213 by electrocrystallization? ESR and x-ray identification of (ET)213 crystals which are mixtures of phases and observation of high Tc states of (ET)213, ranging from 2.5–6.9K. *Synth. Met.*, **27**, A195–A207. [79]

Mooij, W. T. M. (2000). *Ab initio* prediction of crystal structures. *Ph. D. Thesis*, University of Utrecht, Utrecht, The Netherlands. [182]

Mooij, W. T. M., van Eijck, B. P., Price, S. L., Verwer, P. and Kroon, J. (1998). Crystal structure prediction for acetic acids. *J. Comput. Chem.*, **19**, 459–74. [183, 185, 186]

Moorthy, J. N. and Venkatesan, K. (1994). Photobehavior of crystalline 4-styrylcoumarin dimorphs: Structure-reactivity correlations. *Bull. Chem. Soc. Jpn.*, **67**, 1–6. [236f, 236]

Morel, D. L., Stogryn, E. L., Ghosh, A. K., Feng, T., Purin, P. E., Shaw, R. F., Fishman, C., Bird, G. R. and Piechowsky, A. P. (1984). Organic photovoltaic cells. Correlations between cell performance and molecular structure. *J. Phys. Chem.*, **88**, 923–33. [204]

Morris, K. R., Griesser, U. J., Eckhardt, C. J. and Stowell, J. G. (2001). Theoretical approaches to physical transformations of active pharmaceutical ingredients during manufacturing processes. *Adv. Drug Deliv. Reviews*, **48**, 91–114. [256]

Morris, K. R., Nail, S. L., Peck, G. E., Byrn, S. R., Griesser, U. J., Stowell, J. G., Hwang, S. J. and Park, K. (1998). Advances in pharmaceutical material and processing. *Pharm. Sci. Technol. Today*, **1**, 235–45. [249]

Morrison, P., Morrison, P. and the Office of Charles and Ray Eames (1982). *Powers of ten*, Scientific American, New York. [91]

Moser, F. H. and Thomas, A. L. (1963). *Phthalocyanine compounds*. Reinhold Publishing Co., New York. [265]

Moser, F. H. and Thomas, A. L. (1983). *The phthalocyanines*, Vols. I and II, CRC Press, Boca Raton, FL. [265, 266]

Motherwell, W. D. S. (1997). Distribution of molecular centers in crystallographic unit cells. *Acta Crystallogr. B*, **53**, 726–36. [183]

Motherwell, W. D. S. (1999). Crystal structure prediction and the Cambridge structural database. *Nova Acta Leopold, NF*, **79**, 89–98. [57]

Motherwell, W. D. S. (2001). Crystal structure prediction and the Cambridge Structural Database. *Mol. Cryst. Liq. Cryst.*, **356**, 559–67. [264]

Motherwell, W. D. S., Shields, G. P. and Allen, F. H. (1999). Visualization and characterization of non-covalent networks in molecular crystals: automated assignment of graph-set descriptors for asymmetric molecules. *Acta Crystallogr. B*, **55**, 1044–56. [57]

Moustafa, M. A., Ebian, A. R., Khalil, S. A. and Motawi, M. M. (1971). Sulfamethoxydiazine. *J. Pharm. Pharmacol.*, **23**, 868–74. [246]

Müller, A. H. R. (1914). Über total instabile Formen. *Z. Phys. Chem.*, **86**, 177–242. [98]

Mullin, J. W. (1993). *Crystallization*. 3rd edn, Butterworth-Heinemann Ltd., Oxford. [67, 91]

Munn, R. W. and Ironside, C. N. (eds.) (1993). *Principles and applications of nonlinear optical materials*, Blackie A & P. Chapman and Hall, Glasgow. [207]

Murrell, J. N. and Tanaka, J. (1964). The theory of the electronic spectra of aromatic hydrocarbon dimers. *Mol. Phys.*, **7**, 363–80. [231]

Nagai, Y., Goto, N. and Nishi, H. (1965). *Yuki Gosei Kagaku Kyokai Shi*, **23**, 318; *Chem. Abst.*, **62**, 16221. [262]

Nagashima, N., Sano, C., Kishimoto, S. and Iitaka, Y. (1987). The characterization of various crystalline forms of aspartame (a dipeptide sweetener). *Acta Crystallogr. A*, **43**, C54. [305]

Nakano, K., Sada, K. and Miyata, M. (1996). Novel additive effect of inclusion crystals on polymorphs of cholic acid crystals having different hydrogen-bonded networks with the same organic guest. *Chem. Commun.*, 989–90. [58]

Nakasuji, K., Sasaki, M., Kotani, T., Murata, I., Enoki, T., Imaeda, K., Inokuchi, H., Kaawamoto, A. and Tanaka, J. (1987). Methylthio- and ethanediyldithio-substituted 1,6-dithiapyrene and their charge-transfer complexes: new organic molecular metals. *J. Am. Chem. Soc.*, **109**, 6970–5. [191]

Nakatsu, K., Yoshie, N., Yoshioka, H., Nogami, T., Shirota, Y., Shimizu, Y., Uemiya, T. and Yasuda, N. (1990). Polymorphism and the molecular and crystal structures of a 2nd-order nonlinear optical-compound containing a 1,3-dithiole ring. *Mol. Cryst. Liq. Cryst.*, **182**, 59–69. [213]

Nakazawa, Y., Tamura, M., Shirakawa, N., Shiomi, D., Yakahishi, M., Kinoshita, M. and Ishikawa, M. (1992). Low-temperature magnetic properties of the ferromagnetic organic radical, *p*-nitrophenyl nitronyl nitroxide. *Phys. Rev. B*, **46**, 8906–14. [200]

Nash, C. P., Olmstead, M. M., Weiss-Lopez, B., Musker, W. K., Ramasubbu, N. and Parthasarathy, R. (1985). Structures and Raman spectra of two crystalline modifications of dithioglycolic acid. *J. Am. Chem. Soc.*, **107**, 7194–5. [225]

Nass, K. K. (1991). Process implication of polymorphism in organic compounds. In *A. I. Che. E Symp. Ser. No. 284* (eds. R. Ramanarayanan, W. Kern, M. Larson and S. Sikdar), Vol. 87, American Institute of Chemical Engineers, New York, pp. 72–81. [27, 258]

Nassau, K. (1983). *The physics and chemistry of color. Fifteen causes of color.* Wiley-Interscience, New York. [213, 215, 217, 229, 257]

Näther, C., Bock, H. and Claridge, R. F. C. (1996a). Solvent-shared radical ion pairs [pyrene⁻Na⁺O(C₂H₅)₂]∞: ESR evidence for two different aggregates in solution, room temperature crystallization, and structural proof of another polymorphic modification. *Helv. Chim. Acta*, **79**, 84–91. [69]

Näther, C., Nagel, N., Bock, H., Seitz, W. and Havlas, Z. (1996b). Structural, kinetic and thermodynamic aspects of the conformational dimorphism of diethyl 3,6-dibromo-2,5-dihydroxyterephthlate. *Acta Crystallogr. B*, **52**, 697–706. [69, 170, 178, 179f]

Navon, O., Bernstein, J. and Khodorkovsky, V. (1997). Chains, ladders and two-dimensional sheets with halogen⋯halogen and halogen⋯hydrogen interactions. *Angew. Chem., Int. Ed. Engl.*, **36**, 601–3. [55]

Navon, O., Bernstein, J., MacDonald, J. C. and Reutzel, S. (2001). The polymorphism of a barbiturate: 5-5′ diethylbarbituric acid. In preparation. [139f]

Nedelko, V. V., Chukanov, N. V., Raevskii, A. V., Korsounskii, B. L., Larikova, T. S. and Kolesova, O. I. (2000). Comparative investigation of thermal decomposition of various modifications of hexanetrohexaazaisowurtzitane (CL-20). *Propell. Explos. Pyrot.*, **25**, 255–9. [282, 283]

Neidle, S., Kuhlbrandt, W. and Achari, A. (1976). The crystal structure of an orthorhombic form of adenosine-5′-monophosphate. *Acta Crystallogr. B*, **32**, 1850–5. [159]

Neville, G. A., Beckstead, H. D. and Shurvell, H. F. (1992). Utility of Fourier transform Raman and Fourier transform infrared diffuse reflectance spectroscopy for differentiation of polymorphic spirolactone samples. *J. Pharm. Sci.*, **81**, 1141–6. [129, 132]

Newman, A. W., Stephens, P. W., Morrison, H. G., Andres, M. C., Shatly, G. P. and Thomas, A. S. (1999). Quantitation of two polymorphic forms using Rietveld analysis, synchrotron XPRD and traditional XPRD. Presentation notes, pp. 85–96. Pharmaceutical Powder X-ray Diffraction Symposium, organized by International Centre for Diddraction Data, 27–30 Sep., 1999. [122, 123f, 124f]

Ng, W. L. (1990). A study of the kinetics of nucleation in a palm oil melt. *J. Am. Oil Chem. Soc.*, **67**, 879–82. [28]

Nguyen, N. A. T., Ghosh, S., Gatlin, L. A. and Grant, D. J. W. (1994). Physicochemical characterization of the various solid forms of carbovir, an antiviral nucleoside. *J. Pharm. Sci.*, **83**, 1116–23. [5]

Nichols, G. (1998). Optical properties of polymorphic forms I and II of paracetamol. *Microscope*, **46**, 117–22. [100f]

Nichols, G. (1999). Thermodynamic and kinetic control of conformational polymorphism: a case study. 1st International symposium on aspects of polymorphism & crystallization—chemical development issues, pp. 199–209, Scientific Update, Mayfield, East Sussex, United Kingdom. [100f, 144f]

Nichols, G. and Frampton, C. S. (1998). Physicochemical characterization of the orthorhombic polymorph of paracetamol crystallized from solution. *J. Pharm. Sci.*, **87**, 684–93. [113, 144f]

Nielsen, A. T., Chafin, A. P., Christian, S. L., Moore, D. W., Nadler, M. P., Nissan, R. A., Vanderah, D. J., Gilardi, R. D., George, C. F. and Flippen-Anderson, J. L. (1998). Synthesis of polyazapolycyclic caged polynitramines. *Tetrahedron*, **54**, 11793–812. [282]

Niementowski, S. (1896). Über das Chinakridin. *Berichte*, **29**, 76–83. [259]

Niementowski, S. (1906). Oxy-chinakridin und phlorchinyl. *Berichte*, **39**, 385–92. [259]

Niggli, P. (1924). *Lehrbuch der Mineralogie*, Zweite Auflage, Verlag von Gebrüder Borntraeger, Berlin. [26]

Ninomiya, M., Komai, A., Sbirane, N., Ito, Y. and Terui (1979). Nippon Shokubai Kagaku Kogyo Co. Patent JP 53-118427 [*Chem. Abstr.*, **90**, 56431s]; Patent JP 53-36036 [*Chem. Abstr.*, **90**, 105642x]; Patent JP 54-10331 [*Chem. Abstr.*, **90**, 170162e]; Patent JP 54-11135 [*Chem. Abstr.*, **91**, 92990c]; Patent JP 54-11136 [*Chem. Abstr.*, **90**, 205787s]. [267]

NIST (2001). *NIST crystal data*. US National Institute of Standards and Technology. Gaithersburg, Maryland (http://www.nist.gov/srd/nistz.htm). [14]

Norris, T., Aldridge, P. K. and Sekulic, S. S. (1997). Near-infrared spectroscopy. *Analyst*, **122**, 549–52. [125]

Nyburg, S. C. (1974). Some uses of a best molecular fit routine. *Acta Crystallogr. B*, **30**, 251–3. [161]

Nyvlt, J., Söhnel, O., Matuchová, M. and Broul, M. (1985). *The kinetics of industrial crystallization*. Academia, Prague. [68]

O'Connor, R. T. (1960). X-ray diffraction and polymorphism. In *Fatty acids*, Part I, 2nd edn., ed. K. S. Marklay, pp. 285–378. Interscience, New York, U. S. A. [28]

Ogawa, T., Kuwamoto, K., Isoda, S., Kobayashi, T. and Karl, N. (1999). 3,4 : 9,10-perylenetetracarboxylic acid dianhydride (PTCDA) by electron crystallography. *Acta Crystallogr. B*, **55**, 123–30. [260t, 263]

Ogawa, Y. and Tatsumi, M. (1978). Raman spectroscopic studies on conformational polymorphism of 1-bromopentane. *Chem. Lett.*, 947–50. [158]

Ogawa, Y. and Tatsumi, M. (1979). Raman and infrared spectroscopic studies on conformational polymorphism of *n*-propyl acetate. *Chem. Lett.*, 1411–16. [158]

Ojala, C. R., Ojala, W. H., Pennamon, S. Y. and Gleason, W. B. (1998). Conformational polymorphs of acetone tosylhydrazone. *Acta Crystallogr. C*, **54**, 57–60. [83]

Ojala, W. H. and Etter, M. C. (1992). Polymorphism in anthranilic acid: a reexamination of the phase transitions. *J. Am. Chem. Soc.*, **114**, 10288–10293. [59, 181]

Oliver, S. N., Pantelis, P. and Dunn, P. L. (1990). Polymorphism and crystal–crystal transformations of the highly optically nonlinear organic compound α-[(4'-methoxyphenyl)methylene]-4-nitro-benzene-acetonitrile. *Appl. Phys. Lett.*, **56**, 307–9. [213]

Orpen, A. G., Brammer, L., Allen, F. H., Kennard, O., Watson, D. G. and Taylor, R. (1989). Tables of bond lengths determined by X-ray and neutron diffraction. Part 2. Organometallic compounds and coordination complexes of the d- and f-block metals. *J. Chem. Soc., Dalton Trans.*, S1–83. [151]

Ostwald, W. (1902). *Lehrbuch der Allgemein Chemie.* Vol. II, 2, Engelmann, Leipzig, pp. 383ff, 710ff, 773ff. [67]

Ostwald, W. F. (1897). Studien über die bildung und umwandlung fester körper. Studies on formation and transformation of solid materials. *Z. Phys. Chem.*, **22**, 289–330. [8, 23, 44]

Oyumi, Y. and Brill, T. B. (1985). Thermal decomposition of energetic materials. 6. Solid-phase transitions and the decomposition of 1,2,3-triaminoguanidinium nitrate. *J. Phys. Chem.*, **89**, 4325–9. [284]

Oyumi, Y., Brill, T. B. and Rheingold, A. L. (1986*a*). Thermal decomposition of energetic materials. 9. Polymorphism, crystal structures, and thermal decomposition of polynitroazabicyclo [3.3.1]nonanes. *J. Phys. Chem.*, **90**, 2526–33. [286]

Oyumi, Y., Rheingold, A. L. and Brill, T. B. (1986*b*). Thermal decomposition of energetic materials. 16. Solid-phase structural analysis and the thermolysis of 1,4-dinitrofurzano[3,4-*b*]piperazine. *J. Phys. Chem.*, **90**, 4686–90. [294]

Oyumi, Y., Rheingold, A. L. and Brill, T. B. (1987*a*). Thermal decomposition of energetic materials. 19. Unusual condensed-phase and thermolysis properties of a mixed azidomethyl nitramine: 1,7-diazido-2,4,6-trinitro-2,4,6-triazaheptane. *J. Phys. Chem.*, **91**, 920–5. [285]

Oyumi, Y., Brill, T. B. and Rheingold, A. L. (1987*b*) Thermal decomposition of energetic materials. A comparison of energetic materials and thermal reactivity of an acyclic and cyclic tetramethylenetetranitramine pair. *Thermochim. Acta*, **114**, 209–25. [285]

Ozawa, Y., Pressprich, M. R. and Coppens, P. (1998). On the analysis of reversible light-induced changes in molecular crystals. *J. Appl. Crystallogr.*, **31**, 128–35. [234]

Pace, M. D. (1991). EPR-spectra of photochemical NO_2 formation in monocyclic nitramines and hexanitrohexaazaisowurtzitane. *J. Phys. Chem.*, **95**, 5858–64. [283]

Pace, M. D. (1992). Free radical mechanisms in high-density nitrocompounds— hexanitrohexaazaisowurtzitane, a new high energy nitramine. *Mol. Cryst. Liq. Cryst.*, **219**, 139–48. [283]

Padden, B. E., Zell, M. T., Dong, Z., Schroeder, S. A., Grant, D. J. W. and Munson, E. J. (1999). Comparison of solid-state ^{13}C NMR spectroscopy and powder X-ray diffraction for analyzing mixtures of polymorphs of neotame. *Anal. Chem.*, **71**, 3325–31. [141]

Padmaja, N., Ramakumar, S. and Viswamitra, M. A. (1990). Space-group frequencies of proteins and of organic compounds with more than one formula unit in the asymmetric unit. *Acta Crystallogr. A*, **46**, 725–30. [178]

Palacio, F., Antorenna, G., Casatro, M., Burriel, R., Rawson, J., Smith, J. N. B., Bricklebank, N., Novoa, J. J. and Ritter, C. (1997*a*). High temperature magnetic ordering in a new organic magnet. *Phys. Rev. Lett.*, **79**, 2336–9. [200]

Palacio, F., Antorenna, G., Casatro, M., Burriel, R., Rawson, J., Smith, J. N. B., Bricklebank, N., Novoa, J. J. and Ritter, C. (1997b). Spontaneous magnetization at 36K in a sulfur–nitrogen radical. *Mol. Cryst. Liq. Cryst., Sci. Technol. A*, **306**, 293–300. [200]

Palmore, G. T. R., Luo, T. J., Martin, T. L., McBride-Wieser, M. T. and Voong, N. T. (1998). Using the atomic force microscope to study the assembly of molecular solids. *Trans. Am. Crystallogr. Assoc.*, **33**, 45–57. [146]

Panagiotopoulos, N. C., Jeffrey, G. A., LaPlaca, S. J. and Hamilton, W. C. (1974). The A and B forms of potassium D-Gluconate monohydrate. *Acta Crystallogr. B*, **30**, 1421–30. [2, 155, 157]

Panagopoulou-Kaplani, A. and Malamataris, S. (2000). Preparation and characterisation of a new insoluble polymorphic form of glibenclamide. *Int. J. Pharm.*, **195**, 239–46. [249]

Pandarese, F., Ungaretti, L. and Coda, A. (1975). The crystal structure of a monoclinic phase of *m*-nitrophenol. *Acta Crystallogr. B*, **31**, 2671–5. [81, 212]

Partington, J. R. (1952). *An advanced treaty on physical chemistry*. Vol. 3, *The properties of solids*, pp. 512–13. Longmans, Green and Co., London. [20]

Patel, A. D., Luner, P. E. and Kemper, M. S. (2001) Low-level determination of polymorph composition in physical mixtures by near-infrared reflectance spectroscopy. *J. Pharm. Sci.*, **90**, 360–70. [129]

Patil, D. G. and Brill, T. B. (1993). Thermal-decomposition of energetic materials. 59. Characterization of the residue of hexanitrohexaazaisowurtzitane. *Combust. Flame*, **92**, 456–8. [283]

Paul, I. C. and Curtin, D. Y. (1973). Thermally induced organic reactions in the solid state. *Acc. Chem. Res.*, **6**, 217–25. [1, 217, 223, 237]

Paul, I. C. and Curtin, D. Y. (1975). Reactions of organic crystals with gases. *Science*, **187**, 19–26. [1]

Paul, I. C. and Curtin, D. Y. (1987). Gas-solid reactions and polar crystals. In *Organic solid state chemistry*, Vol. 32, *Studies in organic chemistry* (ed. G. Desiraju), pp. 331–70. Elsevier, Amsterdam. [237]

Paul, I. C. and Go, K. T. (1969). The crystal and molecular structure of 5-methyl-1-thia-5-azacyclooctane 1-oxide perchlorate. *J. Chem. Soc. B*, 33–42. [50, 54f]

Pauling, L. (1960). *The nature of the chemical bond*, 3rd edn. Cornell University Press, Ithaca, New York. [182]

Paulus, E. F., Dietz, E., Kroh, A., Prokschy, F. and Lincke, G. (1989). Crystal structure of quinacridones. In *Collected abstracts of twelfth European crystallographic meeting, Moscow, Z. Kristallogr.*, **2** (Suppl 2). [263, 273]

Pavia, D. L., Lampman, G. L. and Kriz, G. S. (1988) *Introduction to organic laboratory techniques*, 3rd edn. Saunder College Publishing, Philadelphia. [90]

Payne, R. S., Roberts, R. J., Rowe, R. C. and Docherty, R. (1998a). Generation of crystal structures of acetic acid and its halogenated analogs. *J. Comput. Chem.*, **19**, 1–20. [185]

Payne, R. S., Rowe, R. C., Roberts, R. J., Charlton, M. H. and Docherty, R. (1998b). Potential polymorphs of aspirin. *J. Comput. Chem.*, **20**, 262–73. [185]

Payne, R. S., Roberts, R. J., Rowe, R. C. and Docherty, R. (1999). Examples of successful crystal prediction: polymorphs of primidone and progesterone. *Int. J. Pharm.*, **177**, 231–45. [185]

Pearson, J. T. and Varney, G. (1973). The anomalous behaviour of some oxyclozanide polymorphs. *J. Pharm. Pharmaceut.*, **25**, (Suppl.) 62–70P. [255]

Pella, P. A. (1976). Generator for producing trace vapor concentrations of 2,4,6-trinitrotoluene,2,4-dinitrotoluene, and ethylene glycol dinitrate for calibrating explosives vapor detectors. *Anal. Chem.*, **48**, 1634–7. [291]

Pella, P. A. (1977). Measurements of the vapor pressures of TNT, 2,4-DNT, 2,6-DNT and EGDN. *J. Chem. Thermodyn.*, **9**, 301–5. [291]

Perlstein, J. (1992). Molecular self-assemblies: Monte Carlo predictions for the structure of the one-dimensional translation aggregate. *J. Am. Chem. Soc.*, **114**, 1955–63. [183]

Perlstein, J. (1994*a*). Molecular self-assemblies: 2. A computational method for the prediction of the structure of one-dimensional screw, glide and inversion molecular aggregates and implications for the packing of molecules in monolayers and crystals. *J. Am. Chem. Soc.*, **116**, 455–70. [183]

Perlstein, J. (1994*b*). Molecular self-assemblies: 4. Using Kitaigorodskii's aufbau principle for quantitatively predicting the packing geometry of semiflexible organic molecules in translation monolayer aggregates. *J. Am. Chem. Soc.*, **116**, 11420–32. [183]

Perlstein, J. (1996). Molecular self-assemblies: 5. Analysis of the vector properties of hydrogen bonding in crystal engineering. *J. Am. Chem. Soc.*, **118**, 8433–43. [183]

Perlstein, J. (1999). Introduction to packing patterns and packing energetics for crystalline self-assembled structures. In *From molecules and crystals to materials* (ed. D. Braga, F. Grepioni and A. G. Orpen), Vol. C538 of *NATO ASI Series*, pp. 23–42. Kluwer Academic Publishers, Dordrecht. [183]

Perrenot, B. and Widmann, G. (1994). Polymorphism by differential scanning calorimetry. *Thermochim. Acta*, **234**, 31–9. [5, 105, 106, 107f]

Perrin, R. and Lamartine, R. (1990). Organic reactions in the solid state. In *Structure and properties of molecular crystals—studies in physical and theoretical chemistry*, Vol. 69, pp. 107–59. Elsevier, Amsterdam. [238]

Pertsin, A. J. and Kitaigorodskii, A. I. (1987). *The atom–atom potential method. Applications to organic molecular solids.* Springer-Verlag, Berlin. [153, 166]

Petit, S., Coquerel, G. and Perez, G. (1994). Influence of water molecules on the nucleation rate of polymorphis complexes with different conformations in solution. In *Hydrogen bond networks* (ed. M.-C. Bellissent-Funel and J. C. Dore), pp. 255–9. Kluwer Academic Publishers, Dordrecht, The Netherlands. [69]

Petty, C. J., Bugay, D. E., Findlay, W. P. and Rodriguez, C. (1996). Applications of FT-Raman spectroscopy in the pharmaceutical industry. *Spectroscopy*, **11**, 41–5. [132]

Pfeiffer, F. L. (1962). Pigmentary copper phthalocyanine in the 'R' form and its preparation. American Cyanamid Patent US 3 051 721. [266]

Pfeiffer, P. (1922). *Organischer Molekülverbindungen* Organic Molecular Compounds. Ferdinand Enke, Stuttgart. [15]

Phillips, J. W. C. and Mumford, A. (1934). Dimorphism of certain aliphatic compounds. V. Primary alcohols and their acetates. *J. Chem. Soc.*, 1657–65. [98]

Pierce-Butler, M. A. (1982). Structures of the barium salt of 2,4,6-trinitro-1,3-benzenediol monohydrate and the isomorphous lead salt (β-polymorph). *Acta Crystallogr. B*, **38**, 3100–4. [294]

Pierce-Butler, M. A. (1984). The structure of the lead salt of 2,4,6-trinitro-1,3-benzenediol monohydrate (α-polymorph), α-Pb$_2^+$C$_6$HN$_3$O$_8^{2-}$ · H$_2$O. *Acta Crystallogr. C*, **40**, 63–5. [294]

Piermani, G. J., Mighell, A. D., Weir, C. E. and Block, S. (1969). Crystal structure of benzene II at 25 kilobars. *Science*, **165**, 1250–5. [238]

Pillardy, J., Wawak, R. J., Arnautova, Y. A., Czaplewski, C. and Scheraga, H. A. (2000). Crystal structure prediction by global optimization as a tool for evaluating potentials: role of the dipole moment correction term in successful predictions. *J. Am. Chem. Soc.*, **122**, 907–21. [182, 185]

Pimentel, G. C. and McClellan, A. L. (1960). *The hydrogen bond*. Freeman, San Francisco, CA. [53]

PLUTO Plotting Module, Cambridge Structural Database, April 1998. Release: Cambridge Crystallographic Data Centre, Cambridge.

Poole, J. W. and Bahal, C. K. (1968). Dissolution behavior and solubility of anhydrous and trihydrate forms of ampicillin. *J. Pharm. Sci.*, **57**, 1945–8. [245]

Poole, J. W., Owen, G., Silverio, J., Freyhof, J. N. and Rosenman, S. B. (1968). Physicochemical factors influencing the absorption of the anhydrous and trihydrate forms of ampicillin. *Curr. Ther. Res. Clin. Experimentat.*, **10**, 292–303. [245]

Popelier, P., Lenstra, A. T. H., van Alsenoy, C. and Geise, H. J. (1989). An *ab initio* study of crystal field effects: solid state and gas phase geometry of acetamide. *J. Am. Chem. Soc.*, **111**, 5658–60. [154]

Pople, A. J. (1999). Quantum chemical models (Noble lecture). *Angew. Chem., Int. Ed. Engl.*, **38**, 1894–902. [165]

Popov, E. V., Shvets, V. I., Shalimova, G. V. and Shtanov, N. P. (1981). Effects of crystal structure on coloristic properties of vat and disperse dyes. III. Morphological features of such vat dyes as violanthrone derivatives. *J. Appl. Chem., USSR* **54**, 2090–4; English translation of *Zh. Prikl. Khim.* (Leningrad) **54**, 2362–6. [262t]

Popovitz-Biro, R., Addadi, L., Leiserowitz, L. and Lahav, M. (1991). Strategies for the design of solids with polar arrangement. *ACS Symp. Ser.*, **455**, 472–83. [209]

Porter, M. W. and Codd, L. W. (1963). Crystals of anorthic system. In *The Barker index of crystals. A method for the identification of crystalline substances*, Vol. 3, W. Heffer & Sons, Cambridge. [14]

Porter, M. W. and Spiller, R. C. (1951). Crystals of the tetragonal, hexagonal, trigonal and orthorhombic systems. In *The Barker index of crystals. A method for the identification of crystalline substances*, Vol. 1, W. Heffer & Sons, Cambridge. [14, 95]

Porter, M. W. and Spiller, R. C. (1956). Crystals of monoclinic system. In *The Barker index of crystals. A method for the identification of crystalline substances*, Vol. 2, W. Heffer & Sons, Cambridge. [14]

Potter, B. S., Palmer, R. A., Withnall, R., Chowdhry, B. Z. and Price, S. L. (1999). Aza analogues of nucleic acid bases: experimental determination and computational prediction of the crystal structure of anhydrous 5-azauracil. *J. Mol. Struct.*, **486**, 349–61. [184]

Potts, G. D., Jones, W., Bullock, J. F., Andrews, S. J. and Maginn, S. J. (1994). The crystal structure of quinacridone: an archetypal pigment. *J. Chem. Soc., Chem. Commun.*, 2565–6. [115, 262, 263]

Power, L. F., Turner, K. E. and Moore, F. H. (1976). The crystal and molecular structure of α-glycine by neutron diffraction—a comparison. *Acta Crystallogr. B*, **32**, 11–6. [56]

Powers, H. E. C. (1971). Nucleation and the sugar industry. *Zeitschrift Zukerindustrie*, **21**, 272–7. [69, 70]

Price, B. J., Clitherow, J. W., Bradshaw, J. W. (1978). Aminoalkyl furan derivatives. *US Patent* 4,128,568. [298]

Price, S. L. (2000). Toward more accurate model of intermolecular potentials for organic molecules. *Rev. Comput. Chem.*, **14**, 225–89. [166]

Price, S. L. and Stone, A. J. (1992). Electrostatic models for polypeptides: Can we assume transferability? *J. Chem. Soc., Faraday Trans.*, **88**, 1755–63. [166]

Price, S. L. and Wiley K. S. (1997). Predictions of crystal packings for uracil, 6-azauracil and allopurinol: the interplay between hydrogen bonding and close packing. *J. Phys. Chem. A*, **101**, 2198–206. [184]

Prodic-Kojic, B., Kajfes, F., Belin, B., Toso, R. and Sunjuc, V. (1979). Study of crystalline forms of *N*-cyano-*N'*-methyl-*N''*-2-[[(4-methyl-1*H*-imidazol-5-yl)methyl]thio]guanidine (cimetidine). *Gazz. Chim. Ital.*, **109**, 535–9. [73]

Prokop'eva, I. A. and Davidov, A. A. (1975). *Akad. Nauk SSSR, Ural Nauk Is.* 34. [203]

Prusiner, P. and Sundaralingam, M. (1976). The crystal and molecular structures of two polymorphic crystalline forms of virazole (1-β-D-ribofuranosyl-1,2,4-triazole-3-carboxamide). A new synthetic broad spectrum antiviral agent. *Acta Crystallogr. B*, **32**, 419–26. [159]

Pushkina, L. L. and Shelyapin, O. P. (1988). Study of preparation of a pigment form of *cis*-naphthoylenebisbenze. *J. Appl. Chem. USSR*, **61**, 2296–301; English translation of *Zh. Prikl. Khim.* (Leningrad) **61**, 2515–9. [261t]

Putz, H., Schon, J. C. and Jansen, M. (1999). Combined method for ab initio structure solution from powder diffraction data. *J. Appl. Crystallogr.*, **32**, 864–70. [111]

Ramamurthy, V. and Venkatesan, K. (1987). Photochemical reactions of organic crystals. *Chem. Rev.*, **87**, 433–81. [236]

Raman, C. V. and Krishnan, K. S. (1928). A new type of secondary radiation. *Nature*, **121**, 501–2. [131]

Ramdas, S., Thomas, J. M., Jordan, M. E. and Eckhardt, C. J. (1981). Enantiomeric intergrowths in hexahelicenes. *J. Phys. Chem.*, **85**, 2421–5. [30]

Rao, C. N. R. (1984). Phase transitions and the chemistry of solids. *Acc. Chem. Res.*, **17**, 83–9. [188]

Rao, C. N. R. (1987). Phase transitions in organic solids. In *Studies in organic chemistry*, Vol. 32, *Organic solid-state chemistry* (ed. G. Desiraju), pp. 371–432. Elsevier, Amsterdam. [188]

Rao, C. N. R. and Gopalakrishnan, J. (1997). *New directions in solid state chemistry*, 2nd. edn., pp. 168–228. Cambridge University Press, Cambridge. [188]

Rao, C. N. R. and Rao, K. J. (1978). *Phase transitions in solids*, McGraw-Hill, New York. [223]

Rastogi, S. and Kurelec, L. (2000). Polymorphism in polymers; its implications for polymer crystallization. *J. Mater. Sci.*, **35**, 5121–38. [28]

Readings, M. (1993). Modulated differential scanning calorimetry—a new way forward in material characterization. *Trends Polym. Sci.*, **1**, 248–53. [108]

Reed, S. M., Weakley, T. J. R. and Hutchison, J. E. (2000). Polymorphism in a conformationally flexible substituted anthraquinone; a crystallographic, thermodynamic, and molecular modeling study. *Cryst. Eng.*, **3**, 85–99. [75, 162, 182, 214]

Reetz, M. T., Höger, S. and Harms, K. (1994). Proton-transfer-dependent reversible phase changes in the 4,4′-bipyridinium salt of squaric acid. *Angew. Chem., Int. Ed. Engl.*, **33**, 181–3. [214, 218]

Reilly, J. and Rae, W. N. (1954). *Physicochemical methods*, 5th edn., Vol. 1, pp. 507–608. van Nostrand, New York, U. S. A. [41]

Reinke, H., Dehne, H. and Hans, M. (1993). A discussion of the term 'polymorphism'. *J. Chem. Educ.*, **70**, 101. [4, 17]

Reznichenko, V. V., Chesnovskaya, E. S., Podrezova, T. N. and Ezhkova, Z. I. (1984). Effect of the method of synthesis on allotropic properties of [2,2]-bi(naphtho[2,1-b]thiophenylidene]-1,1′-dione. *J. Appl. Chem. USSR*, **57**, 194–6; English translation of (1982). *Zhurnal Praktische Khimie* (Leningrad) **57**, 205–7. [267]

Ribka, J. (1961) German patent 1,208,435. [261t]

Ribka, J. (1969) German patent 1,287,731. [261t]

Ribka, J. (1970) German patent 2,043,482. [261t]

Richards, F. M. and Lindley, P. F. (1999). Determination of the density of solids. In *International tables for crystallography* (eds. A. J. C. Wilson and E. Prince), Vol. C, 2nd edn., pp. 156–9. Kluwer Academic Publishers, Dordrecht, The Netherlands. [147]

Richardson, M. R., Yang, Q., Novotny-Bregger, E. and Dunitz, D. J. (1990). Conformational polymorphism of dimethyl 3,6-dichloro-2,5-dihydroxyterphthalate. II. Structural, thermodynamic, kinetic and mechanistic aspects of phase transformations among the three crystal forms. *Acta Crystallogr. B*, **46**, 653–60. [178, 215]

Rieper, W. and Baier, E. (1992). Neue Kristallmodifikationen von C.I. Pigment Yellow 16. European Patent EP 054072. [261t]

Riley, P. E. and Davis, R. E. (1976). Crystal and molecular structures at 35 °C of two crystal forms of bis(2,6-dimethylpyridine)chromium, a bis heterocyclic sandwich complex. *J. Inorg. Chem.*, **15**, 2735–40. [162]

Ripmeester, J. A. (1980). Application of ^{13}C NMR to the study of polymorphs, clathrates and complexes. *Chem. Phys. Lett.*, **74**, 536–8. [134]

Roberts, R. M. (1989). *Serendipity. Accidental discoveries in science.* Wiley, New York. [299]

Robertson, J. M. (1934). An X-ray study of the structure phthalocyanines. I. The metal-free, nickel, copper and platinum compounds. *J. Chem. Soc.*, 615–21.

Robertson, J. M. (1935). An X-ray study of the structure of the phthalocyanines. I. The metal-free, nickel, copper and platinum compounds. *J. Chem. Soc.*, 615–21.

Robertson, J. M. (1936). An X-ray study of the phthalocyanines. II. Structure determination of the metal-free compound. *J. Chem. Soc.*, 1195–209. [266, 270]

Robertson, J. M. (1935). X-ray analysis of the structures of bibenzyl. *Proc. R. Soc. (London)*, **A150**, 348–62. [266, 270]

Robertson, J. M. (1953). *Organic crystals and molecules*, Cornell University Press, Ithaca, New York.

Robertson, J. M. and Woodward, I. (1937). An X-ray study of the phthalocyanines. Part III. Quantitative structure determination of nickel phthalocyanines. *J. Chem. Soc.*, 219–30. [266]

Robles, L., Mondieig, D., Haget, Y. and Cuevas-Diarte, M. A. (1998). Review on the energetic and crystallographic behaviour of n-alkanes. II. Series from $C_{22}H_{46}$ to $C_{27}H_{56}$. *J. Chim. Phys. Phys.-Chim. Biol.*, **95**, 92–111. [28]

Rockley, N. L., Woodard, M. K. and Rockley, M. G. (1984). The effect of particle size on FT-IR-PAS spectra. *Appl. Spectrosc.*, **38**, 329–34. [129]

Rodriguez-Hornedo, N. and Murphy, D. (1999). Significance of controlling crystallization mechanisms and kinetics in pharmaceutical systems. *J. Pharm. Sci.*, **88**, 651–60. [146f, 256]

Romé de l'Isle J. B. L. (1783). *Cristallographie ou description des formes propres a tous les corps de Règne Minéral dans l'Etat dede Combinasion Saline, Pierreuse ou Metallique* (4 vols). Imprimerie de Monsieur: Paris. [19]

Roozeboom, H. W. B. (1911). Die Heterogenen Gleichgewichte von Standpunkte der Phasen lehre. Braunschweig: FriedrichVieweg & Son. [29]

Rosenstein, S. and Lamy, P. P. (1969). Polymorphism. *Am. J. Hos. Pharm.*, **26**, 598–601. [3]

Roston, D. A., Walters, M. C., Rhineberger, R. R. and Ferro, L. J. (1993). Characterization of polymorphs of a new antiinflammatory drug. *J. Pharm. Biomed. Anal.*, **11**, 293–300. [129]

Rovira, C. and Novoa, J. J. (2001). A first principles computation of the low-energy polymorphic form of the acetic acid crystal. A test of the atom–atom force field predictions. *J. Phys. Chem. B*, **105**, 1710–19. [184]

Row, T. N. G. (1999). Hydrogen and fluroine in crystal engineering: systematics from crystallographic studies of hydrogen bonded tartrate-amine complexes and flurosubstituted coumarins, styrylcoumarins and butadienes. *Coord. Chem. Rev.*, **183**, 81–100. [236]

Ruch, T. and Wallquist, O. (1997). Crystal modification of a diketopyrrolopyrrole pigment. European Patent 0825234. [261t]

Rudel, P., Odiot, S., Mutin, J. C. and Peyrard, M. (1990). Crystal-structure and explosive power of molecular crystals. *J. Chim. Phys. Phys.-Chim. Biol.*, **87**, 1307– 44. [275]

Russell, T. P., Miller, P. J., Piermani, G. J. and Block, S. (1992). High pressure phase transition in γ-hexanitrohexaazaisowurtzitane. *J. Phys. Chem.*, **96**, 5509–12. [282]

Russell, T. P., Miller, P. J., Piermani, G. J. and Block, S. (1993). Pressure/temperature phase diagram of hexanitrohexaazaisowurtzitane. *J. Phys. Chem.*, **97**, 1993–7. [283]

Russell, T. P., Piermani, G. J., Block, S. and Miller, P. J. (1996). Pressure, temperature reaction phase diagram ammonium dinitramide. *J. Phys. Chem.*, **100**, 3248–51. [287]

Russell, T. P., Piermani, G. J. and Miller, P. J. (1997). Pressure/temperature and reaction phase diagram for dinitro azetidinium dinitramide. *J. Phys. Chem. B*, **101**, 3566–70. [286, 287]

Rustichelli, C., Gamberini, G., Ferioli, V., Gamberini, M. C., Ficarra, R. and Tommasini, S. (2000). Solid state study of polymorphic drugs: carbamazepine. *J. Pharm. Biomed. Anal.*, **23**, 41–54. [148, 250]

Ryzhkov, L. R. and McBride, J. M. (1996). Low-temperature reactions in single crystals of two polymorphs of the polycyclic nitramine N-15-HNIW. *J. Phys. Chem.*, **100**, 163–9. [283]

Saindon, P. J., Cauchon, N. S., Sutton, P. A., Chang, C. J., Peck, G. E. and Byrn, S. R. (1993). Solid-state nuclear-magnetic-resonance (NMR) spectra of pharmaceutical dosage forms. *Pharm. Res.*, **10**, 197–203. [243]

Saito, G. (1997). Organic superconducting solids. In *Organic molecular solids. Properties and applications* (ed. W. Jones), pp. 309–40. CRC Press, Boca Raton, FL. [191]

Sakaguchi, I. and Hayashi, Y. (1981). Sunmitomo Chemical Co. Ltd. Patent DE 3 031 444 A1. [260t]

Sakaguchi, I. and Hayashi, Y. (1982). Sunmitomo Chemical Co. Ltd. Patent DE 3 211 607 A1. [260t]

Saklatvala, R., Royall, P. G. and Craig, D. Q. M. (1999). The detection of amorphous material in a nominally crystalline drug using modulated temperature DSC—a case study. *Int. J. Pharm.*, **192**, 55–62. [255]

Salje, E. K. H. (1990). *Phase transitions in ferroelastic and co-elastic crystals*, Cambridge University Press, Cambridge. [223]

Samsonov, G. V. (ed.) (1976). *Physicheskie Svoistva Elementov*. Spravochnik, Moscow. [18]

Sanchis, M. J., Marthe, S., Diaz Calleja, R., Sanchez, Martinez, E., Epple, M. and Klar, G. (1997). The thermally induced phase transition in 2,3,7,8-tetramethoxythianthrene. *Ber. Bunsen-Ges. Phys. Chem.*, **101**, 1889–95. [237]

Sandefur, C. W. and Thomas, T. J. (1984). Modified acid dyestuff. Mobay Chemical Corp. Patent US 4 474 577. [260t]

Sano, N. and Tanaka, J. (1986). Electronic spectra of two polymorphs of (5-dimethylamino-2,4-pentadienylidene)dimethylammonium perchlorate. *Bull. Chem. Soc. Jpn.*, **59**, 843–51. [231]

Sappok, R. (1978). Recent progress in the physical chemistry of organic pigments with special reference to phthalocyanines. *J. Oil Colour Chem. Assoc.*, **61**, 299–308. [206f]

Sarma, J. A. R. P. and Desiraju, G. R. (1985). The chloro-substituent as a steering group: a comparative study of non-bonded interactions and hydrogen bonding in crystalline chloroaromatics. *Chem. Phys. Lett.*, **117**, 160–4. [166]

Sato, H., Suzuki, K., Okada, M. and Garti, N. (1985). Solvent effects on kinetics of solution-mediated transition of stearic acid polymorphs. *J. Cryst. Growth*, **72**, 699–704. [28]

Sato, K, and Suzuki, M. (1986). Solvent crystallization of α, β, and γ polymorphs of oleic acid. *J. Am. Oil Chem. Soc.*, **63**, 1356–9. [28]

Sato, K. (1999). Solidification and phase transformation behavior of food fats—a review. *Fett/Lipid*, **101**, 467–74. [28, 111]

Sato, K. and Boistelle, R. (1984). Stability and occurrence of polymorphic modifications of stearic acid in polar and nonpolar solutions. *J. Cryst. Growth*, **66**, 441–50. [67, 71]

Sawada, K., Hashizume, D., Sekine, A., Uekusa, H., Kato, K., Ohashi, Y., Kakinuma, K. and Ohgo, Y. (1996). Four polymorphs of a cobaloxime complex with different solid-state photoisomerization rates. *Acta Crystallogr. B*, **52**, 303–13. [237]

Schaefer, J. and Stejskal, E. O. (1976). Carbon-13 nuclear magnetic resonance of polymers spinning at the magic angle. *J. Am. Chem. Soc.*, **98**, 1031–2. [133]

Schaeffer, W. P. and Marsh, R. E. (1969). Oxygen-carrying cobalt compounds. I. Bis(salicylaldehyde)ethylenediiminecobalt(II) monochloroformate. *Acta Crystallogr. B*, **25**, 1675–2. [238]

Scheidt, W. R., Geiger, D. K. Hayes, R. G. and Lang, G. (1983). Control of spin state in (porphinato)iron(III) complexes. Axial ligand orientation effect leading to an intermediate-spin complex. Molecular structure and physical characterization of the monoclinic form of bis(3-chloropyridine)(octaethylporphinato)iron(III) perchlorate. *J. Am. Chem. Soc.*, **105**, 2625–32. [198]

Schinzer, W. C., Bergren, M. S., Aldrich, D. S., Chao, R. S., Dunn, M. J., Jeganathan, A. and Madden, L. M. (1997). Characterization and interconversion of polymorphs of premafloxacin, a new quinoline antibiotic. *J. Pharm. Sci.*, **86**, 1426–31. [249]

Schlueter, J. A., Carlson, K. D., Williams, J. M., Geiser, U., Wang, H. H., Welp, U., Kowk, W.-K., Fendrich, J. A., Dudek, J. D., Achenbach, C. A., Keane, P. M., Komosa, A. S., Naumann, D., Roy, T., Schirber, J. E. and Bayless, W. R. (1994). A new 9K superconducting organic salt composed of the bis(ethylthio)fulvalene (ET) electron-donor molecule and tetrakis(trifluoromethyl) cuprate(III) anion, [Cu(CF3)4]. *Physica C*, **230**, 378–84. [191]

Schmid, S., Muller-Goymann, C. C. and Schmidt, P. C. (2000). Interactions during aqueous film coating of ibuprofen with Aquacoat ECD. *Int. J. Pharm.*, **197**, 35–9. [243]

Schmidt, A., Kababya, Appel, M., Khatib, S., Botoshansky, M. and Eichen, Y. (1999). Measuring the temperature width of a first-order single crystal to single crystal phase transition using solid-state NMR: Application to the polymorphism of 2-(2,4-dinitrobenzyl)-3-methylpyridine. *J. Am. Chem. Soc.*, **121**, 11291–9. [140]

Schmidt, G. M. J. (1964). Topochemistry. Part III. The crystal chemistry of some *trans*-cinnamic acids. *J. Chem. Soc.*, 2014–21. [22, 234]

Schmidt, G. M. J. (1971). Photodimerization in the solid state. *Pure Appl. Chem.*, **27**, 647–78. [1, 235]

Schmidt, M. U. (1995). *Ph.D. Dissertation.* Technische Hochschule Aachen, Aachen. [185]

Schmidt, M. U. (1999). Process for preparing new crystalline modifications of C.I. Pigment Red 53:2. European Patent EP 1010732 A1. [140]

Schmidt, M. U. (2000). Process for preparing new crystalline modifications of C. I. Pigment Red 53:2. Clariant GMBH Patent DE 19858853 A1, EP 1010732 A1. [260t]

Schmidt, M. U. and Englert, U. (1996). Prediction of crystal structures. *J. Chem. Soc., Dalton Trans.*, 2077–82. [185]

Schmidt, M. U. and Metz, H. J. (1998*a*). Novel crystal modification of C.I. Pigment Red 53:2 (gamma-phase) German Patent DE 19827272.3. [260t]

Schmidt, M. U. and Metz, H. J. (1998*b*). Novel crystal modification of C.I. Pigment Red 53:2 (delta-phase) German Patent DE 19827273.1. [260t]

Schmidt, M. U. and Metz, H. J. (1999*a*). Novel crystal modification of C.I. Pigment Red 53:2 (gamma-phase) European Patent EP 965617 A1. [260t]

Schmidt, M. U. and Metz, H. J. (1999*b*). Novel crystal modification of C.I. Pigment Red 53:2 (delta-phase) European Patent EP 965616 A1. [260t]

Schorlemmer, C. (1874). *The carbon compounds or organic chemistry.* MacMillan & Co., London.

Schui, F., Deubel, R. and Wester, N. (1988). β-Modified crystalline red pigment. Hoechst AG Patent DE 3 2223 888; US 4 719 292. [260t]

Schwarz, E. and de Buhr, J. (1998). Collected applications. Thermal analysis. Pharmaceuticals, Mettler-Toledo GmbH, Schwerzenbach. [105f, 106f, 110f]

Schwarz, E. and Pfeffer, S. (1997). Use of subambient DSC for liquid and semi solid dosage forms—Pharmaceutical product development and quality control. *J. Therm. Anal.*, **48**, 557–67. [251]

Schwiebert, K. E., Chin, D. N., MacDonald, J. C. and Whitesides, G. M. (1995). Engineering the solid state with 2-benzimidazolones. *J. Am. Chem. Soc.*, **118**, 4018–29. [185]

Scottish Dyes (1929*a*). Patent GB 322 169. [266]

Scottish Dyes (1929*b*). Patent DE 586 906. [266]

Selig, W. (1982). New adducts of octahydro-1,3,5,7-tetranitro-1,3,5,7-tetrazocine (HMX). *Propell. Explos.*, **7**, 70–7. [278]

Senechal, M. (1990). Brief history of geometrical crystallography. In *Historical atlas of crystallography* (ed. J. Lima-de- Faria), pp. 43–59. Kluwer Academic Publishers, Dordrecht, The Netherlands. [10]

Serajuddin, A. T. M., Thakur, A. B., Ghoshal, R. N., Fakes, M. G., Ranadive, S. A., Morris, K. R. and Varia, S. A. (1999). Selection of solid dosage form composition through drug-excipient compatibility testing. *J. Pharm. Sci.*, **88**, 696–704. [243]

Serbotuviez, C., Nicoud, J.-F., Fischer, J., Ledoux, I. and Zyss, J. (1994). Crystalline zwitterionic stilbazolium derivatives with large quadratic optical nonlinearities. *Chem. Mater.*, **6**, 1358–68. [213]

Sergio, S. T. (1978). *Studies of the polymorphs of RDXI*, M.Sc. Thesis, University of Delaware, Newark, Delaware. [283]

Seyer, J. J., Luner, P. E. and Kemper, M. S. (2000). Application of diffuse reflectance near infrared spectroscopy for determination of crystallinity. *J. Pharm. Sci.*, **89**, 1305–16. [254]

Shah, J. C., Chen, J. R. and Chow, D. (1999). Metastable polymorph of etoposide with higher dissolution rate. *Drug Dev. Ind. Pharm.*, **25**, 63–67. [244, 252]

Shaik, S. S. (1982). On the stability and properties of organic metals and their isomeric charge-transfer complexes. *J. Am. Chem. Soc.*, **104**, 5328–34. [190]

Shalaev, E. Y. and Zografi, G. (1996). Interrelationships between phase transformation and organic chemical reactivity in the solid state. *J. Phys. Org. Chem.*, **9**, 729–38. [223]

Shankland, K., David, W. I. F. and Sivia, D. S. (1997). Routine *ab initio* structure determination of chlorothiazole by X-ray powder diffraction using optimized data collection and analysis strategies. *J. Mater. Chem.*, **7**, 569–72. [111]

Sharma, B. D. (1987). Allotropes and polymorphs. *J. Chem. Educ.*, **64**, 404–7. [4, 17]

Sharma, S. and Radhakrishnan, T. P. (2000). Modeling polymorphism-solvated supramolecular clusters reveal the solvent selection of SHG active and inactive dimorphs. *J. Phys. Chem. B*, **104**, 10191–5. [213]

Sharp, J. H., Miller, R. L. and Lardon (1972). M.A. Xerox Corporation Patent US 3 862 127; DE-OS 1 944 021 (1970); GB 1 268 422 (1972); FR 2 016 641 (1972). [266]

Shekunov, B. Y. and York, P. (2000). Crystallization processes in pharmaceutical technology and drug delivery design. *J. Cryst. Growth*, **211**, 122–36. [254]

Shelyapin, O. P., Pushkina, L. L., Kaugina, T. I., Tonchilova, V. F., Yaroshevich, G. and Peskova, V. I. (1987). cis- Naphthoylenebisbenzimidazole in a β-modification. Patent SU 1 310 415; *Chem. Abstr.*, **107**, 219146*p*. [261t]

Shenmin, Z., Dongzhi, L. and Shengwu, R. (1992). A study of the synergism and crystal form of some dichlorobenzidine disazo yellow pigments. *Dyes and Pigments*, **18**, 137–49. [261t, 272f, 273f, 273]

Sherwood, J. N. and Gallagher, H. G. (1984). The influence of lattice imperfections on the chemical reactivity of solids. *Final Technical Report* European Research Office of the U. S. Army, Contract Number DAJA-37-81-M-0395. Avail. NTVS Gov. Rep. Announce. Index (NS) 1984, **84**, 83. [287]

Shibaeva, R. P., Yagubskii, E. B., Laukhina, E. E. and Laukhin, V. N. (1990). Organic conductors and superconductors based on (BEDT-TTF)-polyiodides In *The physics and chemistry of organic superconductors* (ed. G. Saito and S. Kagoshima), pp. 342–8. Springer-Verlag, Berlin. [79]

Shibata, M., Kokubo, H., Morimoto, K., Morisaka, K., Ishida, T. and Inoue, M. (1983). X-ray structural studies and physicochemical properties of cimetidine polymorphism. *J. Pharm. Sci.*, **17**, 1436–42. [73]

Shiftan, D., Ravenelle, F., Mateescu, M. A. and Marchessault, R. H. (2000). Change in the V/B polymorph ratio and T-1 relaxation of epichlorohydrin crosslinked high amylose starch excipient. *Starch-Starke*, **52**, 186–95. [243]

Shigorin, V. D. and Shipulo, G. P. (1972). Use of the effect of second harmonic generation during a determination of crystal symmetry. *Kvantovaya Electron. (Moscow)*, **4**, 116–8. [212]

Shimura, K., Tada, K. and Imai, H. (1988). Crystal transformation of benzanthrone compound—shown by X-ray diffraction pattern, giving good dyeing properties. Nippon Kayaku, Co. Ltd. Patent JP 63010672; *Chem. Abstr.*, **109**, 8033c. [261t]

Shriner, R. L., Hermann, C. K. F., Morrill, T. C., Curtin, D. Y. and Fuson, R. C. (1997). *The systematic identification of organic compounds*, 7th edn. John Wiley & Sons, New York. [90]

Shtanov, N. P., Moroz, V. A. and Tikhonov V. I. (1980). Identification of polymorphic forms of the pigment bordeaux anthraquinone. *Khim. Tekhnol.* (Kiev) **1980**, 23–5; *Chem. Abstr.*, **93**, 96873 f. [261t]

Shtanov, N. P., Moroz, V. A., Tikhonov V. I. and Rogovik, V. I. (1981). Relation of structure to the coloring properties of the pigment perylene scarlet. *Khim. Tekhnol.* (Kiev) **1981**, 19–20; *Chem. Abstr.*, **94**, 193745v. [261t]

Silvestri, H. H. and David A. Johnson v. Norman H. Grant and Harvey Alburn (1974). U. S. Court of Customs and Patent Appeals, Patent Appeal No. 8978; 496 F.2d 593. [299]

Singh, N. B., Singh, R. J. and Singh, N. P. (1994). Organic solid-state reactivity. *Tetrahedron*, **50**, 6441–93. [236]

Sirota, N. N. (1982). Certain problems of polymorphism (I). *Cryst. Res. Technol.*, **17**, 661–91. [9, 18]

Sitzmann, M. E., Gilardi, R., Butcher, R. J., Koppes, K. M., Stern, A. G., Thrasher, J. S., Trivedi, N. J. and Yang, Z. Y. (2000). Pentafluorosulfanylnitramide salts. *Inorg. Chem.*, **39**, 843–50. [286]

Smith, D. A. (1993). *Being an effective expert witness*. Thames Publishing, London. [298]

Smith, D. L. (1974). Structure of dyes and aggregates. Evidence from crystal structure analysis. *Photogr. Sci. Eng.*, **18**, 309–22. [229]

Smith, F. J., Armstrong, A. T. and McGlynn, S. P. (1966). Energy of excimer luminescence. V. Excimer fluorescence of naphthalene and methylnaphthalenes. *J. Chem. Phys.*, **44**, 442–8. [231]

Smith, G. D. and Bharadwaj, R. K. (1999). Quantum chemistry based force field for simulations of HMX. *J. Phys. Chem. B*, **103**, 3570–5. [281]

Smith, J., MacNamara, E., Raftery, D., Borchardt, T. and Byrn, S. (1998). Application of two-dimensional ^{13}C solid-state NMR to the study of conformational polymorphism. *J. Am. Chem. Soc.*, **120**, 11710–13. [158]

Smithells, C. J. (ed.) (1976). *Metals reference book*, 5th edn. Butterworths, London. [18]

Sommer, K. and Kruse, H. (1979). Modification of monoazo dispersion dye of benzene series-with improved dyeing stability. Hoechst AG Patent DE 2 835 544. [260t]

Sommer, R., Schulze, J. and Wolfrum, G. (1974). Exhaust dyeing of (semi) synthetic fibers—uniform yellow shades with a modified diazo dye from organic solvents. Bayer AG Patent DE 2 313 356. [260t]

Sorescu, D. C., Rice, B. M. and Thompson, D. L. (1997). Intermolecular potential for the hexahydro-1,3,5,7-trinitro-1,3,5-s-triazine crystal: a crystal packing, Monte Carlo, and molecular dynamics study. *J. Phys. Chem. B*, **101**, 798–808. [283]

Sorescu, D. C., Rice, B. M. and Thompson, D. L. (1998a). Molecular packing and NPT-molecular dynamics investigation of the transferability of the RDX intermolecular potential to 2,4,6,8,10,12-hexanitrohexaazaisowurtzitane. *J. Phys. Chem. B*, **102**, 948–52. [278, 282, 283]

Sorescu, D. C., Rice, B. M. and Thompson, D. L. (1998b). Isothermal–isobaric molecular dynamics simulations of 1,3,5,7-tetranitro-1,3,5,7-tetraazacyclooctane (HMX) crystals. *J. Phys. Chem. B*, **102**, 6692–5. [281, 283]

Sorescu, D. C., Rice, B. M. and Thompson, D. L. (1999). Theoretical studies of the hydrostatic compression of RDX, HMX, HNIW, and PETN crystals. *J. Phys. Chem. B*, **103**, 6783–90. [284]

Spek, A. L. (1990). Platon, an integrated tool for the analysis of the results of a single crystal structure determination. *Acta Crystallogr. A*, **46**, C34. [184]

Spietschta, E. and Tröster, H. (1988a). Mix-crystal pigments based on perylenetetracarbamides, process for preparing and their use. Hoechst AG Patent DE 3 436 206; US 4 769 460. [261t]

Spietschta, E. and Tröster, H. (1988b). Mix-crystal pigments based on perylenetetracarbamides, process for their preparation, and their use. Hoechst AG Patent DE 3 436 209; US 4 742 170. [261t]

Stam, C. H. (1972). Crystal structure of a monoclinic modification and the refinement of a triclinic modification of Vitamin A acid (retinoic acid), $C_{20}H_{28}O_2$. *Acta Crystallogr. B*, **28**, 2936–45. [162]

Stam, C. H. and MacGillavry, C. H. (1963). The crystal structure of the triclinic modification of Vitamin A acid. *Acta Crystallogr.*, **16**, 62–8. [162]

Stanley-Wood, N. G. and Riley, G. S. (1972). The effect of temperature on the physical nature of phenobarbitone produced by acid–base precipitation. *Pharm. Acta Helv.*, **47**, 58–64. [255]

Starbuck, J. Docherty, R., Charlton, M. H. and Buttar, D. (1999). A theoretical investigation of conformational aspects of polymorphism. Part 2. Diarylamines. *J. Chem. Soc., Perkin Trans. 2*, 677–91. [154, 168, 180]

Starr, T. L. and Williams, D. E. (1977a). Comparison of models for molecular hydrogen–molecular hydrogen and molecular hydrogen–helium anisotropic intermolecular repulsions. *J. Chem. Phys.*, **66**, 2054–7. [153]

Starr, T. L. and Williams, D. E. (1977b). Coulombic nonbonded interatomic potential functions derived from crystal-lattice vibrational frequencies in hydrocarbons. *Acta Crystallogr. A*, **33**, 771–6. [153]

Stawasz, M. E., Sampson, D. L. and Parkinson, B. A. (2000). Scanning Tunneling Microscopy investigation of the ordered structures of dialkylamino hydroxylated squaraines absorbed on highly oriented pyrolytic graphite. *Langmuir*, **16**, 2326–42. [205]

Steiner, T., Hinrichs, W., Saenger, W. and Gigg, R. (1993). Jumping crystals—X-ray structures of the 3 crystalline phases of (+/−)-3,4-di-O-acetyl-1,2,5,6-tetra-O-benzyl-*myo*-inositol. *Acta Crystallogr. C*, **49**, 708–19. [219, 223]

Steinmetz, H. (1915). Crystallographic study of some nitro derivatives of benzene. *Zeitschrift*, **54**, 467–97. [212]

Stephenson, G. A., Forbes, R. A. and Reutzel-Edens, S. M. (2001). Characterization of the solid state: quantitative issues. *Adv. Drug Deliv. Revs.*, **48**, 67–90. [117, 129, 142, 254]

Stezowski, J. J., Biedermann, P. U., Hildenbrand, T., Dorsch, J. A., Eckhardt, C. J. and Agranat, I. (1993). Overcrowded enes of the tricycloindane-1,3-dione series: Interplay of twisting, folding and pyramidalization. *J. Chem. Soc., Chem. Commun.*, 213–5. [220]

Stobbe, H. and Lehfeldt, A. (1925). Polymerization and depolymerization by light of different wavelengths. II. α and β-*trans* cinnamic acids, allo-cinnamic acids and their dimers. *Berichte*, **58B**, 2415–27. [234]

Stobbe, H. and Steinberger, F. K. (1922). Light reactions of the *cis*- and *trans*-cinnamic acids. *Berichte*, **55B**, 2225–45. [234]

Stockton, G. W., Godfrey, R., Hitchcock, P., Mendelsohn, R., Mowery, P. C., Rajan, S. and Walker, A. F. (1998). Crystal polymorphism in pendimathalin herbicide is driven by electronic delocalization in intramolecular hydrogen bonding. A crystallographic, spectroscopic and computational study. *J. Chem. Soc., Perkin Trans. 2*, 2061–71. [142, 154, 170, 178]

Störmer, H. and Laage, E. (1921). Truxillic acids. (III). Natural and artificial truxillic and truxinic acids. *Berichte*, **54B**, 77–84. [234]

Störmer, R. and Förster, G. (1919). The truxillic acids and truxones. *Berichte*, **52B**, 1255–72. [234]

Stoltz, M., Oliver, D. W., Wessels, P. L. and Chalmers, A. A. (1991). High resolution solid state ^{13}C NMR spectra of mofebutazone, phenylbutazone and oxyphenbutazone in relation to X-ray crystallographic data. *J. Pharm. Sci.*, **80**, 357–62. [140]

Stott, P. E., McCausland, C. W. and Parish, W. W. (1979). Polymorphic behavior and melting points of certain benzo crown ether compounds. *J. Heterocycl. Chem.*, **16**, 453–5. [158]

Stout, G. and Jensen, L. H. (1989). *X-ray structure determination. A practical guide*, p. 91. John Wiley & Sons, New York, U. S. A. [41, 46]

Stowell, G. W. (2001). Form of fluoxetine hydrochloride. *US patent* 6,316,672. [117]

Stranski, I. N. and Totomanov, D. (1933). Rate of formation of (crystal) nuclei and the Ostwald step rule. *Z. Phys. Chem.*, **163**, 399–408. [44]

Streng, W. H. (1997). Physical chemical characterization of drug substances. *Drug Disc. Today*, **2**, 415–26. [95, 240]

Strickland-Constable, R. F. (1968). *Kinetics and mechanism of crystallization*. Academic Press, London. [68]

Strohmeier, M., Orendt, A. M., Alderman, D. W. and Grant, D. M. (2001). Investigation of the polymorphs of dimethyl-3,6-dichloro-2,5-dihydroxyterephthalate by ^{13}C solid state NMR. *J. Am. Chem. Soc.*, **123**, 1713–22. [133]

Struve, W. S. (1959). U. S. Patent 2,844,485. [262]

Stull, D. R., Westrum, E. F. Jr. and Sinke, G. C. (1969). *The thermodynamics of organic compounds*. J. Wiley and Sons, New York. [30]

Sudo, S., Sato, K. and Harano, Y. (1991). Growth and solvent-mediated phase transition of cimetidine polymorphic forms A and B. *J. Chem. Eng. Jpn.*, **24**, 628–32. [70, 73, 255]

Sugawara, T., Matsushita, M. M., Izuoka, A., Wada, N., Takeda, N. and Ishikawa, M. (1994). An organic ferromagnet: α-phase crystals of 2-(2′5′-dihydroxyphenyl)-4,4,5,5-tetramethyl-4,5- dihydro-1*H*-imidazolyl-1-oxy-3-oxide (α-HQNN). *J. Chem. Soc., Chem. Commun.*, 1723–4. [201]

Sugeta, H. (1975). Normal vibrations and molecular conformations dialkyl disulfides. *Spectrochim. Acta, Part A*, **31A**, 1729–37. [225]

Sugeta, H., Go, A. and Miyazawa, T. (1972). S–S and C–S stretching vibrations and molecular conformations of dialkyl disulfides and cystine. *Chem. Lett.*, 83–6. [225]

Sugeta, H., Go, A. and Miyazawa, T. (1973). Vibrational spectra and molecular conformations of dialkyl disulfides. *Bull. Chem. Soc. Jpn.*, **46**, 3407–11. [225]

Sugita, Y.-H. (1988). Polymorphism of L-glutamic acid crystals and inhibitory substance for β-transition in beet molasses. *Agric. Biol. Chem.*, **52**, 3081–5. [27]

Suihko, E., Poso, A., Korhonen, O., Gynther, J., Ketolainen, J. and Paronen, P. (2000). Deformation behaviors of tolbutamide, hydroxypropyl-beta-cyclodextrin, and their dispersions. *Pharm. Res.*, **17**, 942–8. [256]

Suito, E. and Uyeda, N. (1963). Transformation and growth of a copper phthalocyanine crystal in organic suspension. *Kolloid Z. Z. Polym.*, **193**, 97–111. [268]

Suito, E. and Uyeda, N. (1965). Transformation and growth of a copper phthalocyanine crystal in organic suspensions. *Nippon Kagaku Zasshi*, **86**, 970–7, and references therein. [268]

Suito, E. and Uyeda, N. (1974). *Kolloid Z. Z. Polym.*, **19**, 77. [268]

Summers, M. P., Enever, R. P., Carless, J. E. (1977). Influence of crystal form on tensile strength of compacts of pharmaceutical materials. *J. Pharm. Sci.*, **66**, 1172–5. [256]

Sun, E. (1998). Excerpts from news conference, 15.10.98. IAPAC, International Association of Physicians in AIDS Care, 225 West Washington St., Suite 2200, Chicago, IL 60606 (iapac@iapac.org; http://www.iapac.org). [240]

Sun, C. Q. and Grant, D. W. J. (2001). Influence of crystal structure on the tableting properties of sulfamerazine polymorphs. *Pharm. Res.*, **18**, 274–80. [256]

Suryanarayanan, R. and Wiedmann, T. S. (1990). Quantitation of the relative amounts of anhydrous carbamazepine ($C_{15}H_{12}N_2O$) and carbamazepine dihydrate ($C_{15}H_{12}N_2O \cdot H_2O$) in a mixture by solid-state nuclear magnetic resonance (NMR). *Pharm. Res.*, **7**, 184–7. [142]

Susich, G. (1950). Identification of organic dyestuffs by X-ray powder diffraction. *Anal. Chem.*, **22**, 425–30. [258, 261t, 262t, 266, 270]

Süsse, P. and Wolf, A. (1980). A new crystalline phase of indigo. *Naturwissenschaften*, **67**, 453. [262t]

Suzuki, E. Shirotani, K.-I., Tsuda, Y. and Sekiguchi, K. (1985). Water content and dehydration behavior of crystalline caffeine hydrate. *Chem. Pharm. Bull.*, **33**, 5028–35. [115, 125]

Suzuki, M. and Ogaki, T. (1986). Crystallization and transformation mechanisms of α-, β- and γ-polymorphs of ultra-pure oleic acid. *J. Am. Oil Chem. Soc.*, **62**, 1602–4. [28]

Suzuki, Y., Fujita, T., Hayashi, Y. and Okayasu, H. (1989*a*). New crystal type copper phthalocyanine—is useful in ink, paint and coloring material for resins as high temperature use organic pigment. Sumitomo Chemical Co. Ltd. Patent JP 1153758; *Chem. Abstr.*, **111**, 235058z. [267]

Suzuki, Y., Fujita, T., Hayashi, Y. and Okayasu, H. (1989*b*). Novel crystal type copper phthalocyanine blue pigment—obtained from phthalic anhydride, urea, cuprous chloride, titanium tetrachloride and sulphophthalic acid. Sumitomo Chemical Co. Ltd. Patent JP 1153756. *Chem. Abstr.*, **111**, 235057y. [267]

Sweeting, L., Rheingold, A. L., Gingerich, J. M., Rutter, A. W., Spence, R. A., Cox, C. D. and Kim, T. J. (1997). Crystal structure and triboluminescence. 2,9-anthracenecarboxylic acid and its esters. *Chem. Mater.* **9**, 1103–15. [222]

SYBIL Molecular Modeling Software, Tripos, Inc. 1699 Hanley Road, St. Louis, MO, U. S. A. [181]

Szafranski, M. and Katrusiak, A. (2000). Phase transitions in the layered structure of diguanidinium tetraiodoplumbate. *Phys. Rev. B*, **61**, 1026–35. [238]

Szafranski, M., Czarnecki, P., Katrusiak, A. and Habrylo, S. (1992). DTA investigation of phase-transitions in 1,3-cyclohexanedione under high-pressures. *Solid State Commun.*, **82**, 277–81. [239]

Szelagiewicz, M., Marcoli, C., Cianferani, S., Hard, A. P., Vit, A., Burkhard, A., von Raumer, M., Hofmeier, U. C., Zilian, A., Francotte, E. and Schenker, R. (1999). *In situ* characterization of polymorphic forms the potential of Raman techniques. *J. Thermal Anal. Calorim.*, **57**, 23–43. [250]

Szeverenyi, N. M., Sullivan, M. J. and Maciel, G. E. (1982). Observation of spin exchange by two-dimensional Fourier transform carbon-13 cross- polarization— magic angle spinning. *J. Magn. Reson.*, **47**, 462–75. [141]

Tabei, H. Kurihara, T. and Kaino, T. (1987). Recrystallization solvent effects on second-order nonlinear optical organic materials. *Appl. Phys. Lett.*, **50**, 1855–7. [213]

Takezawa, H., Ohba, S. and Saito, Y. (1986). Structure of monoclinic *o*-aminobenzoic acid. *Acta Crystallogr. C*, **42**, 1880–1. [59]

Tammann, G. (1903). *Kristallisieren und Schmelzen*. Verlag von Johann Ambrosius Barth, Leipzig. [25]

Tammann, G. (1926). *The States of Aggregation* (trans. F. F. Mehl), pp. 116–57. Constable and Company, Ltd., London. [25, 29, 38]

Tammann, G., Elsner, H. and von Gronow, H. E. (1931). Über die spontane Kristallisation unter kühlter schmelzen, und übersättiger Lösungen. *Z. Anorg. Allg. Chem.*, **200**, 57–73. [67]

Tamura, M., Nakazawa, Y., Shiomi, D., Nozawa, K., Hosokoshi, Y., Ishikawa, M., Takahashi, M. and Kinoshita, M. (1991). Bulk ferromagnetism in the β-phase crystal of the *p*-nitrophenyl nitronyl nitroxide radical. *Chem. Phys. Lett.*, **186**, 401–4. [201]

Tamura, M., Shiomi, D., Hosokoshi, Y., Iwasawa, N., Nozawa, K., Kinoshita, M., Sawa, H. and Kato, R. (1993). Magnetic properties of phenyl nitronyl nitroxide. *Mol. Cryst. Liq. Cryst. Sci. Technol. A*, **232**, 45–52. [201]

Tanninem, V. P. and Ylirussi, J. (1992). X-ray powder diffraction profile fitting in quantitative determination of two polymorphs from their powder mixture. *Int. J. Pharmacol.*, **81**, 169–77. [124]

Tarling, S. E. (1992). Declaration to European Patent Office regarding action of the Opposition Division dated 23rd April 1992 in relation to EP 0 091 787. [305]

Tarling, S.E. (2001). Private communication to author; see also http://www.vino.demon.co.uk/ppxrd/cefad.html. [303, 306f]

Tausen (1935). U.S. Patent 2 020 665. [294]

Taylor, J. W. and Crookes, R. J. (1976). Vapor pressure and enthalpy of sublimation of 1,3,5,7-tetranitro-1,3,5,7-tetraazacyclooctane. *J. Chem. Soc., Faraday Trans. 1*, 723–30. [279t]

Taylor, L. S. and Zografi, G. (1998). The quantitative analysis of crystallinity using FT-Raman spectroscopy. *Pharm. Res.*, **15**, 755–61. [254]

Teetsov, A. S. and McCrone, W. C. (1965). The microscopical study of polymorph stability diagrams. *Micros. Cryst. Front*, **15**, 13–29. [279t, 280]

Terao, H., Itoh, Y., Ohno, K., Isogai, M., Kakuta, A. and Mukoh, A. (1990). 2nd order nonlinear optical properties and polymorphism of benzophenone derivatives. *Opt. Commun.*, **74**, 451–3. [213]

Terol, A., Cassanas, G., Nurit, J., Pauvet, B., Bouassab, A., Rambaud, J. and Chevallet, P. (1994). Infrared, Raman and ^{13}C NMR spectra of two crystalline forms of (1R,3S)-3-(*p*-thioanisoyl)-1m2m2-tritrimethylcyclopentanecarboxylic acid. *J. Pharm. Sci.*, **83**, 1437–42. [132]

Terpstra, P. and Codd, L. W. (1961). *Crystallometry*. Academic Press, New York. [95]

ThaiHVAC (2001). http://www.thaihvac.com/knowledge/cleanroom/cleanroo1.htm [91]

Thakar, A. L., Hirsch, C. A. and Page, J. G. (1977). Solid dispersion approach for overcoming bioavailability problems due to polymorphism of nabilone, a cannabinoid derivative. *J. Pharmacol. Pharm.*, **29**, 783–4. [245]

Thenard, L. J. and Biot, J. B. (1809). Mémoire sur l'analyse comparée de l'arragonite, et du carbonate de chaux rhomboidal. *Chem. Phys. II, Soc. D' Arcueil*, **2**, 176–206. [20]

Theocaris, C. R. and Jones, W. (1984). The thermally induced phase transition of crystalline 9-cyanoanthracene dimer: A single crystal study. *J. Chem. Soc., Chem. Commun.*, 369–84. [234, 237]

Theocharis, C. R., Jones, W. and Rao, C. N. R. (1984). An unusual photo-induced conformational polymorphism: A crystallographic study of bis(*p*-methoxy)-*trans*-stilbene. *J. Chem. Soc., Chem. Commun.*, 1291–3. [234]

Thomas, J. M. (1979). Organic reactions in the solid state: accident and design. *Pure Appl. Chem.*, **51**, 1065–82. [236]

Thomas, J. M., Morsi, S. E. and Desvergne, J. P. (1977). Topochemical phenomena in organic solid-state chemistry. *Adv. Phys. Org. Chem.*, **15**, 63–151. [235]

Thompson, K. C. (2000). Pharmaceutical applications of calorimetric measurements in the new millenium. *Thermochim. Acta*, **355**, 83–7. [250]

Threlfall, T. (1995). Analysis of organic polymorphs. A review. *The Analyst*, **120**, 2435–60. [4, 5, 7, 8, 85, 94, 105, 106, 128, 129, 134, 240, 253]

Threlfall, T. L. (1999). Developments in analysis of polymorphs and solvates 1st International symposium on aspects of polymorphism & crystallization—chemical development issues, Scientific Update, Mayfield, East Sussex, United Kingdom, Hinckley, UK, 23–25 June 1999 (http://www.scientificupdate.co.uk). [118f, 128, 139f]

Threlfall, T. L. (2000). Crystallization of polymorphs: Thermodynamic insight into the role of the solvent. *Org. Process Res. Dev.*, **4**, 384–90. [2, 92, 252, 255]

Threlfall, T. and Hursthouse, M. Private communication to author.

Tilley, R. (1999). *Colour and optical properties of materials*. John Wiley & Sons, Chichester. [257]

Timken, M. D., Chen, J. K. and Brill, T. B. (1990). Thermal decomposition of energetic materials. 37. SMATCH FT-IR (simultaneous mass and temperature-change FT-IR) spectroscopy. *Applied Spectrosc.*, **44**, 701–6. [148]

Timofeeva, T. V., Chernikova, N. Yu. and Zorkii, P. M. (1980). Theoretical calculation of the spatial distribution of molecules in crystals. *Russ. Chem. Rev.*, **49**, 509–25. (Translated from *Uspekhi Khimii*, **49**, 966–97 (1980).) [182]

Timofeeva, T. V., Nesterov, V. N., Dolgushin, F. M., Zubavichus, Y. V., Goldshtein, J. T., Sammeth, D. M., Clark, R. D., Penn, B. and Antipin, M. Yu. (2000). One-pot polymorphism of nonlinear optical materials. First example of organic polytypes. *Cryst. Eng.*, **3**, 263–88. [213]

Tipson, R. S. (1956). Crystallization and recrystallization. In *Techniques of organic chemistry* (ed. A. Weissberger) Vol. III, 2nd edn., pp. 395–561. Interscience, New York, U. S. A. [47, 67]

Tokura, Y., Kaneko, Y., Okamoto, H., Tanuma, S., Koda, T., Mitani, T. and Saito, G. (1985). Spectroscopic study of the neutral-to-ionic phase transition in TTF- chloranil. *Mol. Cryst. Liq. Cryst.*, **125**, 71–80. [196]

Toma, P. H., Kelley, M. P., Borchardt, T. B., Byrn, S. R. and Kahr, B. (1994). Chloroi-somers and polymorphs of 9-phenylacridinium hydrogen sulfate. *Chem. Mater.*, **6**, 1317–24. [214]

Tomita, Y., Ando, T. and Ueno, K. (1965). Three crystalline forms of iminodiacetic acid. *Bull. Chem. Soc. Jpn.*, **38**, 138–9. [158]

Torrance, J. B., Vasquez, J. E., Mayerle, J. J. and Lee, V. Y. (1981). Discovery of a neutral-to-ionic phase transition in organic materials. *Phys. Rev. Lett.*, **46**, 253–7. [195]

Toscani, S. (1998) An up-to-date approach to drug polymorphism. *Thermochim. Acta*, **321**, 73–9. [252]

Tremayne, M., Kariuki, B. and Harris, K. D. M. (1997). Structure determination of a complex organic solid from X-ray powder diffraction data by a generalized Monte Carlo method: the crystal structure of red fluorescein. *Angew. Chem., Int. Ed. Engl.*, **36**, 770–2. [111]

Trevor, J. E. (1902). The nomenclature of variance. *J. Phys. Chem.*, **6**, 136–7. [31]

Tristani-Kendra, M. and Eckhardt, C. J. (1984). Influence of crystal fields on the quasimetallic reflection spectra of crystals: optical spectra of polymorphs of a squarylium dye. *J. Chem. Phys.*, **90**, 1160–73. [205, 231f, 231, 232f]

Tristani-Kendra, M., Eckhardt, C. J., Bernstein, J. and Goldstein, E. (1983). Strong coupling in the optical spectra of a squarylium dye. *Chem. Phys. Lett.*, **98**, 57–61. [230]

Trotter, J. (1963). *Acta Crystallogr.*, **16**, 1091–4. [284]

Trovão, M. C. N., Cavaleiro, A. M. V. and de Jesus, J. D. P. (1998). Preparation of poly-morphic crystalline phases of *D*-mannitol: influence of Keggin heteropolyanions. *Carbohydrate Res.*, **309**, 363–6. [141, 142f]

Tsuboi, M., Ueda, T. and Ikeda, T. (1991). Use of a Raman microscope in conforma-tional analysis of a peptide with polymorphism. In *Proc. Fourth Eur. Con. Spectros. Biol. Mole.* (ed. R. E. Hester and R. B. Girling), pp. 29–30. Special Publication #94, Royal Society of Chemistry, London. [305]

Tuck, B., Stirling, J. A., Farnoochi, C. J. and McKay, R. B. (1997). Polymorph of a yellow diarylide pigment. European Patent EP 0790282. [261t]

Tudor, A. M., Davies, M. C., Melia, C. D., Lee, D. C., Mitchell, R. C., Hendra, P. J. and Church, S. J. (1991). The applications of near-infrared Fourier trans-form Raman spectroscopy to the analysis of polymorphic forms of cimetidine. *Spectrochim. Acta Part A*, **47**, 1389–93. [132]

Tudor, A. M., Church, S. J., Hendra, P. J., Davies, M. C. and Melia, C. D. (1993). The qualitative and quantitative analysis of chlorpropanide polymorphic mixtures by near-infrared Fourier transform Raman spectroscopy. *Pharm. Res.*, **10**, 1772–6. [130, 132]

Turek, P., Nozawa, K., Shiomi, D., Awaga, K., Inabe, T., Maruyama, M. and Kinoshita, M. (1991). Ferromagnetic coupling in a new phase of the *p*-nitrophenyl nitronyl nitroxide radical. *Chem. Phys. Lett.*, **180**, 327–31. [200]

Tutton, A. E. H. (1911a). The work of Eilhardt Mitscherlich and his discovery of isomorphism. In *Crystals*, pp. 70–97. Kegan Paul, Trench, Trubner & Co. Ltd., London. [19,22, 25]

Tutton, A. E. H. (1911b). *Crystals*, p. 140, Kegan Paul, Trench, Trubner & Co. Ltd., London. [19, 22, 25]

Tutton, A. E. H. (1922). *Crystallography and practical crystal measurement*, Vol. 1, pp. 625–39. MacMillan, London. [41, 148]

Ullman, F. and Maag, R. (1906). Ueber Chinacridon. *Berichte*, **39**, 1693–6. [259]

Vachon M. G. and Grant, D. J. W. (1987). Enthalpy–entropy compensation in pharmaceutical solids. *Int. J. Pharm.*, **40**, 1–14. [243]

van der Sluis, P. and Kroon, J. (1989). Solvents and X-ray crystallography. *J. Cryst. Growth*, **97**, 645–56. [5]

van Eijck, B. and Kroon, J. (1999). UPACK program package for crystal structure prediction: force field and crystal structure generation for small carbohydrate molecules. *J. Comput. Chem.*, **20**, 799–812. [183, 184, 185, 186]

van Eijck, B., Mooij, W. T. M. and Kroon, J. (1995). Attempted prediction of the crystal structures of six mono-saccharides. *Acta Crystallogr. B*, **51**, 99–103. [183, 185]

van Eijck, B., Spek, A. L., Mooij, W. T. M. and Kroon, J. (1998). Hypothetical crystal structures of benzene at 0 and 30 kbar. *Acta Crystallogr. B*, **54**, 291–9. [185]

Van Hook, A. (1961). *Crystallization in theory and practice*. Reinhold, New York, and references therein. [67, 72]

van Putte, K. P. A. M. and Bakker, B. H. (1987). Crystallization kinetics of palm oil. *J. Am. Oil Chem. Soc.*, **64**, 1138–43. [72]

Van Wart, H. E. and Scheraga, H. A. (1976a). Raman spectra of cystine-related disulfides. Effect of rotational isomerism about carbon–sulfur bonds on sulfur–sulfur stretching frequencies. *J. Phys. Chem.*, **80**, 1812–22. [225]

Van Wart, H. E. and Scheraga, H. A. (1976b). Raman spectra of strained disulfides. Effect of rotation about sulfur-sulfur bonds on sulfur-sulfur stretching frequencies. *J. Phys. Chem.*, **80**, 1823–32. [225]

Van Wart, H. E. and Scheraga, H. A. (1977). Stable conformations of aliphatic disulfides: influence of 1,4 interactions involving sulfur atoms. *Proc. Nat. Acad. Sci. U. S. A.*, **74**, 13–17. [225]

Van Wart, H. E. and Scheraga, H. A. (1986). Agreement with the disulfide stretching frequency-conformation correlation of Sugeta, Go, and Miyazawa. *Proc. Nat. Acad. Sci. U. S. A.*, **83**, 3064–7. [225]

Veciana, J., Cirujeda, J., Rovira, C., Molins, E. and Novoa, J. J. (1996). Organic ferromagnets. Hydrogen bonded supramolecular magnetic organizations derived from hydroxylated phenyl α-nitronyl nitroxide radicals. *J. Phys. I Fr.*, **6**, 1967–86. [201]

Venkatesan, K. and Ramamurthy, V. (1991). Bimolecular photoreactions in crystals. In *Photochemistry in organized and constrained media* (ed. V. Ramamurthy), pp. 133–84. VCH Publishers, Weinheim. [236]

Veregin, R. P., Fyfe, C. A., Marchessault, R. H. and Taylor, M. G. (1986). Characterization of the crystalline A and N starch polymorphs and investigation of starch crystallization by high-resolution ^{13}C CP/MAS NMR. *Macromolecules*, **19**, 1030–4. [141]

Verma, A. R. and Krishna, P. (1966). *Polytypism and polymorphism in crystals*. Wiley, New York, U. S. A. [1, 19, 188, 213]

Verwer, P. and Leusen, F. J. J. (1998). Computer simulations to predict possible crystal polymorphs. In *Reviews in computational chemistry* (ed. K. B. Lipkowitz and D. B. Boyd), Vol. 12 pp. 327–65. Wiley-VCH, New York, U. S. A. [182, 186]

Vidine, D. W. (1982). In Fourier transform infrared spectroscopy (ed. J. R. Ferraro and L. J. Basile), Vol. 3, Academic Press, New York. [129]

Vippagunta, S. R., Brittain, H. G. and Grant, D. J. W. (2001). Crystalline Solids. *Advanced Drug Deliv. Sys.*, **48**, 3–26. [105, 240]

Vishnimurthy, K., Row, T. N. G. and Venkatesan, K. (1996). Studies in crystal engineering: effect of fluorine substitution in crystal packiing and topological photodimerization of styryl coumarins in the solid state. *J. Chem. Soc., Perkin Trans. 2*, 1475–8. [236]

Voght, L. H. Jr, Faigenbaum, H. H. and Wiberly, S. E. (1963). Synthetic reversible oxygen-carrying chelates. *Chem. Rev.*, **63**, 269–77. [238]

Voigt-Martin, I. G., Yan, D. H., Yakimansky, A., Schollmeyer, D., Gilmore, C. J. and Bricogne, G. (1995). Structure determination by electron crystallography using both maximum-entropy and simulation approaches. *Acta Crystallogr.*, **51**, 849–68. [112]

Volmer, M. (1939). *Kinetic der Phasenbildung*. Steinkopf, Leipzig. [43]

von Eller, H. (1955). Structure de colorants indigoïdes. III. Structure cristalline de l'indigo. *Bull. Soc. Chim. Fr.*, **106**, 1433–8. [262t]

von Rambach, L., Daubach, E. and Honigman, B. (1974). Crystalline modification of monoazo dye—from diazotised 2-cyan-4-nitro-aniline coupled with *N*-ethyl *N*-cyanethyl aniline, with improved stability. BASF AG Patent DE 2 249 739. [260t]

von Stackelber, M. (1947). X-ray investigations of inner-complex copper salts. *Z. Anorg. Allg. Chem.*, **253**, 136–60. [163]

Wagener, A. P. and Meisters, G. J. (1970*a*). 4,11-dichloroquinacridone pigments. Sherwin-Williams Co. Patent US 3 524 856. [261t]

Wagener, A. P. and Meisters, G. J. (1970*b*). Sherwin-Williams Co. Patent US 3 547 927-8. [261t]

Wagner, T. and Englert, U. (1997). Packing effects in organometallic compounds: a study of five mixed crystals. *Struct. Chem.*, **8**, 357–65. [154, 163]

Wahlstrom, E. E. (1969). *Optical crystallography*, 4th edn., John Wiley & Sons, New York, U. S. A. [95]

Waki, S. J. (1970). *Lipid metabolism*. Academic Press, New York. [247]

Wang, H. H., Ferraro, J. K., Carlson, K. D., Montgomery, L. K., Geiser, U., Williams, J. M., Whitworth, J. R., Schlueter, J. A., Hill, S., Whangbo, M.-H., Evain, M. and Novoa, J. J. (1989). Electron spin resonance, infrared spectroscopic, and molecular packing studies of the thermally induced conversion of semiconducting α-to semiconducting α_t-(BEDT-TTF)$_2$I$_3$. *Inorg. Chem.*, **28**, 2267–71. [78, 192]

Ward, M. D. (1997). Organic crystal surfaces: structure, properties and reactivity. *Curr. Opin. Colloid Interf. Sci.*, **2**, 51–64. [72]

Wardle, R. B., Hinshaw, J. C. and Hajik, R. M. (1990). Gas generating compositions containing nitrotriazalone. U. S. Patent 4 931 112. [285]

Webb, J. and Anderson, B. (1978). Problems with crystals. *J. Chem. Educ.*, **55**, 644. [20, 91]

Weber, C., Rustemeyer, F. and Durr, H. (1998). A light-driven switch based on photochromic dihydroindolizines. *Adv. Mater.*, **10**, 1348–51. [158, 216]

Webster, S., Smith, D. A. and Batchelder, D. N. (1998). Raman microscopy using a scanning near-field optical probe. *Vib. Spectrosc.*, **18**, 51–9. [132]

Weil, J. A. and Anderson, J. K. (1965). The determination and reaction of 2,2-diphenyl-1-picryl-hydrazyl with thiosalicylic acid. *J. Chem. Soc.*, 5567–70. [203]

Weissbuch, I., Addadi, L., Lahav, M. and Leiserowitz, L. (1991). Molecular recognition at crystal interfaces. *Science*, **253**, 637–45. [93, 252]

Weissbuch, I., Leiserowitz, L. and Lahav, M. (1994). Tailor-made charge-transfer auxiliaries for the control of the crystal polymorphism of glycine. *Adv. Mater.*, **6**, 952–6. [93]

Weissbuch, I., Popovitz-Biro, R., Lahav, M. and Leiserowitz, L. (1995). Understanding and control of nucleation, growth, habit, dissolution and structure of two- and three-dimensional crystals using 'tailor-made' auxilliaries. *Acta Crystallogr. B*, **51**, 115–48. [30, 252, 253]

Wells, A. F. (1946). Crystal habit and internal structure. I. *Philos. Mag.*, **37**, 184–99. [48]

Wells, A. F. (1989). *Structural inorganic chemistry*, 5th edn., pp. 294–315. Clarendon Press, Oxford. [1, 18, 55]

Westrum, E. F. Jr and McCullough, J. P. (1963). Thermodynamics of crystals. In *Physics and chemistry of the organic solid state* (ed. D. Fox, M. M. Labes and A. Weissberger), Vol. I pp. 1–178. Interscience, New York, U. S. A. [29, 38, 39f]

Weygand, C. and Baumgärtel, H. (1929). Isomerism relations in the chalcone series. VII. A natural system of the polymorphous forms of *p'*-methylchalcone. *Liebigs Annalen*, **469**, 225–56. [255]

Wheeler, I. (1979). Copper phthalocyanine pigment production—by reacting compound forming phthalocyanine ring system, benzophenone-tetracarboxylic acid, copper compound, catalyst and nitrogen source. Ciba-Geigy Patent GB 1 544 171. [266, 267]

Whitaker, A. (1977*a*). Fresh X-ray data for two forms of linear *trans*-quinacridone, C. I. Pigment Violet 19. *J. Soc. Dyers Colourists*, **93**, 15–17. [262]

Whitaker, A. (1977*b*). Chapter. 10 of *The analytical chemistry of synthetic dyes* (ed. K. Venkatraman), pp. 269–98. Wiley-Interscience, New York. [266, 270]

Whitaker, A. (1979). Crystal data for a second polymorph (*β*) of C. I. Pigment Red 1,1-[4-nitrophenyl)azo]-2-naphthol. *J. Appl. Crystallogr.*, **12**, 626–7. [260t]

Whitaker, A. (1980*a*). The crystal structure of a second polymorph(*β*) of C. I. Pigment Red 1,1-[(4-nitrophenyl)azo]-2-naphthol. *Z. Kristallogr.*, **152**, 227–37. [260t]

Whitaker, A. (1980*b*). Crystal data for a third polymorph (γ) of C. I. Pigment Red 1,1-[(4-nitrophenyl)azo]-2-naphthol. *J. Appl. Crystallogr.*, **13**, 458–9. [260t]

Whitaker, A. (1981). The crystal structure of a third polymorph (γ) of C. I. Pigment Red 1,1-[(4-nitrophenyl)azo]-2-naphthol. *Z. Kristallogr.*, **156**, 125–36. [260t]

Whitaker, A. (1982). The polymorphism of C. I. Pigment Red 1. *J. Soc. Dyers Colourists*, **98**, 436–9. [260t]

Whitaker, A. (1985*a*). The polymorphism of C. I. Pigment Yellow 5. *J. Soc. Dyers Colourists*, **101**, 21–4. [261t]

Whitaker, A. (1985*b*). The crystal structure of aceto-acetanilide azo pigments VII. A polymorph α) of C. I. Pigment Yellow 5, α -(1-hydroxyethylidene)acetanilide-α-azo-(2'nitrobenzene). *Z. Kristallogr.*, **171**, 17–22. [261t]

Whitaker, A. (1990). Crystal data for Solvent Yellow 18. *J. Soc. Dyers Colourists*, **106**, 108–9. [261t]

Whitaker, A. (1992). The polymorphism of C.I. Solvent Yellow 18. *J. Soc. Dyers Colourists*, **108**, 282–4. [261t]

Whitaker, A. (1995). X-ray powder diffraction of synthetic organic colorants. In *Analytical chemistry of synthetic colorants* (ed. A. T. Peters and H. S. Freeman), pp. 1–48, Blackie, Glasgow. [8, 259, 262t, 266, 270]

Whitesell, J. K., Davis, R. E., Saunders, L. L., Wilson, R. J. and Feagins, J. P. (1991). Influence of molecular dipole interactions on solid state organization. *J. Am. Chem. Soc.*, **113**, 3267–70. [176]

Willcock, J. D., Price, S. L., Leslie, M. and Catlow, C. R. A. (1995). The relaxation of molecular crystal structures using a distributed multipole electrostatic model. *J. Comput. Chem.*, **16**, 628. [183]

Williams, D. E. (1965). Crystallographic data for 2,2-diphenyl-1-picrylhydrazyl. *J. Chem. Soc.*, 7535–6. [203]

Williams, D. E. (1967). Crystal structure of 2,2-diphenyl-1-picrylhydrazyl free radical. *J. Am. Chem. Soc.*, **89**, 4280–7. [203]

Williams, D. E. (1974). Coulombic interactions in crystalline hydrocarbons. *Acta Crystallogr. A*, **30**, 71–7. [154, 166]

Williams, D. E. (1983). *PCK83*. QCPE Program 548. Quantum Chemistry Program Exchange, Chemistry Department, Indiana University, Bloomington, Indiana, U. S. A. [183]

Williams, D. E. (1996). *Ab initio* molecular packing analysis. *Acta Crystallogr. A*, **52**, 326–8. [185, 186]

Williams, D. E. and Hsu, L. -Y. (1985). Transferability of nonbonded Cl $\cdots$ Cl potential energy function to crystalline chlorine. *Acta Crystallogr. A*, **41**, 296–301. [166]

Williams, G. K. and Brill, T. B. (1995). Thermal-decompsition of energetic materials. 68. Decomposition and sublimation kinetics of NTO and evaluation of prior kinetic data. *J. Phys. Chem.*, **99**, 12536–9. [285]

Williams, J. M., Beno, M. A., Wang, H. H., Leung, P. C. W., Enge, Y. J., Geiser, U. and Carlson, K. D. (1985). Organic superconductors: structural aspects and design of new materials. *Acc. Chem. Res.*, **18**, 261–7. [189]

Williams, J. M., Schultz, A. J., Geiser, U., Carlson, K. D., Kini, A. M., Wang, H. H., Kwok, W. -K., Whangbo, M.-H. and Schirber, J. E. R. (1991). Organic super-conductors—new benchmarks. *Science*, **252**, 1501–8. [79, 191, 193]

Williams, K. P. J., Pitt, G. D., Batchelder, D. N. and Kip, B. J. (1994*a*). Confocal Raman microspectrometry using a stigmatic spectrograph and CCD detector. *Appl. Spectros.*, **48**, 232–5. [132]

Williams, K. P. J., Pitt, G. D., Smith, B. J. E., Whitley, A., Batchelder, D. N. and Hayward, I. P. (1994*b*). Use of rapid scanning stigmatic Raman imaging spectrograph in the industrial environment. *J. Raman Spectrosc.*, **25**, 131–8. [133]

Willis, M. R. (1986). Talk at British Crystallographic Association annual meeting, Herriot-Watt University, Edinburgh, Scotland. [188]

Winchell, A. N. (1943). *The optical properties of organic compounds*. University of Wisconsin Press, Madison. [14, 15f, 95]

Winchell, A. N. (1987). *The optical properties of organic compounds*. McCrone Research Institute, Chicago. [14, 15f]

Wingard, R. E. (1982). *IEEE Trans. Ind. Appl.*, **37**, 1251.

Winter, G. (1999). Polymorphs and solvates of molecular solids in the pharmaceutical industry. In *Reactivity of molecular solids, the molecular solid state* (ed. E. Boldyreva and V. Boldyrev), Vol. 3, pp. 241–70, John Wiley & Sons, Ltd., Chichester. [240]

Wirth, D. D. and Stephenson, G. A. (1997). Purification of dirithromycin. Impurity reduction and polymorph manipulation. *Org. Process Res. Dev.*, **1**, 55–60. [256]

Woehrle, D. (1989) in Leznoff, C. C. and Lever, A. B. P. (eds.). *Phthalocyanines: properties and applications*, VCH Verlagsgesellschaft, Weinheim. [265]

Wohler, F. and Liebig, J. (1832). Unter suchengen über das Radikal der Benzosäure. *Ann. Pharm.*, **3**, 249–82. [75]

Wojcik, G. and Marqueton, Y. (1989). The phase transition of *m*-nitrophenol. *Mol. Cryst. Liq. Cryst.*, **168**, 247–54. [81, 212]

Wojtyk, J., McKerrow, A., Kazmaier, P. and Buncel, E. (1999). Quantitative investigations of the aggregation behavior of hydrophobic anilino squaraine dyes through UV/vis spectroscopy and dynamic light scattering. *Can. J. Chem.—Rev. Can. Chem.*, **77**, 903–12. [88, 205]

Wolf, F., Koch, U., Heitrich, W. and Modrow, H.-W. (1986*a*). Martin-Luther Uniersität Halle-Wittenberg Patent DD 236 543. [260t]

Wolf, F., Koch, U. and Modrow, H.-W. (1986*b*). Martin-Luther Universität, Halle-Wittenberg Patent DD 236 544. [260t]

Wolff, J. J. (1996). Crystal packing and molecular geometry. *Angew. Chem., Int. Ed. Engl.*, **35**, 2195–7. [154]

Wollmann, H. and Braun, V. (1983). Die Anwendung der differenzthermoanlyse in der pharmazie. *Pharmazie*, **38**, 5–20. [250]

Wood, W. M. L. (1997). Crystal science techniques in the manufacture of chiral compounds. In *Chirality in industry II. Developments in the commercial manufacture and applications of optically active compounds* (ed. A. N. Collins, G. N. Sheldrake and J. Crosby), pp. 119–56. John Wiley & Sons, New York, U. S. A. [4, 47]

Woodard, G. D. (1970). Calibration of the Mettler FP2 hot stage. *Microscope*, **18**, 105–8. [95]

Woodard, G. D. and McCrone, W. C. (1975). Unusual crystallization behavior. *J. Appl. Cryst.*, **8**, 342. [91]

Wrede, F. (1841). Anwendung des polarisirten Lichtes bei mikroskop. Untersuchungen Jahresbericht uber der Fortschritte und Mineralogie (Berzelius). *Tubingen*, **20**, 11–17. [21]

Wudl, F. (1984). From organic metals to superconductors: managing conducting electrons in organic solids. *Acc. Chem. Res.*, **17**, 227–32. [189]

Wyrouboff, M. G. (1890). Recherches sur le polymorphisme et la pseudosymétrie. *Bull. Soc. Mineral. Fr.*, **13**, 277–319. [22]

Yagishita, T., Ikegami, K., Narusawa, T. and Okuyama, H. (1984). Photoconduction of copper phthalocyanine-binder photoconductor sensitized with poly-N-vinylcarbazole. *IEEE Trans. Ind. Appl.*, **IA-20**, 1642–6. [207]

Yakobson, B. I., Boldyreva, E. V. and Sidelnikov, A. A. (1989). Bending of needle crystals caused by a photochemical-reaction—quantitative description. *Izv. Sibirsk. Ptdel. Akad. Nauk SSSR Ser. Khim. Nauk* 6–10. [285]

Yamamoto, H., Katogi, S., Watanabe, T., Sato, H., Miyata, S. and Hosomi, T. (1992). New molecular design approach for non-centrosymmetric crystal structures - $\wedge$-shaped molecules for frequency doubling. *Appl. Phys. Lett.*, **60**, 935–7. [213]

Yang, Q.-C., Richardson, M. F. and Dunitz, J. D. (1989). Conformational polymorphism of dimethyl 3,6-dichloro-2,5- dihydroxyterephthalate. I. Structures and atomic displacement parameters between 100 and 350 K for three crystal forms. *Acta Crystallogr. B*, **15**, 312–23. [215]

Yannoni, C. S. (1982). High resolution NMR in solids: the CPMAS experiment. *Acc. Chem. Res.*, **15**, 201–8. [133]

Yano, J., Ueno, S., Sato, K., Arishimia, T., Sagi, N., Kaneko, F. and Kobayashi, M. (1993). Polymorphic transformations in SOS, POP and POS. *J. Phys. Chem.*, **97**, 12967–73. [129]

Yano, K., Takamatsu, N., Yamazaki, S., Sako, K., Nagura, S., Tomizawa, S., Shimaya, J. and Yamamoto, K. (1996). Crystal forms, improvements of dissolution and absorption of poorly water-soluble (R)-1-[2,3-dihydro-1-(2'-methylphenacyl)-2-oxo-5- phenyl-1H-1,4-benzodiazepin-3-yl]-3-(3-methylphenyl)urea (YM022). *Yakugaku* Zasshi—J. Pharm. Soc. Jpn., **116**, 639–46. [245]

Yazawa, H. and Momonaga, M. (1994). Reaction crystallisation with additive agents. *Pharm. Manuf. Int.*, 107–10. [256]

Yevich, J. P. (1991). Drug development: from discovery to marketing. In *A textbook of drug design and development* (ed. P. Krogsgaard-Larsen and H. Bundgaard). Harwood Academic Publishers, Chur. [241]

Yokota, M., Mochizuki, M., Saito, K., Sato, A. and Kubota, N. (1999). Simple batch operation for selective crystallization of metastable crystalline phase. *Chem. Eng. Commun.*, **174**, 243–56. [253]

Yokoyama, T., Umeda, T., Kuroda, K., Nagafuku, T., Yamamoto, T. and Asada, S. (1979). Studies on drug nonequivalence. IX. Relationship between polymorphism and rectal absorption of indomethacin. *Yakugaku Zasshi—J. Pharm. Soc. Jpn.*, **99**, 837–42. [245]

Yokoyama, T., Umeda, T., Kuroda, K., Kuroda, T. and Asada, S. (1980). Studies on drug nonequivalence. X. Bioavailability of 6-mercaptopurine polymorphs. *Chem. Pharm. Bull.*, **29**, 194–9. [245]

Yordanov, N. D. and Christova, A. G. (1997). Quantitative spectrophotometric and EPR-determination of 1,1-diphenyl-2-picryl-hydrazyl (DPPH). *Fresenius Z. Anal. Chemie*, **358**, 610–13. [203]

Yoshimoto, N., Takayuki, S., Gamachi, H. and Yoshizawa, M. (1999) Polymorphism and crystal growth of organic conductor (BEDT-TTF)(2)I-3. *Mol. Cryst. Liq. Cryst., Sci. Technol. A—Mol. Cryst. Liq. Cryst.*, **327**, 233–6. [192]

Youming, C., Shengqing, L. and Renyuan, Q. (1986). Polymorphism and electronic properties of 3-ethyl-5-[2-(3-ethyl-2-benzthiazolinylidene)-ethylidene]-rhodamine. *Phys. Status Solidi. A: Appl. Res.*, **98**, 37–42. [204]

Young, R. A. (ed.) (1993). *The Rietveld method.* International Union of Crystallography, Oxford Science Publications, Oxford. [111, 122]

Young, R. H., Marchetti, A. P. and Newhouse, E. I. P. (1989). Optical spectra of a planar aggregate and of a lamellar crystal containing it—how are they related? *J. Chem. Phys.*, **91**, 5743–55. [205]

Young, S. W. (1911). Mechanical stimulus to crystallization in supercooled liquids. *J. Am. Chem. Soc.*, **33**, 148–62. [70]

Yu, L. (1995). Inferring thermodynamic stability relationship of polymorphs from melting data. *J. Pharm. Sci.*, **84**, 966–74. [108, 171]

Yu, L. (2001). Amorphous pharmaceutical solids: preparation, characterization and stabilization. *Adv. Drug. Revs.*, **48**, 27–42. [253, 254]

Yu, L., Reutzel, S. M. and Stephenson, G. A. (1998). Physical characterization of polymorphic drugs: an integrated characterization strategy. *Pharm. Sci. Technol. Today*, **1**, 118–27. [149, 150t, 224, 240]

Yu, L., Stephenson, G. A., Mitchell, C. A., Bunnell, C. A., Snorek, S. V., Bowyer, J. J., Borchardt, T. B., Stowell, J. G. and Byrn, S. R. (2000). Thermochemistry and conformational polymorphism of a hexamorphic crystal system. *J. Am. Chem. Soc.*, **122**, 585–91. [98, 107, 170, 171f, 172f, 173f, 174f, 175f, 176f, 177f, 280]

Yu, R. C., Yakimansky, A V., Kothe, H., Voigt-Martin, I. G., Schollmeyer, D., Jansen, J., Zandbergen, H. and Tenkovtsev, A. V. (2000). Strategies for structure solution and refinement of small organic molecules from electron diffraction data and limitations of the simulation approach. *Acta Crystallogr. A*, **56**, 436–50.

Zamir, S., Bernstein, J. and Greenwood, D. J. (1994). A single crystal to single crystal reversible phase transition which exhibits the 'hopping effect'. *Mol. Cryst. Liq. Cryst.*, **242**, 193–200. [219, 223]

Zannikos, P. N., Li, W.-I., Drennen, J. K. and Lodder, R. A. (1991). Spectrophotometric prediction of the dissolution rate of carbamazepine tablets. *Pharm. Res.*, **8**, 974–8. [244]

Zell, M. T., Padden, B. E., Grant, D. J. W., Chapeau, M.-C., Prakash, I. and Munson, E. J. (1999). Two-dimensional high-speed CP/MAS NMR spectroscopy of polymorphs. 1. Uniformly ^{13}C-labeled aspartame. *J. Am. Chem. Soc.*, **121**, 1372–8. [141, 305]

Zeman, S. (1980). The relationship between differential thermal analysis data and the detonation characteristics of polynitroaromatic compounds. *Thermochim. Acta*, **41**, 199–212. [275]

Zenith Laboratories, Inc. v. Bristol-Meyers Squibb Co. (1992). Civ. A. No. 91-3423; 1991 WL 267892 (D. N. J.) [303]

Zenith Laboratories, Inc. v. Bristol-Meyers Squibb Co. (1994). No. 92-1527; 19 F.3d 1418. [130, 297]

Zepharovich, V. V. (1888). *Z. Kristallogr.*, **13**, 145–9. [80]

Zerkowski, J. A., MacDonald, J. C. and Whitesides, G. M. (1997). Polymorphic packing arrangements in a class of engineered organic crystals. *Chem. Mater.*, **9**, 1933–41. [65f, 65]

Zevin, L. S. and Kimmel, G. (1995). *Quantitative X-ray diffracometry*. Springer, New York. [117, 119, 254]

Zhang, Y., Wu, G., Wenner, B. R., Bright, F. V. and Coppens, P. (1999). Engineering crystals for excited-state studies: Two polymorphs of 4,4'-dihydroxybenzophenon/4,13-diaza-18-crown-6. Abstracts of the American Crystallographic Association Meeting, Abstract 03.02.09, p. 48. [234]

Zhitomirskaya, N. G., Eremenko, L. T., Golovina, N. I. and Atovmayan, L. O. (1987). Structural and electronic parameters of certain cyclic nitramines. *Izv. Akad. Nauk SSSR, Ser. Khim.*, 576–80. [279t]

Zink, J. I. (1978). Triboluminescence. *Acc. Chem. Res.*, **11**, 289–95. [221]

Zollinger, H. (1991). *Color Chemistry: Synthesis, Properties and applications of organic dyes and pigments*, 2nd edn., VCH, Weinheim. [258]

Zorky, P. M. and Kuleshova, L. N. (1980). Comparison of the hydrogen bonds in polymorphic modifications of organic substances. *Zh. Strukt. Khim.*, **22**, 153–6. [55]

Zupan, J. and Gasteiger, J. (1991). Neural networks: A new method for solving chemical problems or just a passing phase? *Anal. Chim. Acta*, **248**, 1–30. [124]

Zyss, J. (ed.) (1994). *Molecular nonlinear optics: materials, physics and devices*. Academic Press, Boston. [207]

Index